34/95

St. Olaf College

AUG 19 1985

Science Library

Brownian Motion
and
Martingales in Analysis

The Wadsworth Mathematics Series

Series Editors
Raoul H. Bott, Harvard University
David Eisenbud, Brandeis University
Hugh L. Montgomery, University of Michigan
Paul J. Sally, Jr., University of Chicago
Barry Simon, California Institute of Technology
Richard P. Stanley, Massachusetts Institute of Technology

W. Beckner, A. Calderon, R. Fefferman, P. Jones, *Conference on Harmonic Analysis in Honor of Antoni Zygmund*
M. Behzad, G. Chartrand, L. Lesniak-Foster, *Graphs and Digraphs*
J. Cochran, *Applied Mathematics: Principles, Techniques, and Applications*
W. Derrick, *Complex Analysis and Applications*, Second Edition
R. Durrett, *Brownian Motion and Martingales in Analysis*
A. Garsia, *Topics in Almost Everywhere Convergence*
K. Stromberg, *An Introduction to Classical Real Analysis*
R. Salem, *Algebraic Numbers and Fourier Analysis*, and L. Carleson, *Selected Problems on Exceptional Sets*

Brownian Motion and Martingales in Analysis

Richard Durrett
University of California, Los Angeles

Wadsworth Advanced Books & Software
Belmont, California
A Division of Wadsworth, Inc.

Acquisitions Editor: John Kimmel
Production Editor: Marta Kongsle
Designer: Janet Bollow
Copy Editor: Mary Roybal
Technical Illustrator: Brown & Sullivan

© 1984 by Wadsworth, Inc. All rights reserved. No part of this book may be reproduced, stored in a retrieval system, or transcribed, in any form or by any means, electronic, mechanical, photocopying, recording, or otherwise, without the prior written permission of the publisher, Wadsworth Advanced Books & Software, Belmont, California 94002, a division of Wadsworth, Inc.

Printed in the United States of America

1 2 3 4 5 6 7 8 9 10—88 87 86 85 84

ISBN 0-534-03065-3

Library of Congress Cataloging in Publication Data
Durrett, Richard, 1951–
   Brownian motion and martingales in analysis.

   (The Wadsworth mathematics series)
   Bibliography: p. 313
   Includes index.
   1. Brownian motion processes. 2. Martingales (Mathematics)
I. Title. II. Series.
QA274.75.D87 1984   515   84-7230
ISBN 0-534-03065-3

Preface

In the years that have passed since the pioneering work of Kakutani, Kac, and Doob, it has been shown that Brownian motion can be used to prove many results in classical analysis, primarily concerning the behavior of harmonic and analytic functions and the solutions of certain partial differential equations. In spite of the many pages that have been written on this subject, the results in this area are not widely known, primarily because they appear in articles that are scattered throughout the literature and are written in a style appropriate for technical journals. The purpose of this book, then, is to bring some of these results together and to explain them as simply and as clearly as we can.

In Chapters 1 and 2, we introduce the two objects that will be the cornerstones for our later developments—Brownian motion and the stochastic integral. This material is necessary for all that follows, but after digesting Chapters 1 and 2, readers can turn to their favorite applications. The remaining seven chapters fall into four almost independent groups.

In Chapters 3 and 4, we will study the boundary limits of functions that are harmonic in the upper half space $H = \{x \in R^d : x_d > 0\}$. Chapter 3 is devoted to developing the relevant probabilistic machinery, that is, we define conditioned Brownian motions (or h-transforms) and give their basic properties. In Chapter 4, we apply these results to the study of harmonic functions; to be precise, we show that (modulo null sets) nontangential convergence, nontangential boundedness, and finiteness of the "area function" are equivalent, and we investigate the relationship between these notions and their probabilistic counterparts.

In Chapter 5, we turn our attention to analytic functions and use Brownian motion to prove results about their boundary limits and mapping properties (e.g., Picard's theorem). The first results are, I think, some of the most striking applications of Brownian motion. By observing that a complex Brownian motion never visits 0 at any (positive) time, we can make Privalov's theorem "obvious," and we can take $\log f$ unambiguously along Brownian paths to prove that functions in the Nevanlinna class have nontangential limits, without relying on factorization theorems to remove the zeros.

In Chapters 6 and 7, we use Brownian motion to study the classical Hardy spaces $H^p, p > 0$, that is, the set of functions that are analytic in $D = \{z\colon |z| < 1\}$ and have

$$\sup_{r<1} \int |f(re^{i\theta})|^p \, d\theta < \infty.$$

The first task is to establish the equivalence of H^p to a subspace of a space of martingales. After this is done, we can prove results in analysis by proving their probabilistic analogues. Some of the topics we investigate in this way are: (i) boundary limits of functions in H^p, (ii) the relationship between the L^p norms of conjugate functions, (iii) Fefferman's duality theorem $(H^1)^* = BMO$, and (iv) properties of functions in BMO.

In Chapters 8 and 9, we investigate the relationship between Brownian motion and partial differential equations. The key words associated with the developments in Chapter 8 are heat equation, Feynman-Kac formula, Cameron-Martin transformation, Dirichlet problem, Poisson's equation, and eigenvalues of Schrödinger operators. Although the middle four of these topics have been treated in many books, our approach, which is based on stochastic integration, is five-sixths new and allows us to treat all the equations in the same manner. One slightly kinky aspect of our development is that the words *semigroup* and *generator* do not appear in the book (well, only once). We picked up this idea ("all you need is Itô's formula") and many others in the book from conversations with Mike Harrison.

In all the equations in Chapter 8, the highest-order term is of the form $\frac{1}{2}\Delta$. If we want to replace this term with something more general, then we have to replace our Brownian motion with a more general diffusion. In Chapter 9, we take a feeble first step in this direction. We introduce stochastic differential equations and solve them to get processes that we can run to solve second-order parabolic equations. In many respects, this is a strange place to stop, since at this point there are obviously a number of other things to talk about, such as elliptic equations, Neumann boundary conditions, small time asymptotics for diffusions, large λ asymptotics for eigenvalues, Malliavin calculus, and so on. However, the book had to end somewhere.

Since this is a book about the relationship between probability and analysis, we have tried to write it so that it could be read by someone in either field. In keeping with this aim, the only formal prerequisite for this book is a familiarity with measure theory as usually developed in the first quarter of a graduate course in analysis or probability. We have included an appendix that describes all the results we need from a first-year graduate course in probability theory (chiefly the basics of results of martingale theory). Anyone who has had a graduate course in probability can safely skip the appendix, and more advanced readers can start with Chapter 2 or turn to their favorite application. Readers who skip the appendix should keep in mind, however, that when we call something a "standard result" it can (usually) be found in the appendix.

In addition to trying to keep the book self-contained, we have tried to do several things to make the book easy to read. (a) The first two chapters contain a number of exercises to help the reader who is learning the material for the first time. (b) We are content in most cases to give the reader a taste of an area by confining our attention to a special situation or proving somewhat less than the best possible result and referring the reader to the literature for more refined results. (c) Last but not least, we have tried to keep attention focused on the "ideas behind the proofs" and the most important techniques by making remarks about the how and why and, occasionally, by reviewing at the end of a proof the key steps that led to the conclusion. The last point, along with the decision to supply details usually left to the reader, may cause the pace to be too slow at times, but as Stroock and Varadhan said in their preface, we believe "it is preferable to bore than to batter."

The idea of writing this book was born while I was visiting Stanford during the fall and winter quarters of 1979–1980. When I arrived, my academic godfather, Kai Lai Chung, suggested that we write a book on Brownian motion. Several months later, when it became clear that the book would not be finished before my visit was over, he changed his mind, but I was on my way. Returning to UCLA in the spring, I taught the first of three seminars that led to this book, the other two being at Cornell in the spring semester of 1981 and at UCLA in the fall quarter of 1982. Between seminars, I revised and expanded my notes, and finally, on a leave of absence in the winter quarter of 1983, I (almost) finished writing this book.

A number of people read the manuscript or heard lectures from it at various stages of its development and made suggestions for improving the exposition. In this regard I would especially like to thank Donald Burkholder, René Carmona, Burgess Davis, Richard Dudley, Daniel Revuz, Barry Simon, and Ruth Williams. Other people who provided advice on specific points are thanked at appropriate places in the text.

From time to time during the writing of this book, I was supported by funds from the National Science Foundation (at Cornell and during four summers at UCLA) and the Sloan Foundation (for two months in Paris and for one quarter while I was finishing the book). I am very grateful for the support of these foundations, but by far the most important member of the supporting cast is my wife, Susan, who has patiently endured the writing of this book.

Contents

1 Brownian Motion   1
1.1 Definition and Construction   1
1.2 The Markov Property   7
1.3 The Right Continuous Filtration, Blumenthal's 0-1 Law   11
1.4 Stopping Times   17
1.5 The Strong Markov Property   21
1.6 Martingale Properties of Brownian Motion   25
1.7 Hitting Probabilities, Recurrence, and Transience   27
1.8 The Potential Kernels   30
1.9 Brownian Motion in a Half Space   32
1.10 Exit Distributions for the Sphere   36
1.11 Occupation Times for the Sphere   39
  Notes on Chapter 1   43

2 Stochastic Integration   44
2.1 Integration w.r.t. Brownian Motion   44
2.2 Integration w.r.t. Discrete Martingales   48
2.3 The Basic Ingredients for Our Stochastic Integral   50
2.4 The Variance and Covariance of Continuous Local Martingales   52
2.5 Integration w.r.t. Continuous Local Martingales   55
2.6 The Kunita-Watanabe Inequality   59
2.7 Stochastic Differentials, the Associative Law   62
2.8 Change of Variables, Itô's Formula   64
2.9 Extension to Functions of Several Semimartingales   67
2.10 Applications of Itô's Formula   70
2.11 Change of Time, Lévy's Theorem   75
2.12 Conformal Invariance in $d \geq 2$, Kelvin's Transformations   78
2.13 Change of Measure, Girsanov's Formula   82
2.14 Martingales Adapted to Brownian Filtrations   85
  Notes on Chapter 2   89
  A Word about the Notes   90

3 Conditioned Brownian Motions 91

- 3.1 Warm-Up: Conditioned Random Walks 91
- 3.2 Brownian Motion Conditioned to Exit $H = R^{d-1} \times (0, \infty)$ at 0 94
- 3.3 Other Conditioned Processes in H 97
- 3.4 Inversion in $d \geq 3$, B_t Conditioned to Converge to 0 as $t \to \infty$ 100
- 3.5 A Zero-One Law for Conditioned Processes 102

4 Boundary Limits of Harmonic Functions 105

- 4.1 Probabilistic Analogues of the Theorems of Privalov and Spencer 105
- 4.2 Probability Is Less Stringent than Analysis 108
- 4.3 Equivalence of Brownian and Nontangential Convergence in $d = 2$ 113
- 4.4 Burkholder and Gundy's Counterexample ($d = 3$) 116
- 4.5 With a Little Help from Analysis, Probability Works in $d \geq 3$: Brossard's Proof of Calderon's Theorem 119

5 Complex Brownian Motion and Analytic Functions 123

- 5.1 Conformal Invariance, Applications to Brownian Motion 123
- 5.2 Nontangential Convergence in D 126
- 5.3 Boundary Limits of Functions in the Nevanlinna Class N 128
- 5.4 Two Special Properties of Boundary Limits of Analytic Functions 132
- 5.5 Winding of Brownian Motion in $C - \{0\}$ (Spitzer's Theorem) 134
- 5.6 Tangling of Brownian Motion in $C - \{-1, 1\}$ (Picard's Theorem) 139

6 Hardy Spaces and Related Spaces of Martingales 144

- 6.1 Definition of H^p, an Important Example 144
- 6.2 First Definition of $\mathcal{M}^p$, Differences Between $p > 1$ and $p = 1$ 146
- 6.3 A Second Definition of $\mathcal{M}^p$ 152
- 6.4 Equivalence of H^p to a Subspace of $\mathcal{M}^p$ 155
- 6.5 Boundary Limits and Representation of Functions in H^p 158
- 6.6 Martingale Transforms 162
- 6.7 Janson's Characterization of $\mathcal{M}^1$ 166
- 6.8 Inequalities for Conjugate Harmonic Functions 170
- 6.9 Conjugate Functions of Indicators and Singular Measures 180

7 H^1 and BMO, $\mathcal{M}^1$ and $\mathcal{BMO}$ 184

- 7.1 The Duality Theorem for $\mathcal{M}^1$ 184
- 7.2 A Second Proof of $(\mathcal{M}^1)^* = \mathcal{BMO}$ 188
- 7.3 Equivalence of BMO to a Subspace of $\mathcal{BMO}$ 192
- 7.4 The Duality Theorem for H^1, Fefferman-Stein Decomposition 199

7.5	Examples of Martingales in $\mathcal{BMO}$	205
7.6	The John-Nirenberg Inequality	208
7.7	The Garnett-Jones Theorem	211
7.8	A Disappointing Look at $(\mathcal{M}^p)^*$ When $p < 1$	215

8 PDE's That Can Be Solved by Running a Brownian Motion 219

A	**Parabolic Equations** 219	
8.1	The Heat Equation	220
8.2	The Inhomogeneous Equation	223
8.3	The Feynman-Kac Formula	229
8.4	The Cameron-Martin Transformation	234
B	**Elliptic Equations** 245	
8.5	The Dirichlet Problem	246
8.6	Poisson's Equation	251
8.7	The Schrödinger Equation	255
8.8	Eigenvalues of $\Delta + c$	263

9 Stochastic Differential Equations 271

9.1	PDE's That Can Be Solved by Running an SDE	271
9.2	Existence of Solutions to SDE's with Continuous Coefficients	274
9.3	Uniqueness of Solutions to SDE's with Lipschitz Coefficients	278
9.4	Some Examples	283
9.5	Solutions Weak and Strong, Uniqueness Questions	286
9.6	Markov and Feller Properties	288
9.7	Conditions for Smoothness	290
	Notes on Chapter 9	293

Appendix A Primer of Probability Theory 294

A.1	Some Differences in the Language	294
A.2	Independence and Laws of Large Numbers	296
A.3	Conditional Expectation	300
A.4	Martingales	302
A.5	Gambling Systems and the Martingale Convergence Theorem	303
A.6	Doob's Inequality, Convergence in $L^p, p > 1$	306
A.7	Uniform Integrability and Convergence in L^1	307
A.8	Optional Stopping Theorems	309

References 313

Index of Notation 325

Subject Index 327

1 Brownian Motion

1.1 Definition and Construction

A d-dimensional Brownian motion is a process B_t, $t \geq 0$, taking values in R^d, that has the following properties:

(i) if $t_0 < t_1 < \cdots < t_n$, then $B(t_0), B(t_1) - B(t_0), \ldots, B(t_n) - B(t_{n-1})$ are independent

(ii) if $s, t \geq 0$, then
$$P(B(s+t) - B(s) \in A) = \int_A (2\pi t)^{-d/2} e^{-|x|^2/2t}$$

(iii) with probability 1, $t \to B_t$ is continuous.

(i) says that B_t has independent increments. (ii) says that the increment $B_{s+t} - B_s$ has a d-dimensional normal distribution with mean 0 and covariance tI, that is, the coordinates $B^i_{s+t} - B^i_t$ are independent and each one has a normal distribution with mean zero and variance t. (iii) is self-explanatory.

The first question that must be confronted in any discussion of Brownian motion is, "Is there a process with these three properties?" The answer to this question is yes, of course. There are dozens of books about Brownian motion, and there are at least four or five essentially different constructions (one of which we will give below), so there can be no doubt that the process exists. For the moment, however, I want to try to shake your faith in this fact by pointing out that there are two things to worry about:

(a) Are assumptions (i) and (ii) consistent?
(b) If we specify the distribution of B_0, then (i) and (ii) determine the distribution of $(B(t_1), \ldots, B(t_n))$ for any finite set of times. Is (iii) consistent with these distributions?

To build the suspense a little bit, let's fix our attention on the case $d = 1$ and see what happens if we change (ii) to (ii') $P(B_{s+t} - B_s \in A) = \int_A f_t(x)\,dx$

where f_t is a family of probability densities, that is, each $f_t \geq 0$ and $\int f_t = 1$. If $u \geq 0$, then (i) implies

(*) $f_{t+u}(y) = \int f_t(x) f_u(y - x)\,dx$,

or introducing the characteristic function (a.k.a. Fourier transform)

$$\hat{f}_t(\theta) = \int e^{ix\cdot\theta} f(x)\,dx$$

(ˆ) $\hat{f}_{t+u}(\theta) = \hat{f}_t(\theta) \hat{f}_u(\theta)$.

It is easy to show that if $t \to \hat{f}_t(\theta)$ is continuous, then (ˆ) implies $\hat{f}_t(\theta) = \exp(c_\theta t)$ (and if you work hard you can show this is true when $\hat{f}_t(\theta)$ is bounded and measurable), so the distributions $f_t(x)$ are far from arbitrary. When

$$f_t(x) = (2\pi t)^{-1/2} e^{-x^2/2t}$$
$$\hat{f}_t(\theta) = \exp(-t\theta^2/2)$$

so (*) holds, but shifting our attention now to question (b), this is not the only possibility. The Cauchy density with parameter t

$$f_t(x) = \frac{t}{\pi(t^2 + x^2)}$$

has

$$\hat{f}_t(\theta) = \exp(-t|\theta|),$$

so these f_t's are another possible choice, and the process that results is called the Cauchy process, $C_t, t \geq 0$.

It is easy to see from the formulas for the normal and Cauchy densities that

(1) $\qquad B_t \stackrel{d}{=} t^{1/2} B_1 \quad \text{and} \quad C_t \stackrel{d}{=} t C_1,$

so as $t \to 0$, $P(|B_t| > \varepsilon) \to 0$, $P(|C_t| > \varepsilon) \to 0$. On this basis one might naively expect that both processes can be defined in such a way that the paths are continuous. This is true for Brownian motion but false for the Cauchy process, and, in fact, we will see in Section 1.9 that with probability 1 the set of discontinuities of C_t is dense in $[0, \infty)$.

The preceding discussion has hopefully convinced you that the fact that Brownian paths are continuous is not obvious, so we turn now to the somewhat tedious details of the construction of Brownian motion. For pedagogical reasons, we will first pursue an approach that leads us to a dead end, and then we will retreat a little to rectify the difficulty.

Fix $x \in R^d$ and for each $0 < t_1 < \cdots < t_n$ define a measure $\mu_{t_1,\ldots,t_n}$ on $(R^d)^n$ by

1.1 Definition and Construction

$$\mu_{t_1,\ldots,t_n}(A_1 \times \cdots \times A_n) = \int_{A_1} dx_1 \cdots \int_{A_n} dx_n \prod_{i=1}^{n} p_{t_i-t_{i-1}}(x_{i-1}, x_i)$$

where $x_0 = x$, $t_0 = 0$, and

$$p_t(x, y) = (2\mu t)^{-d/2} e^{-|y-x|^2/2t}.$$

In this notation, (*) says

$$\mu_{t,t+u}(R \times A) = \mu_{t+u}(A).$$

This is the first step in showing that the family of μ's is a *consistent set of finite dimensional distributions* (or, for short, f.d.d.'s), that is, if $\{s_1, \ldots, s_{n-1}\} \subset \{t_1, \ldots, t_n\}$ and $t_j \notin \{s_1, \ldots, s_{n-1}\}$, then

$$\mu_{s_1,\ldots,s_{n-1}}(A_1 \times \cdots \times A_{n-1})$$
$$= \mu_{t_1,\ldots,t_n}(A_1 \times \cdots \times A_{j-1} \times R \times A_j \times \cdots \times A_{n-1}).$$

It is easy to check (details are left to the reader) that the measures given above are a consistent set of f.d.d.'s, so we can use Kolmogorov's extension theorem to give our first construction of Brownian motion.

(2) Let $\Omega_0 = \{\text{functions } \omega : [0, \infty) \to R\}$ and $\mathscr{F}_0 = \sigma$-algebra generated by the finite dimensional sets $\{\omega : \omega(t_i) \in A_i \text{ for } 1 \leq i \leq n\}$, where each $A_i \in \mathscr{R}^d$, the set of Borel subsets of R^d. Given a consistent set of f.d.d.'s $\mu_{t_1,\ldots,t_n}$, there is a unique probability measure μ on $(\Omega_0, \mathscr{F}_0)$ so that for all the finite dimensional sets,

$$\mu(\{\omega : \omega_{t_i} \in A_i, 1 \leq i \leq n\}) = \mu_{t_1,\ldots,t_n}(A_1 \times \cdots \times A_n).$$

Proof This result is a consequence of the Caratheodary Extension theorem. For a detailed proof, see Chung (1974), page 60, or Breiman (1968), pages 23–24.

At this point we are at the dead end referred to above. If $C = \{\omega : t \to \omega_t$ is continuous$\}$, then $C \notin \mathscr{F}_0$, that is, C is not a measurable set. The easiest way of proving this is to do the following.

Exercise 1 $A \in \mathscr{F}_0$ if and only if there is a sequence of times $t_1, t_2, \ldots$ and a set $B \in \mathscr{R}^{\{1,2,\ldots\}}$ so that

$$A = \{\omega : (\omega_{t_1}, \omega_{t_2}, \ldots) \in B\}.$$

The problem above is easy to solve. Let $Q_2 = \{m2^{-n} : m, n \geq 0\}$ be the dyadic rationals. Kolmogorov's theorem guarantees that we can define on some probability space $(\Omega, \mathscr{F}, P)$ a family of random variables $\{B_t, t \in Q_2\}$ with the desired joint distributions. To extend B_t to a process defined on $[0, \infty)$, we will show that with probability 1, $t \to B_t$ is uniformly continuous on $Q_2 \cap [0, T]$ for each $T < \infty$, so there is a unique continuous extension to $[0, \infty)$.

To prove the last result, it suffices to show that each coordinate is uniformly continuous on $Q_2 \cap [0, 1]$, that is, we can suppose without loss that $d = 1$ and by scaling that $T=1$. The key to proving the result in this case is the following trivial consequence of (1):

(3) $E|B_t|^4 = Ct^2$ where $C = E|B_1|^4 < \infty$,

and a not-so-trivial computation due to Kolmogorov. Let $\gamma < 1/4$, $\delta > 0$, and observe that $a^4 P(|X| > a) \le E|X|^4$ (Chebyshev's inequality), so

$$P(|B(j2^{-n}) - B(i2^{-n})| > ((j-i)2^{-n})^\gamma \text{ for some } 0 \le i < j \le 2^n, j - i \le 2^{n\delta})$$
$$\le \Sigma((j-i)2^{-n})^{-4\gamma} E|B(j2^{-n}) - B(i2^{-n})|^4$$

where the sum is over the set of (i,j) satisfying the conditions on the left-hand side. Using (3), we see that the right-hand side of the last inequality

$$\le C\Sigma((j-i)2^{-n})^{-4\gamma+2}$$
$$\le C \cdot 2^n \cdot 2^{n\delta} \cdot (2^{n\delta}2^{-n})^{-4\gamma+2},$$

since there are 2^n choices for i, $(0 \le i < 2^n)$, $0 < j - i \le 2^{n\delta}$. The right-hand side is $C2^{-n\varepsilon}$, where $\varepsilon = (1-\delta)(2-4\gamma) - (1+\delta)$. Since $\gamma < 1/4$, $(2-4\gamma) > 1$, and if we pick δ small enough, then $\varepsilon > 0$ and $\Sigma 2^{-n\varepsilon} < \infty$, so the Borel Cantelli lemma implies the following:

(4)   For almost every ω there is an N (which depends on ω) so that for all $n \ge N$,

$$|B(j2^{-n}) - B(i2^{-n})| \le ((j-i)2^{-n})^\gamma$$

for all $0 \le i < j \le 2^n$ that have $j - i \le 2^{n\delta}$.

We will now check that (4) implies that $B_t(\omega)$ is uniformly continuous on $Q_2 \cap [0,1]$. This is not hard, but, as the reader will discover if he or she tries to do the proof alone, it requires some care. Let $q, r \in Q_2 \cap [0,T]$ with $0 < r - q < 2^{-N(1-\delta)}$, pick $m \ge N$ so that

$$2^{-(m+1)(1-\delta)} \le r - q < 2^{-m(1-\delta)},$$

and write

$$q = i2^{-m} - 2^{-q_1} - \cdots - 2^{-q_k}$$
$$r = j2^{-m} + 2^{-r_1} + \cdots + 2^{-r_\ell},$$

where $m < q_1 < \cdots < q_k$ and $m < r_1 < \cdots < r_\ell$ (see Figure 1.1).
Now $r - q < 2^{-m(1-\delta)}$, so $(i - j) < 2^{m\delta}$, and it follows from the choice of m that

$$|B(i2^{-m}) - B(j2^{-m})| \le ((2^{m\delta})2^{-m})^\gamma.$$

$(i-1)2^{-m}$ $i2^{-m}$ $j2^{-m}$ $(j+1)2^{-m}$

Figure 1.1

1.1 Definition and Construction

To estimate $|B(q) - B(i2^{-m})|$ now, we observe that using (4) and the triangle inequality gives

$$|B(q) - B(i2^{-m})| \leq \sum_{i=1}^{k} (2^{-q_i})^\gamma$$

$$\leq \sum_{j=m+1}^{\infty} (2^{-\gamma})^j = C 2^{-\gamma m}$$

and

$$|B(r) - B(j2^{-m})| \leq \sum_{j=m+1}^{\infty} (2^{-\gamma})^j = C 2^{-\gamma m}.$$

Combining the last three inequalities shows

$$|B(q) - B(r)| \leq C 2^{\gamma m (1-\delta)} \leq C' |r - q|^\gamma$$

(since $2^{-m(1-\delta)} \leq 2^{1-\delta}|r-q|$), so B_t is Hölder continuous with exponent γ.

Note: In the last part of the proof we used two conventions that we will use throughout the book (without further notice):

(a) The letter C denotes a constant $\in (0, \infty)$, whose value is unimportant and may change from line to line.
(b) When a constant changes within a string of inequalities, we will use C', C'', ... to alert you and to avoid absurd expressions like $2e^C \leq C$.

Combining the observations above, we see that with probability 1 the process $B_t(\omega)$ initially defined for $t \in Q_2$ has a unique extension so that $t \to B_t(\omega)$ is continuous on $[0, \infty)$. This is our Brownian motion. Now that the birth has been accomplished, we want to tidy it up a little bit. To do this, we observe that the map that takes ω into the trajectory $t \to B_t(\omega)$ maps the original Ω into $C =$ the set of continuous functions from $[0, \infty)$ to R^d, so if we build a Brownian motion starting from $B_0 = x$, the mapping gives us a measure P_x on $(C, \mathscr{C})$ where $\mathscr{C}$ is the smallest σ-algebra that makes all coordinate maps $\omega \to \omega(t)$ measurable, and under P_x the random variables $B_t(\omega) = \omega(t)$ are a Brownian motion starting at x.

The last picture of Brownian motion is the one we will use in what follows, so it is important to get it firmly in mind. We have a special measure space $(C, \mathscr{C})$, one family of random variables $B_t(\omega) = \omega(t)$, and a family of probability measures P_x, so that under P_x, B_t is a Brownian motion with $P_x(B_0 = x) = 1$. This setup gives us all the Brownian motions we will ever need. If we want a Brownian motion in which B_0 is random and has distribution μ, then we let

$$P_\mu(A) = \int \mu(dx) P_x(A),$$

and this gives us a Brownian motion with $P_\mu(B_0 \in A) = \mu(A)$.

The argument above, which led to the continuity of the Brownian path, is an important one, so it is good to stop for a minute and reconsider the details.

Exercise 2 Show that if we replace (3) by

(3′) $\quad E|X_t - X_s|^\beta \leq C|t-s|^{1+\alpha}$

where α and β are positive, then the paths of X are Hölder continuous of order γ for any $\gamma < \alpha/\beta$. (This is Kolmogorov's criterion for a stochastic process to have continuous paths.)

Let B_t be a one-dimensional Brownian motion. Since

$$E|B_t - B_s|^{2m} = C_m|t-s|^m$$

where $C_m = E|B_1|^{2m} < \infty$, applying the result in Exercise 2 and letting $m \to \infty$ shows that Brownian paths are Hölder continuous of order γ for any $\gamma < 1/2$. It is easy to show that Brownian paths are with probability 1 not Lipschitz continuous at any point. Let

$A_n = \{\omega : \text{there is an } s \in [0,1] \text{ such that } |B_t - B_s| \leq C|t-s| \text{ when } |t-s| \leq 2/n\}$

$Y_{k,n} = \max\left\{\left|B\left(\frac{k+j}{n}\right) - B\left(\frac{k+j-1}{n}\right)\right| : j = 0, 1, 2\right\}, \quad 1 \leq k \leq n-2$

$B_n = \left\{\text{at least one of the } Y_{k,n} \leq \frac{4C}{n}\right\}.$

Since $A_n \subset B_n$, using the Brownian scaling relation (2) and the formula for the normal density shows

$$P(A_n) \leq P(B_n) \leq nP\left\{\left|B\left(\frac{1}{n}\right)\right| \leq \frac{4C}{n}\right\}^3$$

$$= nP\left\{|B(1)| \leq \frac{4C}{\sqrt{n}}\right\}^3$$

$$\leq n\left(\frac{4C}{\sqrt{n}} \cdot \frac{1}{\sqrt{2\pi}}\right)^3 \to 0 \text{ as } n \to \infty,$$

which proves our claim, since $n \to A_n$ is increasing and C is arbitrary.

From the last result we immediately get

(5) With probability 1, $t \to B_t$ is not differentiable at any point.

This fact is due to Paley, Wiener, and Zygmund (1933). The simple proof given above is due to Dvoretsky, Erdös, and Kakutani (1961). Since functions of bounded variation are differentiable a.e., (5) has the following corollary:

(6) With probability 1, $t \to B_t$ has unbounded variation in any interval (a, b) with $a < b$.

The next result sharpens and extends (6).

(7)   As $n \to \infty$,
$$\sum_{m=1}^{2^n} |B(tm2^{-n}) - B(t(m-1)2^{-n})|^2 \to t \quad \text{a.s.}$$

Proof Let $\Delta_{m,n} = |B(tm2^{-n}) - B(t(m-1)2^{-n})|^2$. For each $n \Delta_{m,n}$, $m = 1, \ldots, 2^n$ are independent and have $E\Delta_{m,n} = E(B(t2^{-n}))^2 = t2^{-n}$, so
$$E\left(\sum_{m=1}^{2^m} \Delta_{m,n} - t\right)^2 = \sum_{m=1}^{2^n} E(\Delta_{m,n} - t2^{-n})^2 = 2^n t^2 2^{-2n} E(B_1^2 - 1)^2.$$

With this established, the rest is easy. Using Chebyshev's inequality gives
$$P\left(\left(\sum_{m=1}^{2^n} \Delta_{m,n} - t\right)^2 > \varepsilon\right) \leq \varepsilon^{-1} t^2 2^{-n} E(B_1^2 - 1)^2,$$

and then the desired conclusion follows from the Borel Cantelli lemma.

(7) says that the quadratic variation of B_s, $0 \leq s \leq t$ is almost surely t. This shows that (a) locally the Brownian path looks like $t^{1/2}$ and (b) B_t is irregular in a very regular way. The next result is more evidence for (a).

Exercise 3 Generalize the proof of (5) to show that if $\gamma > 1/2$, then with probability 1, $t \to B_t$ is not Hölder continuous of order γ at any point. (The borderline case $\gamma = 1/2$ will be treated in Section 1.3.)

1.2 The Markov Property

Since Brownian motion has independent increments, it is easy (I hope) to believe that

(1)   If we let $s \geq 0$, then $B_{s+t} - B_s$, $t \geq 0$, is a Brownian motion that is independent of what happened before time s.

The last sentence is, I think, the clearest expression of the Markov property of Brownian motion. This section is devoted to formulating and proving a result that makes this precise, but reduces (1) to a rather cryptic formula that requires several definitions to even explain what it means.

Why do I want to do this? It is not for sport or for the love of secret code, but because the "cryptic form" of the Markov property is the most useful for doing computations and proving theorems. It is easy (I am told by my students) to get lost in the measure theoretic details, so before we plunge into them, I will do two sample applications to convince you that the Markov property is easy to understand and use:

Example 1 If you want to compute $P_x(B_t = 0$ for some $t \in [a,b])$ where $0 < a < b$, then it seems reasonable to break things down according to the value of B_a and use the fact that Brownian motion has "independent increments" to conclude that

$$P_x(B_t = 0 \text{ for some } t \in [a,b])$$
$$= \int p_a(x,y) P_y(B_t = 0 \text{ for some } t \leq b-a)\,dy$$

where $p_t(x,y) = (2\pi t)^{-d/2} e^{-|x-y|^2/2t}$.

Example 2 The next level of complication is to compute $P_x(B_t = 0$ for some $t \in [0,a]$ and some $t \in [a,b])$. This time, when we break things up according to B_a the first factor in the answer changes to

$$\bar{p}_a(x,y) \equiv P_x(B_a = y, B_s = 0 \text{ for some } s \in [0,a])$$
$$\leq p_a(x,y),$$

but the second factor stays the same, that is,

$$P_x(B_t = 0 \text{ for some } t \in [0,a] \text{ and some } t \in [a,b])$$
$$= \int \bar{p}_a(x,y) P_y(B_t = 0 \text{ for some } t \leq b-a)\,dy.$$

Intuitively the last equality holds, because if we condition on the value of B_a, then the behavior of B_t for $t \in [a,b]$ is independent of whether or not Brownian motion hit 0 in the interval $[0,a]$. The Markov property below will allow us to prove that this formula is correct.

The first step in giving a precise statement of (1) is to explain what we mean by "what happened before time s." The first thing that comes to mind is

$$\mathscr{F}_s^0 = \sigma(B_r : r \leq s)$$

where the right-hand side denotes the smallest σ-field, which makes all the B_r, $r \leq s$ measurable. Ultimately, for technical reasons, we will trade this in for some σ-fields that are a little large, but for the moment these (organic) σ-fields are fine and we can state and prove our first version of the Markov property.

(2) If we let $s \geq 0$, then for any $t \geq 0$, $B_{s+t} - B_s$ is independent of what happened before time s.

To be precise,

(2) If $s \geq 0$, $t \geq 0$, and f is bounded, then for all $x \in R^d$,

$$E_x(f(B_{t+s} - B_s)|\mathscr{F}_s^0) = E_x f(B_{t+s} - B_s).$$

Proof This is almost an immediate consequence of the definition of Brownian motion. If $s_1 < s_2 < \cdots < s_n \leq s$ and $A = \{\omega : B_{s_i}(\omega) \in C_i, 1 \leq i \leq n\}$ where the $C_i \in \mathscr{R}^d$ (the Borel subsets of R^d), then definition of Brownian motion implies 1_A and $f(B_t - B_s)$ are independent, so in this case

(*) $E_x(f(B_t - B_s); A) = P_x(A) E_x f(B_t - B_s).$

1.2 The Markov Property

To prove (2) we need to show that (∗) holds for all $A \in \mathcal{F}_s^0$. To do this, we will use an extension theorem, commonly called Dynkin's $\pi - \lambda$ theorem, which was tailor-made for situations like this.

(3) Let $\mathcal{A}$ be a collection of subsets of Ω that contains Ω and is closed under intersection. Let $\mathcal{G}$ be a collection of sets that satisfy

(i) if $A, B \in \mathcal{G}$ and $A \supset B$, then $A - B \in \mathcal{G}$
(ii) if $A_n \in \mathcal{G}$ and $A_n \uparrow A$, then $A \in \mathcal{G}$.

If $\mathcal{A} \subset \mathcal{G}$, then the σ-field generated by $\mathcal{A}$, $\sigma(\mathcal{A}) \subset \mathcal{G}$.

The proof of this result is not hard, but as you will discover if you try to prove it for yourself, it is not trivial either. Since the ideas involved in giving an efficient proof of this are not needed in the developments that follow, we will call this a result from measure theory and refer you to page 5 of Blumenthal and Getoor (1968) or page 34 of Billingsley (1979) for a proof. The formulation above is from Billingsley.

With (3) in hand, the proof of (2) is trivial. Let $\mathcal{A} =$ the sets of the form $\{B_{s_i} \in C_i, 1 \le i \le n\}$ where $s_1 < s_2 < \cdots < s_n \le s$ and let $\mathcal{G} =$ the collection of A for which (∗) holds, and observe that $\mathcal{G}$ clearly satisfies (i) and (ii), so $\mathcal{G} \supset \sigma(\mathcal{A}) = \mathcal{F}_s^0$. (The reader should also notice that it is not so easy to show that if $A, B \in \mathcal{G}$, then $A \cap B \in \mathcal{G}$, but the proof of (i) is trivial.)

Our next step toward the Markov property is

(4) If we let $s \ge 0$, then for any $t \ge 0$, $B_{s+t} - B_s$ is independent of what happened before time s and has the same distribution as $B_t - B_0$.

To be precise,

(4) If $s \ge 0, t \ge 0$, and f is bounded, then for all $x \in R^d$,

$$E_x(f(B_{t+s})|\mathcal{F}_s^0) = E_{B(s)}f(B_t)$$

where the right-hand side is the function $\varphi(x) = E_x f(B_t)$ evaluated at $x = B_s$.

Proof To prove (4), we will prove a slightly more general result (let $g(x, y) = f(x + y)$).

(5) If $s \ge 0, t \ge 0$, and g is bounded, then

$$E_x(g(B_s, B_{t+s} - B_s)|\mathcal{F}_s^0) = \varphi_g(B_s)$$

where

$$\varphi_g(x) = \int g(x, y)(2\pi t)^{-d/2} e^{-|y|^2/2t}\, dy.$$

To prove (5), we observe that if $g(x, y) = g_1(x) g_2(y)$, then

$$\varphi_g(x) = g_1(x) \int g_2(y)(2\pi t)^{-d/2} e^{-|y|^2/2t}\, dy$$

and

$$E_x(g(B_s, B_{t+s} - B_s)|\mathscr{F}_s) = g_1(B_s)E_x(g_2(B_{t+s} - B_s)|\mathscr{F}_s)$$
$$= g_1(B_s)E_x(g_2(B_{t+s} - B_s))$$

by (2), so the equality holds in this case.

To extend the last result to all bounded measurable functions, we need an extension theorem for functions. The one we will use is a "monotone class theorem."

(6) Let $\mathscr{A}$ be a collection of subsets of Ω that contains Ω and is closed under intersection. Let $\mathscr{H}$ be a vector space of real-valued functions on Ω satisfying

(i) if $A \in \mathscr{A}$, $1_A \in \mathscr{H}$
(ii) if $f_n \in \mathscr{H}$ are nonnegative and increase to a bounded function f, then $f \in \mathscr{H}$.

Then $\mathscr{H}$ contains all bounded functions on Ω that are measurable with respect to $\sigma(\mathscr{A}) = $ the σ-field generated by $\mathscr{A}$.

As before, we will refer the reader to Blumenthal and Getoor (1968) for the proof (page 6 this time) and we will content ourselves with applying (6) to complete the proof of (5). Let $\mathscr{A} = $ the set of all rectangles ($A \times B$ where A, $B \in \mathscr{R}^d$ and $\mathscr{H} = $ the set of all bounded g for which (5) holds. Taking $g_1 = 1_A$ and $g_2 = 1_B$ and applying our first result shows that (i) holds. It is clear that $\mathscr{H}$ satisfies (ii), so applying (6) now proves (5) and hence (4).

With (4) proved, the last step is to extend the reasoning to a general measurable "function of the path B_{t+s}, $t \geq 0$." To describe this class of functions we need some notation. Recall that our Brownian motions are a family P_x, $x \in R^d$ of measures on $(C, \mathscr{C})$. For $s \geq 0$, define the shift transformation $\theta_s: C \to C$ by setting $(\theta_s \omega)(t) = \omega(s + t)$ for $t \geq 0$. In words, we cut off the part of the path before time s and then shift the time scale so that time s becomes time 0.

With this notation introduced, it is clear how to define a "function of the path B_{t+s}, $t \geq 0$." It is simply $Y \circ \theta_s$ where $Y: C \to R$ is some $\mathscr{C}$ measurable function. We can now finally state the "cryptic form" of the Markov property.

(7) If $s \geq 0$ and Y is bounded and $\mathscr{C}$ measurable, then for all $x \in R^d$,

$$E_x(Y \circ \theta_s | \mathscr{F}_s^0) = E_{B(s)} Y$$

where the right-hand side is the function $\varphi(x) = E_x Y$ evaluated at $x = B_s$.

Proof To prove (7) we start by retracing the steps in the proof of (4). First let $s = t_0 < t_1 < \cdots < t_n$ and $\Delta_i = B(t_i) - B(t_{i-1})$. From the proof of (2), we see that if $f_1, \ldots, f_n$ are bounded,

$$E_x\left(\prod_{i=1}^n f_i(\Delta_i); A\right) = E_x\left(\prod_{i=1}^n f_i(\Delta_i)\right) P_x(A),$$

so repeating the proof of (4) shows that if g is a bounded measurable function on $(R^d)^{n+1}$,

$$E_x(g(B_s, \Delta_1, \ldots, \Delta_n)|\mathscr{F}_s^0) = \psi(B_s)$$

where

$$\psi(y) = \int g(y, z_1, \ldots, z_n) \prod_{i=1}^{n} p_{t_i - t_{i-1}}(z_i)\, dz,$$

proving (7) in the case where Y depends on finitely many coordinates, and applying the monotone class theorem again proves (7).

(7) is a very important formula which will be used many times below. You should try a few of the following exercises to see how (7) is used in computations.

Exercise 1 Let B_t be a one-dimensional Brownian motion. Let $S = \inf\{t > 1 : B_t = 0\}$ and $T = \inf\{t > 0 : B_t = 0\}$. Show that $S = T \circ \theta_1$ and use the Markov property to conclude that if $t \geq 0$,

$$P_x(S > 1 + t) = \int p_1(x, y) P_y(T > t)\, dy.$$

Exercise 2 Let B_t and T be as in Exercise 1 and let $R = \sup\{t < 1 : B_t = 0\}$. Use the Markov property to conclude that if $0 \leq t \leq 1$,

$$P_x(R \leq t) = \int p_t(x, y) P_y(T > 1 - t)\, dy.$$

The last two formulas can be used to compute the distributions of R and S after we find $P_y(T > s)$ in Section 1.5. The next three examples will be important in Chapter 8.

Exercise 3 Let $u(t, x) = E_x f(B_t)$ where f is bounded. Use the Markov property to conclude that if $0 < s < t$,

$$E_x(f(B_t)|\mathscr{F}_s^0) = u(t - s, B_s).$$

Exercise 4 Let $u(t, x) = E_x \int_0^t g(B_r)\, dr$ where g is bounded. Use the Markov property to conclude that if $0 < s < t$,

$$E_x\left(\int_0^t g(B_r)\, dr \,\Big|\, \mathscr{F}_s^0\right) = \int_0^s g(B_r)\, dr + u(t - s, B_s).$$

Exercise 5 Let $c_t = \int_0^t c(B_r)\, dr$ and $u(t, x) = E_x \exp(c_t)$ where c is bounded. Use the Markov property to conclude that if $0 < s < t$,

$$E_x(\exp(c_t)|\mathscr{F}_s^0) = \exp(c_s) u(t - s, B_s).$$

1.3 The Right Continuous Filtration, Blumenthal's 0-1 Law

The first part of this section is devoted to a technical matter. For reasons that will become apparent in Section 1.4, it is convenient to replace the fields $\mathscr{F}_t^0$

defined in the last section by the (slightly larger) fields

$$\mathscr{F}_t^+ = \bigcap_{u>t} \mathscr{F}_u^0,$$

which are nicer because they are right continuous, that is,

(1) $$\bigcap_{t>s} \mathscr{F}_t^+ = \bigcap_{t>s} \left(\bigcap_{u>t} \mathscr{F}_u^0 \right) = \bigcap_{u>s} \mathscr{F}_u^0 = \mathscr{F}_s^+.$$

In words, the $\mathscr{F}_t^+$ allow us "an infinitesimal peek at the future." Intuitively, this should not give us any information that is useful for predicting the value of $B_{t+s} - B_t$. In this section we will show that this intuition is correct, and furthermore that $\mathscr{F}_t^+$ and $\mathscr{F}_t^0$ are equal (modulo null sets).

The first step in proving the last result is to prove that the simplest form of the Markov property proved in Section 1.2 holds when $\mathscr{F}_s^0$ is replaced by $\mathscr{F}_s^+$.

(2) If f is bounded, then for all $s, t \geq 0$ and $x \in R^d$,

$$E_x(f(B_{t+s})|\mathscr{F}_s^+) = E_{B(s)} f(B_t)$$

where the right-hand side is the function $\varphi(x) = E_x f(B_t)$ evaluated at $x = B(s)$.

Proof The result is trivial if $t = 0$, so we will suppose $t > 0$. Let $r \in (s, s+t)$. Applying (4) from Section 1.2 gives

$$E_x(f(B_{t+s})|\mathscr{F}_r^0) = E_{B(r)} f(B_{t+s-r}),$$

so if we let $\varphi(x, u) = E_x f(B_u)$ and integrate over A, we get

$$E_x(f(B_{t+s}); A) = E_x(\varphi(B_r, t+s-r); A).$$

Now if f is bounded and continuous,

$$\varphi(x, t) = \int (2\pi t)^{-d/2} e^{-|y-x|^2/2t} f(y)\, dy$$

is a bounded continuous function of (x, t); so letting $r \downarrow s$, we conclude that

$$E_x(f(B_{t+s}); A) = E_x(\varphi(B_s, t); A)$$

for all bounded continuous functions.

To extend this to all bounded functions, let $\mathscr{H} =$ the set of all bounded functions for which the equality holds, let $\mathscr{A} =$ the set of all rectangles $(a_1, b_1) \times \cdots \times (a_d, b_d)$ where for each $i - \infty \leq a_i < b_i \leq \infty$. $\mathscr{H}$ is clearly a vector space that satisfies the hypotheses of our monotone class theorem ((6) in Section 1.2). As for $\mathscr{A}$, if $A \in \mathscr{A}$, then 1_A is an increasing limit of bounded nonnegative continuous f_n, so $1_A \in \mathscr{H}$.

With (2) established, it is routine to prove that the Markov property holds in the general form:

(3) If $s \geq 0$ and Y is bounded and $\mathscr{C}$ measurable, then for all $x \in R^d$,

1.3 The Right Continuous Filtration, Blumenthal's 0-1 Law

$$E_x(Y \circ \theta_s | \mathscr{F}_s^+) = E_{B(s)} Y$$

where the right-hand side is the function $\varphi(x) = E_x Y$ evaluated at $x = B_s$.

Proof Rereading the proof of (7) in Section 1.2 reveals that all we did was use induction and the monotone class theorem to turn the special case $Y \circ \theta_s = f(B_{t+s})$ into the general case. To help the reader check this, we will repeat the proof with the numbers changed below.

To prove (3), we start by retracing the steps in the proof of (2). First let $s = t_0 < t_1 < \cdots < t_n$ and $\Delta_i = B(t_i) - B(t_{i-1})$. From the proof of (2), we see that if $f_1, \ldots, f_n$ are bounded and continuous,

$$E_x\left(\prod_{i=1}^n f_i(\Delta_i); A\right) = E_x\left(\prod_{i=1}^n f_i(\Delta_i)\right) P_x(A),$$

so repeating the proof of (2) shows that if g is a bounded measurable function on $(R^d)^{n+1}$,

$$E_x(g(B_s, \Delta_1, \ldots, \Delta_n) | \mathscr{F}_s^+) = \psi(B_s)$$

where

$$\psi(y) = \int g(y, z_1, \ldots, z_n) \prod_{i=1}^n p_{t_i - t_{i-1}}(z_i) \, dz,$$

proving (3) in the case where Y depends on finitely many coordinates, and applying the monotone class theorem now proves the desired result.

Combining (3) above with (7) from Section 1.2 gives

$$E_x(Y \circ \theta_s | \mathscr{F}_s^+) = E_{B(s)} Y = E_x(Y \circ \theta_s | \mathscr{F}_s^0).$$

The last equality shows that conditional expectations of functions of the future are the same for $\mathscr{F}_s^+$ and $\mathscr{F}_s^0$. Since it is trivial that $E(X|\mathscr{F}_s^+) = E(X|\mathscr{F}_s^0)$ for $X \in \mathscr{F}_s^0 \subset \mathscr{F}_s^+$, we are led immediately to conclude that

(4) If $Z \in \mathscr{C}$ is bounded, then for all $s \geq 0$ and $x \in R^d$,

$$E_x(Z|\mathscr{F}_s^+) = E_x(Z|\mathscr{F}_s^0).$$

Proof By the monotone class theorem it suffices to prove the result when $Z = \prod_{i=1}^n f_i(B_{t_i})$ where the f_i are bounded and measurable, but in this case $Z = X(Y \circ \theta_t)$ where $X \in \mathscr{F}_t^0$ and $Y \in \mathscr{C}$, so

$$E_x(Z|\mathscr{F}_s^+) = X E_x(Y \circ \theta_s | \mathscr{F}_s^+) = X E_{B(s)} Y \in \mathscr{F}_s^0.$$

Since $\mathscr{F}_s^0 \subset \mathscr{F}_s^+$, it follows (from the definition of conditional expectation) that $E_x(Z|\mathscr{F}_s^+) = E_x(Z|\mathscr{F}_s^0)$.

To steal a line from Chung (1974), page 341, "The reader is urged to ponder over the intuitive meaning of this result and judge for himself whether it is obvious or incredible." Above we have treated the result as a technical necessity (i.e., an obvious result). The rest of this section is devoted to the other viewpoint.

The fun starts when we take $s = 0$ in (4) to get

(5) **Blumenthal's 0-1 Law.** If $A \in \mathscr{F}_0^+$, then for all $x \in R^d$, $P_x(A) \in \{0, 1\}$.

Proof Since $A \in \mathscr{F}_0^+$,

$$1_A = E_x(1_A|\mathscr{F}_0^+).$$

The Markov property (3) implies

$$E_x(1_A|\mathscr{F}_0^+) = E_{B(0)}1_A = E_x 1_A, \quad P_x \text{ a.s.,}$$

so $P_x(A) = E_x 1_A = 1_A(x) \in \{0, 1\}$, proving the desired result.

In words, Blumenthal's 0-1 law says that the "germ field" $\mathscr{F}_0^+ = \bigcap_{\varepsilon > 0} \mathscr{F}_\varepsilon^0$ is trivial. This result is very useful in studying the local behavior of Brownian paths. Let B_t be a one-dimensional Brownian motion.

(6) Let $\tau = \inf\{t \geq 0 : B_t > 0\}$. Then $P_0(\tau = 0) = 1$.

Proof

$$P_0(\tau \leq t) \geq P_0(B_t > 0) = \frac{1}{2}$$

since the normal distribution is symmetric about 0. Letting $t \downarrow 0$, we conclude

$$P_0(\tau = 0) = \lim_{t \downarrow 0} P_0(\tau \leq t) \geq \frac{1}{2},$$

so it follows from (5) that $P_0(\tau = 0) = 1$.

Once Brownian motion must hit $(0, \infty)$ immediately starting from 0, it must also hit $(-\infty, 0)$ immediately, and this forces it to cross 0 infinitely many times in $[0, \varepsilon]$ for any $\varepsilon > 0$.

(7) Let t_n be a sequence of numbers that decreases to 0. Then P_0 almost surely, we have $B_{t_n} > 0$ for infinitely many n and $B_{t_n} < 0$ for infinitely many n.

Proof Let $A_n = \{B_{t_n} > 0\}$. Then

$$P_0(\limsup A_n) = \lim_{N \to \infty} P_0\left(\bigcup_{n=N}^{\infty} A_n\right)$$

$$\geq \limsup_{N \to \infty} P_0(A_N) = \frac{1}{2},$$

so it follows from the 0-1 law that $P_0(\limsup A_n) = 1$.

(7) is a prime example of how the Blumenthal 0-1 law forces the Brownian path to behave erratically. The following is a more humorous example:

(8) If you run a two-dimensional Brownian motion for a positive amount of time, it will write your name an infinite number of times, that is, if $g : [0, 1] \to R^2$ is a continuous function, $\varepsilon > 0$ and $t_n \downarrow 0$, then P_0 almost surely, we will have

1.3 The Right Continuous Filtration, Blumenthal's 0-1 Law

$$\sup_{0 \le \theta \le 1} \left| \frac{B(\theta t_n)}{\sqrt{t_n}} - g(\theta) \right| < \varepsilon$$

for infinitely many n.

The results above are just a few of the many results about the local behavior of Brownian paths that can be studied using Blumenthal's 0-1 law. In the exercises below, we give some more examples. For the rest of the section, we restrict our attention to one-dimensional Brownian motion.

Exercise 1 Let $t_n \downarrow 0$. Show that

$$\limsup_{n \to \infty} B(t_n)/t_n^{1/2} = \infty, \quad P_0 \text{ a.s.},$$

so Brownian paths are not Hölder continuous of order $1/2$.

Exercise 2 Let $t \to f(t)$ be an increasing function with $f(t) > 0$ for $t > 0$. Use Blumenthal's 0-1 law to conclude that

$$\limsup_{t \to 0} B_t/f(t) = c, \quad P_0 \text{ a.s.}$$

where c is a constant $\in [0, \infty]$.

Up to this point all the arguments have been soft. If we want to figure out how fast B_t can grow, we have to do some computations. The first two steps are useful to know, so we have stated them as separate exercises.

Exercise 3 If $S \subset [0, T]$, then

$$P_0 \left(\sup_{t \in S} B_t > x \right) \le 2 P_0(B_T > x).$$

Hint: It suffices to prove the result when S is finite. In this case the result is a consequence of independent increments and the fact that $P_0(B_t > 0) = 1/2$ for all $t > 0$. In Section 1.5 we will show that the inequality is an equality when $S = [0, T]$.

Exercise 4

$$\int_x^\infty e^{-z^2/2} \, dz \le \frac{1}{x} \int_x^\infty z e^{-z^2/2} = \frac{1}{x} e^{-x^2/2}$$

and

$$\int_x^\infty e^{-z^2/2} \, dz \sim \frac{1}{x} e^{-x^2/2} \quad \text{as } x \to \infty.$$

Here and in what follows, $f(x) \sim g(x)$ as $x \to \infty$ means $f(x)/g(x) \to 1$ as $x \to \infty$.

Exercise 5 *Law of the Iterated Logarithm.* Let $\varphi(t) = (2t \log \log(1/t))^{1/2}$. Then

$$\limsup_{t \downarrow 0} B_t/\varphi(t) = 1, \quad P_0 \text{ a.s.}$$

(a) The first step is to explain why this is the right order of magnitude. To do this we observe

$$P(B_t > (ct \log\log(1/t))^{1/2})$$
$$= P\left(\frac{B_t}{\sqrt{t}} > (c \log\log(1/t))^{1/2}\right) \sim \frac{1}{(c \log\log(1/t))^{1/2}} \exp\left(-\frac{c}{2} \log\log(1/t)\right),$$

so if we let $t_n = \alpha^n$ where $\alpha < 1$,

$$P(B(t_n) > (ct_n \log\log(1/t_n))^{1/2})$$
$$\sim (c \log(-n \log \alpha))^{-1/2} \frac{1}{(-n \log \alpha)^{c/2}},$$

so the sum is finite if $c > 2$ and infinite if $c < 2$.

(b) To prove $\limsup B_t/\varphi(t) \leq 1$, we observe that since $t \to \varphi(t)$ is increasing, using the result of Exercise 3 shows

$$P_0\left(\max_{t_{n+1} \leq s \leq t_n} \frac{B_s}{\varphi(s)} > (1 + \varepsilon)\right) \leq P_0\left(\max_{t_{n+1} \leq s \leq t_n} B_s > (1 + \varepsilon)\varphi(t_{n+1})\right)$$
$$\leq 2P_0(B_{t_n} > (1 + \varepsilon)\varphi(t_{n+1}))$$
$$= 2P_0(B_{t_{n+1}} > \sqrt{\alpha}(1 + \varepsilon)\varphi(t_{n+1}))$$

by scaling, so if α is close to 1, then $\sqrt{\alpha}(1 + \varepsilon) > 1$, and the result follows from the Borel Cantelli lemma.

(c) To prove $\liminf B_t/\varphi(t) \geq 1$, observe that

$$P_0(B_{t_n} - B_{t_{n+1}} > (1 - \varepsilon)\varphi(t_n)) = P_0(B_{t_n(1-\alpha)} > (1 - \varepsilon)\varphi(t_n))$$
$$= P_0\left(B_{t_n} > \frac{(1-\varepsilon)}{\sqrt{1-\alpha}} \varphi(t_n)\right),$$

so if α is close to 0, then $(1 - \varepsilon)/\sqrt{1 - \alpha} < 1$, and the other Borel Cantelli lemma implies

$$P_0(B_{t_n} - B_{t_{n+1}} > (1 - \varepsilon)\varphi(t_n) \text{ i.o.}) = 1.$$

An easy argument (see Chung (1974), Theorem 9.5.2, for details) now improves the last result to

$$P_0(B_{t_n} - B_{t_{n+1}} > (1 - \varepsilon)\varphi(t_n) \text{ and } B_{t_{n+1}} > 0 \text{ i.o.}) = 1,$$

which proves the desired result.

The results above concern the behavior of B_t as $t \to 0$. By using a trick, we can use this result to get information about the behavior as $t \to \infty$.

Exercise 6 If B_t is a Brownian motion starting at 0, then so is $X_t = tB(1/t)$, $t > 0$.

Proof It is easy to check that $EX_t = 0$, $EX_t^2 = t$, and $EX_s X_t = s \wedge t$. Since for each $t_1 < t_2 < \cdots < t_n$, $(X(t_1), \ldots, X(t_n))$ has a multivariate normal distribution, it follows that X has the right finite dimensional distributions and hence must be a Brownian motion.

Since a d-dimensional Brownian motion is a vector of d independent one-dimensional Brownian motions, the last result generalizes trivially to $d > 1$. Combining the last result with Blumenthal's 0-1 law leads to a very useful result.

Let $\mathscr{F}_t' = \sigma(B_s : s \geq t)$.
Let $\mathscr{A} = \bigcap_{t \geq 0} \mathscr{F}_t'$.

$\mathscr{A}$ is called the asymptotic σ-field, since it concerns the asymptotic behavior of B_t as $t \to \infty$. As was the case for $\mathscr{F}_0^+$, one can think of a lot of events that are in $\mathscr{A}$, but they are all trivial.

(9) If $A \in \mathscr{A}$, then either $P_x(A) \equiv 0$ or $P_x(A) \equiv 1$.

Proof Since the asymptotic σ-field for B is the same as the germ σ-field for X, we see that $P_0(A) \in \{0, 1\}$. To improve this to the conclusion given, we observe that $A \in \mathscr{F}_1'$, so 1_A can be written as $1_A = 1_B \circ \theta_1$. Applying the Markov property gives

$$P_x(A) = E_x(1_B \circ \theta_1)$$
$$= E_x E_x(1_B \circ \theta_1 | \mathscr{F}_1)$$
$$= E_x E_{B(1)} 1_B$$
$$= \int \frac{1}{(2\pi)^{d/2}} e^{-|x-y|^2/2} P_y(B) \, dy.$$

Taking $x = 0$, we see that if $P_0(A) = 0$, then $P_y(B) = 0$ for a.e. y and hence $P_x(A) = 0$ for every x. Repeating the last argument shows that if $P_0(A) = 1$, then $P_x(A) = 1$ for every x, and proves the desired result.

Remark: The reader should note that the conclusion above is stronger than that in Blumenthal's 0-1 law. In that case, for example, if $A = \{B_0 \in C\}$, then $P_x(A)$ may depend on x.

1.4 Stopping Times

In this section and the next we will develop and explore the most important property of Brownian motion—the strong Markov property. The intuitive idea is simple, but since the rigorous formulation requires a number of definitions that somewhat obscure the intuitive content, we will begin with a simple example.

Let B_t be a one-dimensional Brownian motion, let $a > 0$, and let $T_a = \inf\{t \geq 0 : B_t = a\}$ be the first time Brownian motion hits a. If we know that $T_a = t$, then this says something about the past ($B_s \neq a$ for $s < t$) and the present ($B_t = a$), but nothing about B_u for $u > T_a$, so it seems clear that $B(T_a + t)$, $t \geq 0$, will have the same distribution as a Brownian motion starting at a and that this process will be independent of T_a.

The argument in the last paragraph may be convincing, but it is not rigorous. In Section 1.5 we will prove a result that captures the essence of the reasoning used above and makes it precise. To prepare for this, we need to define and discuss a class of random times S that "say nothing about the behavior of the process after time S."

A random variable S taking values in $[0, \infty]$ is said to be a stopping time if for all $t \geq 0$, $\{S < t\} \in \mathcal{F}_t^+$.

If you think of our Brownian motion B_t as giving the value of a stock or some other commodity, and think of S as the time at which you sell the quantity in question, then the condition $\{S < t\} \in \mathcal{F}_t^+$ has a simple interpretation: The decision to sell at a time $< t$ must be based on the information available at time t, that is, based on $\bigcap_{\varepsilon > 0} \sigma(B_s : s \leq t + \varepsilon)$.

From the definition given above, it should be painfully obvious that we have made a choice between $\{S < t\}$ and $\{S \leq t\}$ and also between $\in \mathcal{F}_t^+$ and $\in \mathcal{F}_t^0$. The quantity we have defined above is what is officially known (see Dellacherie and Meyer (1978), page 115, (49.2)) as a "wide sense stopping time." This name comes from the fact that if $t \to \mathcal{G}_t$ is an increasing family of σ-fields, then $\{S \leq t\} \in \mathcal{G}_t$ implies that

$$\{S < t\} = \bigcup_n \{S \leq t - n^{-1}\} \in \mathcal{G}_t,$$

but $\{S < t\} \in \mathcal{G}_t$ only implies that

$$\{S \leq t\} = \bigcap_n \{S < t + n^{-1}\} \in \mathcal{G}_t^+ = \bigcap_{u > t} \mathcal{G}_t.$$

In general, there can be a difference between the definitions $\{S < t\} \in \mathcal{G}_t$ and $\{S \leq t\} \in \mathcal{G}_t$, but (as we have shown above) there is no difference if $t \to \mathcal{G}_t$ is right continuous. (2) and (3) below show that when checking that something is a stopping time, it is nice to know that the two definitions are equivalent.

(1)   If G is an open set and $T = \inf\{t \geq 0 : B_t \in G\}$, then T is a stopping time.

Proof Since G is open,

$$\{T < t\} = \bigcup_{q < t} \{B_q \in G\} \in \mathcal{F}_t^0$$

where the union is overall rational q, so T is a stopping time.

Remark: The reader should observe that if $t > 0$,

$$\{T = t\} = \{B_s \notin G \text{ for all } s \leq t, \text{ but there are } t_n \downarrow t \text{ so that } B(t_n) \in G\}.$$

Therefore $\{T = t\}$ (and hence $\{T \leq t\}$) $\in \mathcal{F}_t^+$ but is not in $\mathcal{F}_t^0$, although by the results in the last section, the difference is a null set.

1.4 Stopping Times

(2) If T_n is a sequence of stopping times and $T_n \downarrow T$, then T is a stopping time.

Proof $\{T < t\} = \bigcup_{n=1}^{\infty} \{T_n < t\}$.

(3) If T_n is a sequence of stopping times and $T_n \uparrow T$, then T is a stopping time.

Proof $\{T \le t\} = \bigcap_{n=1}^{\infty} \{T_n \le t\}$.

(4) If K is a closed set and $T = \inf\{t \ge 0 : B_t \in K\}$, then T is a stopping time.

Proof Let $G_n = \bigcup \{D(x, 1/n) : x \in K\}$ where $D(x, r) = \{y : |x - y| < r\}$, and let $T_n = \inf\{t \ge 0 : B_t \in G_n\}$. Since G_n is open, it follows from (1) that T is a stopping time. I claim that as $n \to \infty$, $T_n \uparrow T$. To prove this we need to consider two cases.

(i) $T_n \uparrow \infty$. Since $T \ge T_n$ for all n, it follows that $T = \infty$ and hence $T_n \uparrow T$.
(ii) If $T_n \uparrow t < \infty$, then by the argument in (i), $T \ge t$. On the other hand, as $T_n \uparrow t$, $B(T_n) \to B_t$. Since $B(T_n) \in \bar{G}_n$ for all n, it follows that $B_t \in \bigcap \bar{G}_n = K$ and $T \le t$, completing the proof.

One of the reasons we have written out the last proof in such great detail is to show its dependence on the fact that K is closed. (2) implies that if $T_n = \inf\{t \ge 0 : B_t \in A_n\}$ is a stopping time and $A_n \uparrow A$, then $T = \inf\{t \ge 0 : B_t \in A\}$ is also; but to go the other way, that is, let $A_n \downarrow A$ and conclude $T_n \uparrow T$, we need to know that A is closed. The last annoying fact makes it difficult to show that the hitting time of a Borel set is a stopping time, and, in fact, this is not true unless the σ-fields are completed in a suitable way. To remedy this difficulty and to bring ourselves into line with the "usual conditions" (*les conditiones habituelles*) of the general theory of Markov processes, we will now pause to complete our σ-fields.

Let $\mathcal{N} = \{A : P_x(A) = 0 \text{ for all } x \in R^d\}$. A set $A \in \mathcal{N}$ is said to be a null set. Since our Brownian motion is a collection of measures P_x on our probability space $(C, \mathcal{C})$, a set A can be safely ignored only if $P_x(A) = 0$ for all $x \in R^d$.

Let $\mathcal{F}_t = \mathcal{F}_t^+ \vee \mathcal{N}$ (where here, and if we ever use this notation below, $\mathcal{A} \vee \mathcal{B}$ means the smallest σ-field containing $\mathcal{A}$ and $\mathcal{B}$). $\{\mathcal{F}_t, t \ge 0\}$ is called the augmented filtration and is the one which we will use throughout the rest of the book.

It is trivial to see that the Markov property (formula (3) in Section 1.3) remains valid when $\mathcal{F}_t^+$ is replaced by $\mathcal{F}_t$ (the conditional expectation is unchanged because we have only added null sets), so it seems reasonable to use the larger filtration $\mathcal{F}_t$—it allows more things to be measurable and still retains the Markov property.

As we mentioned above, the real reason for wanting to use the completed filtration is that it is needed to make $T_A = \inf\{t \ge 0 : B_t \in A\}$ measurable for every Borel set. Hunt (1957–8) was the first to prove this. The reader can find a discussion of this result in Section 10 of Chapter 1 of Blumenthal and Getoor (1968) or in Chapter 3 of Dellacherie and Meyer (1978). We will not prove it here because the results we have given above imply the following:

(5) If A is a countable union of closed sets and $T = \inf\{t \geq 0 : B_t \in A\}$, then T is a stopping time.

and this result is sufficient for our applications.

In Section 1.5 we will state and prove the strong Markov property. If one does not worry about details, it can be described in one sentence: If S is a stopping time, then formula (3) in Section 1.3 holds if we replace the fixed time s by the stopping time S. To make the resulting formula meaningful, we have to define θ_S and $\mathscr{F}_S$. It is clear how we should define the random shift:

(6) $$(\theta_S \omega)(t) = \begin{cases} \omega(S(\omega) + t) & \text{on } \{S < \infty\} \\ \text{undefined} & \text{on } \{S = \infty\}. \end{cases}$$

The second quantity is a little more subtle. We defined $\mathscr{F}_s = \bigcap_{\varepsilon > 0} \sigma(B_{t \wedge (s+\varepsilon)}, t \geq 0)$, so, by analogy, we should set $\mathscr{F}_S = \bigcap_{\varepsilon > 0} \sigma(B_{t \wedge (S+\varepsilon)}, t \geq 0)$. The definition we will give is less transparent than the last one, but easier to work with.

(7) $\mathscr{F}_S = \{A : A \cap \{S \leq t\} \in \mathscr{F}_t \text{ for all } t \geq 0\}.$

Intuitively, a set $A \in \mathscr{F}_S$ if it depends only upon what happened before time S. To see why (7) says this, let $A = \{\max_{0 \leq s \leq S} |B_s| > 3\}$. A obviously depends only on what happened before time S and is in $\mathscr{F}_S$, since for any $t \geq 0$

$$A \cap \{S \leq t\} = \left\{ \max_{0 \leq s \leq S} |B_s| > 3, S \leq t \right\} \in \mathscr{F}_t.$$

To get a feel for $\mathscr{F}_S$ and stopping times in general, the reader should try a few of the exercises below. Solutions to most of the exercises can be found in Blumenthal and Getoor (1968), Dellacherie and Meyer (1978), and in many other books about Markov processes.

Exercise 1 If S and T are stopping times, then $S \wedge T = \min\{S, T\}$, $S \vee T = \max\{S, T\}$, and $S + T$ are also stopping times. In particular, if $t \geq 0$, then $S \wedge t$, $S \vee t$, and $S + t$ are stopping times.

Exercise 2 If S and T are stopping times, then $T + S \circ \theta_T$ is also.

Exercise 3 Let S be a stopping time, let $A \in \mathscr{F}_S$, and let

$$R = \begin{cases} S & \text{on } A \\ \infty & \text{on } A^c. \end{cases}$$

Show that R is a stopping time.

Exercise 4 Let $S \leq T$ be stopping times. Show that $\mathscr{F}_S \subset \mathscr{F}_T$.

1.5 The Strong Markov Property 21

Exercise 5 Show that the definition of $\mathscr{F}_S$ we have given is equivalent to requiring $A \cap \{S < t\} \in \mathscr{F}_t$ for all t.

Exercise 6

(i) If t is a constant, then $\{S < t\}, \{S = t\},$ and $\{S > t\}$ are in $\mathscr{F}_S$.
(ii) If T is a stopping time, then $\{S < T\}, \{S = T\},$ and $\{S > T\}$ are in $\mathscr{F}_S$ (and also in $\mathscr{F}_T$).

Exercise 7 If $T_n \downarrow T$ in a sequence of stopping times, then $\mathscr{F}_T = \bigcap_n \mathscr{F}_{T_n}$.

1.5 The Strong Markov Property

In this section we will prove the strong Markov property:

(1) If S is a stopping time and Y is bounded and $\mathscr{C}$ measurable, then for all $x \in R^d$,

$$E_x(Y \circ \theta_S | \mathscr{F}_S) = E_{B(S)} Y \quad \text{on} \quad \{S < \infty\}$$

where the right-hand side is the function $\varphi(x) = E_x Y$ evaluated at $x = B(S)$.

Setting $S = T_a$ and $Y(\omega) = f(\omega_t)$ where $t \geq 0$ and f is bounded and using (1) gives that

$$E_0(f(B_{T_a + t}) | \mathscr{F}_{T_a}) = E_a f(B_t) \quad \text{on} \quad \{T_a < \infty\},$$

since $B(T_a) = a$ on $\{T_a < \infty\}$. Extending the argument above to $Y(\omega) = f(\omega(t_1), \ldots, \omega(t_n))$ where $0 \leq t_1 < \cdots < t_n$ and $f: (R^d)^n \to R$ is bounded justifies the remark we made at the beginning of Section 1.4: On $\{T_a < \infty\}$, $B(T_a + t), t \geq 0$, is a Brownian motion that is independent of $\mathscr{F}(T_a)$ and hence of T_a.

For some applications we will need to let the function Y that we apply to the shifted path $\theta_S \omega$ depend on the time S, so we will prove (1) in a more general form that allows for this.

(2) Let $(s, \omega) \to Y_s(\omega)$ be bounded and $\mathscr{R} \times \mathscr{C}$ measurable. If S is a stopping time, then for all $x \in R^d$,

$$E_x(Y_S \circ \theta_S | \mathscr{F}_S) = E_{B(S)} Y_S \quad \text{on} \quad \{S < \infty\}$$

where the right-hand side is the function $\varphi(x, t) = E_x Y_t$ evaluated at $x = B(S)$, $t = S$.

Proof Let $x \in R^d$. We will first prove the result under the assumption that there is a sequence of times $t_n \uparrow \infty$ so that $P_x(S < \infty) = \sum_n P_x(S = t_n)$. If we let $Z_n(\omega) = Y_{t_n}(\omega)$ and $A \in \mathscr{F}_S$, then

$$E_x(Y_S \circ \theta_S ; A \cap \{S < \infty\}) = \sum_{n=1}^{\infty} E_x(Z_n \circ \theta_S ; A \cap \{S = t_n\}).$$

Now if $A \in \mathscr{F}_S$, $A \cap \{S = t_n\} = (A \cap \{S \le t_n\}) - (A \cap \{S \le t_{n-1}\}) \in \mathscr{F}(t_n)$, so it follows from the Markov property that the sum above

$$= \sum_{n=1}^{\infty} E_x(E_{B(t_n)} Z_n ; A \cap \{S = t_n\})$$

$$= \sum_{n=1}^{\infty} E_x(E_{B(S)} Y_S ; A \cap \{S = t_n\})$$

$$= E_x(E_{B(S)} Y_S ; A \cap \{S < \infty\}).$$

To prove the result in general, we let $S_n = ([2^n S] + 1)/2^n$, where $[x] =$ largest integer $\le x$, and let $n \to \infty$. The first thing to observe is that if $t \in (m2^{-n}, (m+1)2^{-n}], \{S_n < t\} = \{S < m2^{-n}\}$, so S_n is a stopping time and $S_n \downarrow S$ as $n \to \infty$. To be able to take the limit, we will first restrict our attention to Y's of the form

$(*)\quad Y_s(\omega) = f_0(s) \prod_{i=1}^{n} f_i(\omega(t_i))$

where $0 = t_0 < t_1 < \cdots < t_n$ and $f_0, \ldots, f_n$ are bounded and continuous. In this case if we let $\varphi(x, s) = E_x Y_s$, then

$$\varphi(x, s) = f_0(s) E_x \prod_{i=1}^{n} f_i(B_{t_i})$$

$$= f_0(s) \int \cdots \int \prod_{i=1}^{n} f_i(x_i)(2\pi\Delta_i)^{-d/2} e^{-|x_i - x_{i-1}|^2/2\Delta_i} dx_1 \cdots dx_n$$

where $\Delta_i = t_i - t_{i-1}$, so the dominated convergence theorem implies that $(x, s) \to \varphi(x, s)$ is bounded and continuous.

Let $A \in \mathscr{F}_S$. If $m2^{-n} \le t < (m+1)2^{-n}$, then $A \cap \{S_n \le t\} = A \cap \{S < m2^{-n}\} \in \mathscr{F}(m2^{-n})$, so $A \in \mathscr{F}(S_n)$. Applying the special case of (2) proved above to S_n and observing that $\{S_n < \infty\} = \{S < \infty\}$ gives

$$E_x(Y_{S_n} \circ \theta_{S_n} ; A \cap \{S < \infty\}) = E_x(\varphi(B(S_n), S_n); A \cap \{S < \infty\}).$$

Now as $n \to \infty$, $Y_{S_n} \circ \theta_{S_n} \to Y_S \circ \theta_S$ and $\varphi(B(S_n), S_n) \to \varphi(B(S), S)$, so the bounded convergence theorem implies that (2) holds when $Y_s(\omega)$ has the form given in $(*)$, and an application of the monotone class theorem completes the proof.

Remark: Later we will want to generalize this result without repeating the argument, so we want the reader to observe that:

(i) in the first part of the proof all we did was dissect Y into the Z_n, apply the ordinary Markov property on each piece $\{S = t_n\}$, and sum things up to get the desired formula;
(ii) in the second part of the argument all we did was use the fact that $\varphi(x, s)$ is bounded and continuous to extend the special case to the general case.

The rest of this section is devoted to examples and exercises that illustrate how the strong Markov property is used to derive formulas for various quantities associated with Brownian motion. Since in most cases the reasoning that

1.5 The Strong Markov Property

is used to discover the formula is much different (and much more important!) from the sequence of steps used to "derive" it, I have tried in the first two examples to explain the intuition behind the result as well as the mechanics of obtaining it from (1).

Example 1 Let G be an open set, let $A \subset \partial G$, let $T = \inf\{t : B_t \notin G\}$, and let $u(x) = P_x(B_T \in A)$. I claim that if we let $\delta > 0$ be chosen so that $D(x, \delta) = \{y : |y - x| < \delta\} \subset G$ and let $S = \inf\{t \geq 0 : B_t \notin D(x, \delta)\}$, then

$$u(x) = E_x u(B_S).$$

Intuitive Proof Since $D(x, \delta) \subset G$, B_t cannot exit G without first exiting $D(x, \delta)$ at some point y, and after this occurs the probability of exiting G in A is $u(y)$ independent of how B got to y, so

$$P_x(B_T \in A) = E_x u(B_S).$$

Proof To make the intuitive proof rigorous, we have to write things down in such a way that we can apply (1). Working back from the answer, we see that we want

$$Y = 1_{(B_T \in A)}.$$

To check that this leads to the right formula, we observe that since $D(x, \delta) \subset G$,

$$T(\theta_S \omega) = \inf\{r \geq 0 : \omega(S + r) \notin G\} = T - S,$$

so

$$B_T \circ \theta_S = B_T$$

and

$$1_{(B_T \in A)} \circ \theta_S = 1_{(B_T \in A)}.$$

(In words, ω and the shifted path $\theta_S \omega$ must exit G in the same place.)
With the last equality in hand, the rest is easy.

$$\begin{aligned} u(x) &= E_x 1_{(B_T \in A)} = E_x(1_{(B_T \in A)} \circ \theta_S) \\ &= E_x E_x(1_{(B_T \in A)} \circ \theta_S | \mathcal{F}_S), \end{aligned}$$

so applying (1) and recalling the definition of u shows

$$u(x) = E_x u(B_S).$$

An example of a situation that requires the generality of (2) is the following:

Example 2 Let B_t be a one-dimensional Brownian motion, let $a > 0$, and $T_a = \inf\{t : B_t = a\}$. Then

(3) $$P_0(T_a \leq t) = 2P_0(B_t \geq a).$$

We will prove the result in words first and then show how it follows from (2).

Intuitive Proof Clearly $\{B_t \geq a\} \subset \{T_a \leq t\}$. To compute the probability of $\{T_a \leq t, B_t < a\} = \{T_a < t, B_t < a\}$, we observe that if B_s hits a at some time $s < t$, then the strong Markov property implies that $B_t - B(T_a)$ is independent of what happened before T_a. The symmetry of the normal distribution implies that for $s < t$, $P(B_t - B_s < 0) = 1/2$, so we have

(4) $$P_0(T_a < t, B_t < a) = \tfrac{1}{2} P_0(T_a < t),$$

and since $P_0(T_a = t) \leq P_0(B_t = a) = 0$, the result follows.

Proof To make the intuitive proof rigorous, we have to prove (4). To deduce this from (2), we let

$$Y_s(\omega) = \begin{cases} 1 & \text{if } s < t, B_{t-s} < a \\ 0 & \text{otherwise.} \end{cases}$$

We do this so that

$$Y_{T_a}(\theta_{T_a}\omega) = \begin{cases} 1 & \text{if } T_a < t \text{ and } B_t < a \\ 0 & \text{otherwise.} \end{cases}$$

If we let $S = T_a$ in (2), we get

$$E_0(Y_{T_a} \circ \theta_{T_a} | \mathcal{F}_{T_a}) = \varphi(B(T_a), T_a)$$

where

$$\varphi(x, s) = \begin{cases} 0 & \text{if } s \geq t \\ 1/2 & \text{if } s < t, x = a, \end{cases}$$

so

$$E_0(Y_{T_a} \circ \theta_{T_a} | \mathcal{F}_{T_a}) = \tfrac{1}{2} 1_{(T_a < t)}.$$

Taking expected values gives

$$P_0(T_a < t, B_t < a) = E_0(Y_{T_a} \circ \theta_{T_a}) = \tfrac{1}{2} P_0(T_a < t)$$

and proves (4).

The reader can now see why I said in Section 1.4 that the rigorous formulation of the Markov property somewhat obscures the intuitive content. To test his understanding of the intuition, the reader should try some of the following exercises. In each case, B_t is a one-dimensional Brownian motion.

Exercise 1 Let $T_a = \inf\{t \geq 0 : B_t = a\}$.

(i) Use the strong Markov property to conclude that under P_0, $\{T_a, a \geq 0\}$ has stationary independent increments, that is, if $a < b$, $T_b - T_a$ is independent of T_a and has the same distribution as T_{b-a}.

(ii) Let $\varphi_a(\lambda) = E_0 \exp(-\lambda T_a)$. By (i), $\varphi_a(\lambda)\varphi_b(\lambda) = \varphi_{a+b}(\lambda)$, and the Brownian scaling relationship implies $T_a \stackrel{d}{=} a^2 T_1$, that is, $\varphi_a(\lambda) = \varphi_1(\lambda a^2)$. Combine the last two observations with an argument from Section 1.1 to conclude that $\varphi_a(\lambda) = e^{-ca\sqrt{\lambda}}$ for some $c \in (0, \infty)$ (actually $c = \sqrt{2}$).

Exercise 2

(i) Let $T_a^+ = \inf\{t \geq 0 : B_t > a\}$. Use the strong Markov property to conclude that $P_0(T_a = T_a^+) = 1$ for all $a \geq 0$, so $a \to T_a$ is with probability 1, continuous for each fixed a.

(ii) Show that if $s < t$, $P_0(T_a$ is discontinuous at some point $a \in [s, t]) > 0$, and then use scaling and independent increments to conclude that the probability must be 1.

Exercise 3 Let $T = \inf\{t : B_t \notin (a, b)\}$ and $\lambda > 0$.

(i) Use the strong Markov property to conclude that if $x \in (a, b)$,
$$E_x e^{-\lambda T_a} = E_x(e^{-\lambda T}; T_a < T_b) + E_x(e^{-\lambda T}; T_b < T_a) E_b(e^{-\lambda T_a}).$$

(ii) Interchanging the roles of a and b in (i) and using the fact that $E_x e^{-\lambda T_y} = e^{-|y-x|\sqrt{2\lambda}}$ gives us two equations in two unknowns that can be solved to yield
$$E_x(e^{-\lambda T}, T_b < T_a) = \frac{\sinh(\sqrt{2\lambda}(x - a))}{\sinh(\sqrt{2\lambda}(b - a))}$$
$$E_x(e^{-\lambda T}, T_a < T_b) = \frac{\sinh(\sqrt{2\lambda}(b - x))}{\sinh(\sqrt{2\lambda}(b - a))}.$$

Exercise 4 Let $R_t = \inf\{u > t : B_u = 0\}$ and let $\tau = R_0$. In Section 1.3 we showed that $P_0(\tau = 0) = 1$. Use this fact and the strong Markov property to conclude that $Z(\omega) = \{t : B_t(\omega) = 0\}$ is a closed set that has no isolated points and hence must be uncountable (for the last step, see Hewitt and Stromberg (1969), page 72).

1.6 Martingale Properties of Brownian Motion

In the first half of this chapter we concentrated on the Markov properties of Brownian motion. In this section we describe some of its martingale properties. We start with the one-dimensional case:

(1) B_t is a martingale.

Note: To be precise we should say that for all $x \in R^d$, B_t (that is, the coordinate maps on $(C, \mathscr{C}, P_x)$) is a martingale w.r.t. $\mathscr{F}_t$ (the σ-fields defined in Section 1.4), but this is too rigorous, so we will stick to the casual statement here and below.

Proof The Markov property implies that

$$E_x(B_t|\mathscr{F}_s) = E_{B(s)}(B_{t-s}) = B_s,$$

since the symmetry of the normal distribution implies that $E_y B_u = y$ for all $u \geq 0$.

The next martingale is not as obvious as the first, but, as we will see in Chapter 2, it is just as important.

(2) $B_t^2 - t$ is a martingale.

The proof is a simple computation:

$$\begin{aligned} E_x(B_t^2|\mathscr{F}_s) &= E_x(B_s^2 + 2B_s(B_t - B_s) + (B_t - B_s)^2|\mathscr{F}_s) \\ &= B_s^2 + 2B_s E_x(B_t - B_s|\mathscr{F}_s) + E_x((B_t - B_s)^2|\mathscr{F}_s) \\ &= B_s^2 + 0 + (t - s), \end{aligned}$$

since $B_t - B_s$ is independent of $\mathscr{F}_s$ and has mean 0 and variance $t - s$.

(1) and (2) generalize immediately to B_t^i, the ith component of a d-dimensional Brownian motion. Repeating the proof of (2) shows

(3) If $i \neq j$, $B_t^i B_t^j$ is a martingale.

Proof

$$E_x(B_t^i B_t^j|\mathscr{F}_s) = B_s^i B_s^j + B_s^i E_x(B_t^j - B_s^j|\mathscr{F}_s) + B_s^j E_x(B_t^i - B_s^i|\mathscr{F}_s) \\ + E_x((B_t^i - B_s^i)(B_t^j - B_s^j)|\mathscr{F}_s).$$

Since $B_t^i - B_s^i$, $B_t^j - B_s^j$, and $\mathscr{F}_s$ are independent, the last three terms are 0.

(1), (2), and (3) are special cases of the following result:

(4) If $f \in C^2$ (i.e., the second-order partial derivatives of f are continuous) and f and Δf are bounded, then

$$f(B_t) - \int_0^t \frac{1}{2}\Delta f(B_s)\,ds \text{ is a martingale.}$$

We will prove this result in Chapter 2, but unfortunately we want to use it in the next section and we do not want to give a direct proof now. Stuck in this position, we will engage in the somewhat undesirable approach of using (4), or more precisely the following corollary of (4), before we prove it.

(5) Let G be a bounded open set and $\tau = \inf\{t : B_t \notin G\}$. If $f \in C^2$ and $\Delta f = 0$ in G, then $f(B_{t \wedge \tau})$ is a martingale.

If the reader wants to insist on a strictly logical development, then he should read Chapter 2 and then come back to read the last part of Chapter 1. I think this is unnecessary, though. It is easy to see that the results in Sections 1.7–1.10 are not dependent on the proof of Itô's formula (which will be given in Section 2.8).

One of the nice things about martingales is that they allow us to compute various quantities associated with Brownian motion. We will see a number of instances of this later, especially in the next section. The exercises below give a number of other applications.

Exercise 1 Let B_t be a d-dimensional Brownian motion and let $T = \inf\{t : |B_t| = r\}$. Use the fact that $|B_t|^2 - td$ is a martingale to conclude that if $|x| < r$, $E_x T = (r^2 - |x|^2)/d$.

Exercise 2 If u is harmonic (i.e., $\Delta u = 0$) and bounded, then (4) implies $u(B_t)$ is a martingale. Combine this observation with the martingale convergence theorem and (9) of Section 1.3 to prove Liouville's theorem: Any bounded harmonic function is constant.

Exercise 3 *The Exponential Martingale.* Let B_t be a one-dimensional Brownian motion.

(i) Use the fact that $E_0(\exp(\theta B_t)) = \exp(\theta^2 t/2)$ to prove that $\exp(\theta B_t - \theta^2 t/2)$ is a martingale.

(ii) Let $a > 0$ and $T = \inf\{t : B_t = a + bt\}$. If $b \le 0$ and $\theta \ge 0$, or if $b > 0$ and $\theta \ge 2b$, then the martingale in (i) is bounded, so the optional stopping theorem can be applied to conclude

$$1 = E_0\left(\exp\left(\theta B_T - \frac{\theta^2 T}{2}\right); T < \infty\right)$$

$$= E_0\left(\exp\left(\theta a + \theta b T - \frac{\theta^2 T}{2}\right); T < \infty\right).$$

If $\lambda > 0$, setting $\theta = b + (b^2 + 2\lambda)^{1/2}$ and solving gives

$$E_0(e^{-\lambda T}) = e^{-a(b+(b^2+2\lambda)^{1/2})}.$$

Letting $\lambda \to 0$ gives

$$P_0(T < \infty) = \begin{cases} 1 & b \le 0 \\ e^{-2ab} & b > 0. \end{cases}$$

1.7 Hitting Probabilities, Recurrence, and Transience

In this section we prove some results concerning the range of Brownian motion $\{B_t : t \ge 0\}$. We start with the one-dimensional case. Let $a < x < b$ and $T = \inf\{t : B_t \notin (a, b)\}$. Since B_t is a martingale, it is easy to guess the distribution of B_T:

(1) $$P_x(B_T = a) = \frac{b - x}{b - a}$$

$$P_x(B_T = b) = \frac{x - a}{b - a}.$$

(This is the only probability distribution with support $\{a, b\}$ and mean x.)

Proof The first step is to show that $T < \infty$, P_x a.s. Observe that if $y \in (a,b)$,
$$P_y(T > 1) \le P_y(B_1 \in (a,b)) \le P_0(|B_1| \le (b-a)) < 1,$$
so $\rho = \sup_y P_y(T > 1) < 1$, and it follows from the Markov property that $P_x(T > n) \le \rho^n \to 0$ as $n \to \infty$.

B_t is a martingale, T is a stopping time, and $B_t \in (a,b)$ for $t < T$, so we can apply the optional stopping theorem to conclude
$$x = E_x B_T = a P_x(B_T = a) + b P_x(B_T = b).$$
Since $P_x(B_T = a) + P_x(B_T = b) = 1$, the expression above can be rewritten as
$$x - a = (-a + b) P_x(B_T = b),$$
so
$$P_x(B_T = b) = \frac{x-a}{b-a},$$
proving the result.

Let $T_x = \inf\{t : B_t = x\}$. From (1), it follows immediately that

(2)   For all x, $P_0(T_x < \infty) = 1$.

Proof Suppose $x > 0$. $P_0(T_x < T_{-Mx}) = M/M+1$, and the right-hand side approaches 1 as $M \to \infty$.

It is trivial to improve (2) to conclude that

(3)   For any $s < \infty$, $P_0(B_t = x \text{ for some } t \ge s) = 1$.

Proof By the Markov property,
$$P_0(B_t = x \text{ for some } t \ge s) = E_0(P_{B(s)}(T_x < \infty)) = 1.$$

The conclusion of (3) implies that with probability 1 there is a sequence of times $t_n \uparrow \infty$ (which will depend on the outcome ω) so that $B_{t_n} = x$ (a conclusion we will hereafter abbreviate as "$B_t = x$ infinitely often" or "$B_t = x$ i.o."), so in the terminology of the theory of Markov processes, one-dimensional Brownian motion is recurrent.

In order to study Brownian motion in $d \ge 2$, we need to generalize (1). In view of (5) in Section 1.6 and the spherical symmetry of Brownian motion, an obvious way to do this is to let $\varphi(x) = f(|x|^2)$ and try to pick f so that $\Delta \varphi = 0$. A little differentiation gives
$$D_i f(|x|^2) = f'(|x|^2) 2x_i$$
$$D_{ii} f(|x|^2) = f''(|x|^2) 4x_i^2 + 2f'(|x|^2).$$
(Now you see why we wrote $f(|x|^2)$.) Therefore, for $\Delta \varphi = 0$ we need
$$0 = \sum_i (f''(|x|^2) 4x_i^2 + 2f'(|x|^2))$$
$$= 4|x|^2 f''(|x|^2) + 2df'(|x|^2).$$

1.7 Hitting Probabilities, Recurrence, and Transience

Letting $y = |x|^2$, we can write the above as
$$4yf''(y) + 2df'(y) = 0,$$
or, if $y > 0$,
$$f''(y) = \frac{-d}{2y} f'(y).$$

From the last equation we see $f'(y) = Cy^{-d/2}$ guarantees $\Delta\varphi = 0$ for $x \neq 0$, so we can let
$$\varphi(x) = \log|x| \quad d = 2$$
$$\varphi(x) = |x|^{2-d} \quad d \geq 3.$$

We are now ready to imitate the proof of (1) in $d \geq 2$. Let $S_r = \inf\{t : |B_t| = r\}$ and $r < R$. Since φ is bounded and has $\Delta\varphi = 0$ in $\{x : r < |x| < R\}$, applying the optional stopping theorem at $T = S_r \wedge S_R$ gives
$$\varphi(x) = E_x \varphi(B_T) = \varphi(r) P(S_r < S_R) + \varphi(R)(1 - P(S_r < S_R)),$$
and solving gives

(4) $$P_x(S_r < S_R) = \frac{\varphi(R) - \varphi(x)}{\varphi(R) - \varphi(r)}.$$

In $d = 2$, the last formula says

(5) $$P_x(S_r < S_R) = \frac{\log R - \log|x|}{\log R - \log r}.$$

If we fix r and let $R \to \infty$ in (5), the right-hand side goes to 1; so $P_x(S_r < \infty) = 1$ for $x, r > 0$, and repeating the proof of (3) shows that two-dimensional Brownian motion is recurrent in the sense that if G is any open set, then $P_x(B_t \in G \text{ i.o.}) \equiv 1$.

If we fix R, let $r \to 0$ in (5), and let $S_0 = \inf\{t > 0 : B_t = 0\}$, then for $x \neq 0$
$$P_x(S_0 < S_R) \leq \lim_{r \to 0} P_x(S_r < S_R) = 0.$$

Since this holds for all R and since the continuity of Brownian paths implies $S_R \uparrow \infty$ as $R \uparrow \infty$, we have $P_x(S_0 < \infty) = 0$ for all $x \neq 0$. The strong Markov property implies
$$P_0(B_t = 0 \text{ for some } t \geq \varepsilon) = E_0[P_{B(\varepsilon)}(T_0 < \infty)] = 0$$
for all $\varepsilon > 0$, so $P_0(B_t = 0 \text{ for some } t > 0) = 0$, and thanks to our definition of S_0 as $\inf\{t > 0 : B_t = 0\}$, we also have $P_x(S_0 < \infty) = 0$ for $x = 0$. In $d \geq 2$, Brownian motion will not hit 0 at a positive time even if it starts there.

For $d \geq 3$, formula (4) says

(6) $$P_x(S_r < S_R) = \frac{R^{2-d} - |x|^{2-d}}{R^{2-d} - r^{2-d}}.$$

If we fix r and let $R \to \infty$ in (6), the right-hand side approaches $(x/r)^{2-d} < 1$; so if $|x| > r$,

(7) $$P_x(S_r < \infty) = (r/|x|)^{d-2} < 1.$$

From the last result it follows easily that for $d \geq 3$, Brownian motion is "transient."

(8) As $t \to \infty$, $|B_t| \to \infty$ a.s.

Proof The strong Markov property implies
$$P_x(|B_t| \leq M^{1/2} \text{ for some } t \geq S_M) = E_x(P_{B(S_M)}(S_{M^{1/2}} < \infty))$$
$$= (M^{1/2}/M)^{d-2} \to 0$$
as $M \to \infty$.

At this point, we have derived the basic facts about the recurrence and transience of Brownian motion. To review for a moment, what we have found is that

(i) $P_x(|B_t| < 1 \text{ for some } t \geq 0) \equiv 1$ iff $d \leq 2$
(ii) $P_x(B_t = 0 \text{ for some } t > 0) \equiv 0$ in $d \geq 2$.

The reader should observe that these facts can be traced to properties of what we have called φ, the (unique up to linear transformations) spherically symmetric function that has $\Delta\varphi(x) = 0$ for all $x \neq 0$, that is:

(9) $$\varphi(x) = \begin{cases} |x| & d = 1 \\ \log|x| & d = 2 \\ |x|^{2-d} & d \geq 3 \end{cases}$$

and the features relevant for (i) and (ii) above are

(i) $\varphi(x) \to \infty$ as $|x| \to \infty$ iff $d \leq 2$
(ii) $\varphi(x) \to \infty$ as $x \to 0$ in $d \geq 2$.

Exercise 1 Generalize the reasoning used in the first part of the proof of (1) to show that if A is a set with $|A| < \infty$ ($|A|$ = Lebesgue measure of A) and $\tau = \inf\{t : B_t \notin A\}$, then there is an $\varepsilon > 0$ (that depends only on $|A|$) so that $E_x \exp(\varepsilon\tau) < \infty$ for all $x \in R^d$.

1.8 The Potential Kernels

If f is a nonnegative function, then
$$E_x \int_0^\infty f(B_t)\,dt = \int_0^\infty E_x f(B_t)\,dt$$
$$= \int_0^\infty \int p_t(x,y) f(y)\,dy\,dt$$
$$= \int \int_0^\infty p_t(x,y)\,dt\, f(y)\,dy$$

1.8 The Potential Kernels

where $p_t(x,y) = (2\pi t)^{-d/2} e^{-|x-y|^2/2t}$ is the transition density for Brownian motion. As $t \to \infty$, $p_t(x,y) \sim (2\pi t)^{-d/2}$, so if $d \le 2$, then $\int p_t(x,y)\,dt = \infty$. When $d \ge 3$, changing variables $t = |x-y|^2/2s$ gives

(1)
$$\int_0^\infty p_t(x,y)\,dt = \int_\infty^0 \left(\frac{s}{\pi|x-y|^2}\right)^{d/2} e^{-s}\left(-\frac{|x-y|^2}{2s^2}\right) ds$$

$$= |x-y|^{2-d}\pi^{-d/2}\frac{1}{2}\int_0^\infty s^{(d/2)-2}e^{-s}\,ds$$

$$= \frac{\Gamma(\frac{d}{2}-1)}{2\pi^{d/2}}|x-y|^{2-d}$$

where $\Gamma(\alpha) = \int_0^\infty s^{\alpha-1}e^{-s}\,ds$ is the usual gamma function, so if we define

$$G(x,y) = \int_0^\infty p_t(x,y)\,dt,$$

then $G(x,y) < \infty$ for $x \ne y$, and

$$E_x \int_0^\infty f(B_t)\,dt = \int G(x,y)f(y)\,dy.$$

We call $G(x,y)$ the potential kernel, because it will turn out (see Section 8.6) that $G(\cdot,y)$ is the potential of a unit charge at y. At the moment we are not prepared to discuss this, so we will only say that (1) is a useful formula for Brownian motion, and with an eye on applications in the next section and later on, we will define the potential kernels for $d \le 2$ by

$$G(x,y) = \int_0^\infty p_t(x,y) - a_t\,dt$$

where the a_t are constants we will choose to make the integral converge (at least when $x \ne y$).

When $d = 1$, we let $a_t = p_t(0,0)$. With this choice,

$$G(x,y) = \frac{1}{\sqrt{2\pi}}\int_0^\infty (e^{-(y-x)^2/2t} - 1)t^{-1/2}\,dt$$

and the integral converges, since the integrand is ≤ 0 and $\sim -(y-x)^2/2t^{3/2}$ as $t \to \infty$. Changing variables $u = (y-x)^2/2t$ gives

(2)
$$G(x,y) = \frac{1}{\sqrt{2\pi}}\frac{-(y-x)^2}{2}\int_\infty^0 (e^{-u}-1)\left(\frac{2u}{(y-x)^2}\right)^{1/2}\frac{du}{u^2}$$

$$= \frac{-|y-x|}{2\sqrt{\pi}}\int_0^\infty \left(\int_0^u e^{-s}\,ds\right) u^{-3/2}\,du$$

$$= \frac{-|y-x|}{\sqrt{\pi}}\int_0^\infty ds\, e^{-s}\int_s^\infty du\,\frac{1}{2}u^{-3/2}$$

$$= \frac{-|y-x|}{\sqrt{\pi}}\int_0^\infty ds\, e^{-s}s^{-1/2} = -|y-x|.$$

The computation is almost the same for $d = 2$. The only thing that changes is the choice of a_t. If we try $a_t = p_t(0,0)$ again, then for $x \ne y$ the integrand $\sim -t^{-1}$ as $t \to 0$ and the integral diverges (for the wrong reason), so we let $a_t = p_t(0, e_1)$ where $e_1 = (1, 0)$. With this choice of a_t, we get

(3)
$$G(x,y) = \frac{1}{2\pi} \int_0^\infty (e^{|x-y|/2t} - e^{-1/2t}) t^{-1} \, dt$$

$$= \frac{1}{2\pi} \int_0^\infty \left(\int_{|x-y|^2/2t}^{1/2t} e^{-s} \, ds \right) t^{-1} \, dt$$

$$= \frac{1}{2\pi} \int_0^\infty ds \, e^{-s} \int_{|x-y|^2/2s}^{1/2s} t^{-1} \, dt$$

$$= \frac{1}{2\pi} \left(\int_0^\infty ds \, e^{-s} \right) (-\log(|x-y|^2)) = \frac{-1}{\pi} \log(|x-y|).$$

To sum up, the potential kernels are given by

$$\frac{\Gamma(\frac{d}{2} - 1)}{2\pi^{d/2}} |x - y|^{2-d} \quad d \geq 3$$

$$-\frac{1}{\pi} \cdot \log(|x - y|) \quad d = 2$$

$$-1 \cdot |x - y| \quad d = 1.$$

The reader should note that in each case, $G(x,y) = C\varphi(|x - y|)$ where φ is the harmonic function we used in Section 1.7. This is, of course, no accident. $x \to G(x, 0)$ is obviously spherically symmetric and, as we will see in Section 8.6, satisfies $\Delta G(x, 0) = 0$ for $x \ne 0$, so the results above imply $G(x, 0) = A + B\varphi(|x|)$.

The formulas above correspond to $A = 0$, which is nice and simple. But what about the weird looking B's? What is special about them? The answer is simple: They are chosen to make $\frac{1}{2} \Delta G(x, 0) = -\delta_0$ (a point mass at 0) in the distributional sense. It is easy to see that this happens in $d = 1$:

$$\varphi'(x) = \begin{cases} -1 & x > 0 \\ 1 & x < 0, \end{cases}$$

so

$$\varphi''(x) = -2\delta_0.$$

More sophisticated readers can easily check that this is also true in $d \geq 2$. (See F. John (1982), pages 96–97.)

1.9 Brownian Motion in a Half Space

Let $H = \{x \in R^d : x_d > 0\}$ be the upper half space and let $\tau = \inf\{t : B_t \notin H\}$. For several applications below, we will need to know about the behavior of

1.9 Brownian Motion in a Half Space

$B(t \wedge \tau)$—Brownian motion "killed when it leaves H." In this section, we will derive some of the basic formulas concerning this process.

Let $x \in R^{d-1}$ and $y > 0$. Since $\tau = \inf\{t > 0 : B_t^d = 0\}$, it follows from (3) of Section 1.5 and obvious symmetries of Brownian motion that

$$P_{(x,y)}(\tau \leq t) = 2P(B_t \geq y) = 2 \int_y^\infty (2\pi t)^{-1/2} e^{-z^2/2t} \, dz.$$

To find the probability density of τ, we change variables $z = (t^{1/2}y)/s^{1/2}$ to obtain

(1)
$$P_{(x,y)}(\tau \leq t) = 2 \int_t^0 (2\pi t)^{-1/2} e^{-y^2/2s} \left(\frac{-t^{1/2}y}{2s^{3/2}} \right) ds$$

$$= \int_0^t (2\pi s^3)^{-1/2} y e^{-y^2/2s} \, ds.$$

Since the exit time depends only on the last coordinate, it is independent of the first $d - 1$ coordinates and we can compute the distribution of B_τ by writing

$$P_{(x,y)}(B_\tau = (\theta, 0)) = \int_0^\infty ds \, P_{(x,y)}(\tau = s)(2\pi s)^{-(d-1)/2} e^{-|x-\theta|^2/2s}$$

$$= \frac{y}{(2\pi)^{d/2}} \int_0^\infty ds \, s^{-(d+2)/2} e^{-(|x-\theta|^2+y^2)/2s}.$$

Changing variables $s = (|x - \theta|^2 + y^2)/2t$ gives

$$\frac{y}{(2\pi)^{d/2}} \int_\infty^0 \frac{-(|x - \theta|^2 + y^2)}{2t^2} dt \left(\frac{2t}{|x - \theta|^2 + y^2} \right)^{(d+2)/2} e^{-t},$$

so we have

(2)
$$P_{(x,y)}(B_\tau = (\theta, 0)) = \frac{y}{(|x - \theta|^2 + y^2)^{d/2}} \frac{\Gamma(d/2)}{\pi^{d/2}},$$

where $\Gamma(\alpha) = \int_0^\infty y^{\alpha - 1} e^{-y} \, dy$ is the usual gamma function.

When $d = 2$ and $x = 0$, probabilists should recognize this as a Cauchy distribution. At first glance, the fact that B_τ has a Cauchy distribution might be surprising, but a moment's thought reveals that this must be true. If we let $B_0 = 0$, $T_s = \inf\{t \geq 0 : B_t^2 = s\}$, and $C_s = B^1(T_s)$, then the strong Markov property and spatial homogeneity of Brownian motion imply that C_s has stationary independent increments. An obvious scaling argument implies $C_s \stackrel{d}{=} sC_1$, and symmetry implies $C_s \stackrel{d}{=} -C_s$, so if we let $\varphi_\theta(s) = E(\exp(i\theta C_s))$, then the observations above imply that $\varphi_\theta(s)\varphi_\theta(t) = \varphi_\theta(s + t)$, $\varphi_\theta(s) = \varphi_{\theta s}(1)$, and $\varphi_\theta(s) = \varphi_{-\theta}(s)$. Since $\theta \to \varphi_\theta(1)$ is continuous, the second equation implies $s \to \varphi_\theta(s) = \varphi_{\theta s}(1)$ is continuous, and a simple argument (details left to the reader) shows that for each θ, $\varphi_\theta(s) = \exp(c_\theta s)$ and the last two equations imply that $c_\theta = -a|\theta|$, so C_s has a Cauchy distribution.

The representation of the Cauchy process given above allows us to justify the remark made in Section 1.1 that the set of discontinuities of C_t is dense in $[0, \infty)$. We will simply outline the steps and leave the details to the reader.

Exercise 1

(i) With probability 1,
$$C = \sup_{0 \le t \le 1} B_t > B_1,$$
so $s \to T_s$ is discontinuous at $s = C$.

(ii) If $a < b$, then by scaling, $P_0(s \to T_s$ is discontinuous in $[a, b])$ is independent of the value of $b - a$ and hence must be $\equiv 1$.

The discussion above has focused on how B_t leaves H. The rest of the section is devoted to studying where it goes before it leaves H. We begin with the case $d = 1$.

(3)   If $x, y > 0$, then
$$P_x(B_t = y, T_0 > t) = p_t(x, y) - p_t(x, -y)$$
where
$$p_t(x, y) = (2\pi t)^{-d/2} e^{|y-x|^2/2t}.$$

Proof The proof is a simple extension of the argument we used in Section 1.5 to prove that $P_0(T_a \le t) = 2P_0(B_t \ge a)$.

Let $f \ge 0$ and $f(x) = 0$ for $x \le 0$. Clearly
$$E_x(f(B_t); T_0 > t) = E_x f(B_t) - E_x(f(B_t); T_0 \le t).$$
If we let $\bar{f}(x) = f(-x)$, then it follows from the strong Markov property and symmetry of Brownian motion that
$$E_x(f(B_t); T_0 \le t) = E_x[E_0 f(B_{t-T_0}); T_0 \le t]$$
$$= E_x[E_0 \bar{f}(B_{t-T_0}); T_0 \le t]$$
$$= E_x[\bar{f}(B_t); T_0 \le t] = E_x(\bar{f}(B_t)),$$
since $\bar{f}(y) = 0$ for $y \ge 0$. Combining this with the first equality shows
$$E_x(f(B_t); T_0 > t) = E_x f(B_t) - E_x \bar{f}(B_t)$$
$$= \int (p_t(x, y) - p_t(x, -y)) f(y) \, dy,$$
proving (3).

The last formula generalizes easily to $d \ge 2$.

(4)   If $x, y \in H$,
$$P_x(B_t = y, \tau > t) = p_t(x, y) - p_t(x, \bar{y})$$

1.9 Brownian Motion in a Half Space

where

$$\bar{y} = (y_1, \ldots, y_{d-1}, -y_d).$$

For some of the developments below, we will need the following formula, which is an easy consequence of (4), and the results in Section 1.8.

(5) If $x \in H$ and G is the potential kernel defined in Section 1.5, then

$$E_x\left(\int_0^\tau f(B_t)\,dt\right) = \int G(x,y)f(y)\,dy - \int G(x,\bar{y})f(y)\,dy$$

whenever the two integrals on the right-hand side are finite.

Proof

$$E_x \int_0^\tau f(B_t)\,dt = \int_0^\infty E_x(f(B_t); \tau > t)\,dt$$

$$= \int_0^\infty \int_H (p_t(x,y) - p_t(x,\bar{y}))f(y)\,dy$$

$$= \int_0^\infty \int_H (p_t(x,y) - a_t)f(y)\,dy - \int_0^\infty \int_H (p_t(x,\bar{y}) - a_t)f(y)\,dy$$

$$= \int G(x,y)f(y)\,dy - \int G(x,\bar{y})f(y)\,dy.$$

(The last two equalities being justified by the fact that $\int |G(x,y)f(y)|\,dy$ and $\int |G(x,\bar{y})f(y)|\,dy$ are finite.)

Remark: The proof given above simplifies considerably in the case $d \geq 3$; however, part of the point of the proof above is that, with the definition we have chosen for G in the recurrent case, the formulas and proofs can be the same for all d.

Let $G_H(x,y) = G(x,y) - G(x,\bar{y})$. We think of $G_H(x,y)$ as the "expected occupation time (density) at y for a Brownian motion starting at x and killed when it leaves H." The rationale for this interpretation is that

$$E_x\left(\int_0^\tau f(B_t)\,dt\right) = \int G_H(x,y)f(y)\,dy$$

whenever the right-hand side exists in the sense specified in (5).

With this interpretation for G_H introduced, the reader should pause for a minute and imagine what $y \to G_H(x,y)$ looks like in one dimension. If you don't already know the answer, you will probably not guess the behavior as $y \to \infty$. So much for small talk. The computation is easier than guessing the answer:

$$G(x,y) = -|x-y|,$$

so

$$G_H(x,y) = -|x-y| + |x+y|.$$

Separating things into cases, we see that

$$G_H(x,y) = \begin{cases} -(x-y) + (x+y) = 2y & \text{when } 0 < y < x \\ -(y-x) + (x+y) = 2x & \text{when } x < y, \end{cases}$$

so we can write

$$G_H(x,y) = 2(x \wedge y) \quad \text{for all } x, y > 0.$$

It is somewhat surprising that $y \to G_H(x,y)$ is constant $= 2x$ for $y \geq x$, that is, all points $y > x$ have the same expected occupation time!

1.10 Exit Distributions for the Sphere

Having dealt with the half space, the next object to contemplate is the sphere. Let $D = \{x : |x| < 1\}$ and $\tau = \inf\{t : B_t \notin D\}$. In the last section we computed the exit distribution and occupation density for the half space. The key to the second computation was reflecting points across ∂H. Eventually (in Section 3.4) we will show there is a reflection function for the sphere (namely, inversion $x \to x/|x|^2$) that allows us to compute things for D as we did for H. At this point this tool is not available, so we will forget about the occupation time density for the moment and we will cheat to find the exit distribution: We will look up the answer in Port and Stone (1978) and verify that their formula is correct.

(1)   If f is bounded, then

$$E_x f(B_\tau) = \int_{\partial D} \frac{1 - |x|^2}{|x - y|^d} f(y) \, d\pi(y)$$

where $\pi =$ surface measure on ∂D normalized to be a probability measure.

Proof To prove that our "guess" is right, we begin by proving a version of (5) in Section 1.6 that is easier to work with.

(2)   Suppose u is C^2 in D and continuous on $\bar{D}$. If $\Delta u = 0$ in D, then $u(x) = E_x u(B_\tau)$.

Proof Let $\tau_n = \inf\{t : B_t \notin D(0, 1 - 1/n)\}$. Since $\tau_n \uparrow$ and $|B(\tau_n)| \geq 1 - 1/n$, it is easy to see that $\tau_n \uparrow \tau$ a.s. From (5) in Section 1.6, it follows that $u(x) = E_x u(B_{\tau_n})$. Letting $n \to \infty$ and using the bounded convergence theorem proves (2).

In light of (2), we will be finished when we verify that the right-hand side of (1) has the indicated properties whenever f is C^∞. The first, somewhat painful, step in doing this is to show that for fixed y,

(3) $k_y(x) = \dfrac{1 - |x|^2}{|x - y|^d}$ is harmonic in D.

1.10 Exit Distributions for the Sphere

To warm up for this, we observe

$$D_i |x-y|^p = D_i \left(\sum_j (x_j - y_j)^2 \right)^{p/2} = p|x-y|^{p-2}(x_i - y_i),$$

so we have

$$D_i k_y(x) = (-2x_i) \cdot \frac{1}{|x-y|^d} + (1-|x|^2) \cdot \frac{-d(x_i - y_i)}{|x-y|^{d+2}}.$$

Differentiating again gives

$$D_{ii} k_y(x) = (-2) \cdot \frac{1}{|x-y|^d} + 2 \cdot (-2x_i) \cdot \frac{-d(x_i - y_i)}{|x-y|^{d+2}}$$

$$+ (1-|x|^2)\left(\frac{d(d+2)(x_i - y_i)}{|x-y|^{d+4}} - \frac{d}{|x-y|^{d+2}} \right).$$

Summing the last expression on i gives

$$\Delta k_y(x) = \frac{-2d}{|x-y|^d} + 4d \frac{|x|^2 - x \cdot y}{|x-y|^{d+2}} + \frac{(d^2 + 2d)(1-|x|^2)}{|x-y|^{d+2}} - \frac{d^2(1-|x|^2)}{|x-y|^{d+2}}$$

$$= \frac{-2d|x-y|^2}{|x-y|^{d+2}} + 4d \frac{|x|^2 - x \cdot y}{|x-y|^{d+2}} + 2d \frac{(1-|x|^2)}{|x-y|^{d+2}},$$

and if we replace 1 by $|y|^2$, the expression collapses

$$\Delta k_y(x) = \frac{2d}{|x-y|^{d+2}}(-|x-y|^2 + 2|x|^2 - 2x \cdot y + |y|^2 - |x|^2) = 0,$$

since $|x-y|^2 = |x|^2 - 2x \cdot y + |y|^2$.

Let $f \in C^\infty$ and let

$$u(x) = \begin{cases} \int_{\partial D} d\pi(y) f(y) k_y(x) & x \in D \\ f(x) & x \in \partial D. \end{cases}$$

In D, u is a linear combination of the k_y, so bringing the differentiation under the integral (and leaving it to the reader to justify this in an exercise below) gives

$$\Delta u(x) = \int_{\partial D} d\pi(y) f(y) \Delta k_y(x) = 0,$$

so u satisfies the conditions of (2) that concern its behavior in D.

To check the behavior of u at ∂D, the first step is to show

$$I(x) = \int_{\partial D} d\pi(y) \frac{1 - |x|^2}{|x-y|^d} \equiv 1.$$

This is just a calculus "exercise," but since it is a rather difficult one, we will use a soft noncomputational approach instead. Applying the results above with $f \equiv 1$, we see $\Delta I = 0$ in D and I is invariant under rotations, so starting a Brownian motion at 0 and applying (5) of Section 1.6 with $G = D(0,r), r < 1$, shows $I(x) = I(0) = 1$ for all $x \in D$.

To show that $u(x) \to f(y)$ as $x \to y \in \partial D$, we observe that if $z \neq y$, $k_z(y) \to 0$, and if $\delta > 0$, the convergence is uniform for $z \in B_0 = \partial D - D(y, \delta)$, so if we let $B_1 = \partial D - B_0$, then

$$\int_{B_1} h_z(x) f(z) \, d\pi(z) \to 0$$

and

$$\int_{B_2} h_z(x) \, d\pi(z) \to 1.$$

from which it follows easily that $u(x) \to f(y)$, and we have established that (3) is correct.

The derivation given above is a little unsatisfying, since it starts with the answer and then verifies it, but it is simpler than messing around with Kelvin's transformations (see pages 100–103 in Port and Stone (1978) or Section 3.4 below), and it also has the merit, I think, of explaining why $k_y(x)$ is the probability density of exiting at y: k_y is a nonnegative harmonic function that has $k_y(0) = 1$ and $k_y(x) \to 0$ when $x \to x \in \partial D$ and $z \neq y$.

Exercise 1 Show that the functions

$$h_\theta(x, y) = \frac{y}{(|x - \theta|^2 + y^2)^{d/2}}$$

are harmonic in $H = \{(x, y) : x \in R^{d-1}, y > 0\}$ and satisfy

(i) $\int dx \, h_\theta(x, y) = 1$
(ii) as $y \to 0$, $\int_{D(\theta, \varepsilon)^c} dx \, h_\theta(x, y) \to 0$ for all $\varepsilon > 0$.

Exercise 2 *Differentiating under the Integral Sign.* It is easiest to do the proof and then decide what we need to assume.
Suppose

$$u(x) = \int_S K(x, y) f(y) \, dm(y)$$

and write

$$u(x + he_i) - u(x) = \int_S (K(x + he_i, y) - K(x, y)) f(y) \, dm(y)$$

$$= \int_S \int_0^h \frac{\partial K}{\partial x_i}(x + \theta e_i, y) f(y) \, dm(y).$$

From the last expression, it follows easily that we have

(4) Suppose $u_i(x) = \int \frac{\partial K}{\partial x_i}(x, y) f(y) \, dm(y)$ is continuous in an open set G and that for some $h > 0$,

$$\int_S \int_0^h \left| \frac{\partial K}{\partial x_i}(x + \theta e_i, y) f(y) \right| d\theta \, dm(y) < \infty.$$

Then $\partial u / \partial x_i$ exists and equals u_i.

1.11 Occupation Times for the Sphere

In the last section we considered the exit distributions, that is, how B_t leaves D. To continue our development of D in parallel with that of H, we now consider where B_t goes before it leaves D. Let $\tau = \inf\{t : B_t \notin D\}$ and f be a bounded function. By analogy with (5) in Section 1.9, we should expect

(1) $$E_x \int_0^\tau f(B_t) \, dt = \int G_D(x, y) f(y) \, dy.$$

In this section we will show that this is correct and find an explicit formula for the "Green's function" G_D in the same way that we found the exit distributions in the last section. We will find a property that characterizes G_D, look up the answer, and verify that it is correct.

For simplicity, we begin with the case $d \geq 3$. In this case, if $G(x, y)$ is the potential kernel defined in Section 1.8 and we suppose $f \equiv 0$ on D^c, then

$$\int G(x, y) |f(y)| \, dy \leq \int_D \frac{c}{|x - y|^d} \|f\|_\infty \, dy < \infty$$

where $c = \Gamma(d/2 - 1)/2\pi^{d/2}$, so Fubini's theorem can be applied to conclude that

$$w(x) \equiv \int G(x, y) f(y) \, dy = E_x \int_0^\infty f(B_t) \, dt.$$

The strong Markov property implies that

$$E_x \int_\tau^\infty f(B_t) \, dt = E_x E_{B(\tau)} \int_0^\infty f(B_t) \, dt = E_x w(B_\tau)$$

$$= \int_{\partial D} \frac{1 - |x|^2}{|x - y|^d} w(y) \, d\pi(y).$$

Recalling the definition of w and using Fubini's theorem gives

$$E_x \int_0^\tau f(B_t) \, dt = w(x) - E_x w(B_\tau)$$

$$= \int \left(G(x, z) - \int \frac{1 - |x|^2}{|x - y|^d} G(y, z) \, d\pi(y) \right) f(z) \, dz,$$

so (1) holds in $d \geq 3$ with

(2) $$G_D(x,z) = G(x,z) - \int \frac{1-|x|^2}{|x-y|^d} G(y,z)\, d\pi(y).$$

At this point, it "only" remains to do the integral. This is easy if $z=0$, for then
$$G(y,z) = \frac{C}{|y|^{d-2}} = C \quad \text{for} \quad y \in \partial D,$$
and we have

(3a) $$G_D(x,0) = \frac{C}{|x|^{d-2}} - C,$$

since, by results in the last section,
$$\int \frac{1-|x|^2}{|x-y|^d}\, d\pi(y) = 1.$$

When $z \neq 0$, however, $G(y,z)$ is not constant on ∂D and we are left with a difficult integral to do. In Section 3.4 we will see that knowing what inversion $x \to x/|x|^2$ does to Brownian motion makes it possible to evaluate this integral, but since that is not available now, we will have to resort to more underhanded means to evaluate the integral. We will look up what we know to be the answer in Folland (1976) and then verify that it is correct. Although this approach is morally reprehensible (we are using analysis to prove something in probability!), it has the advantage of demonstrating that $G_D(x,y)$ is nothing more than the Green's function for D with Dirichlet boundary conditions.

Folland (1976), page 109, defines the Green's function for D to be the function $K(x,y)$ on $D \times D$ determined by the following properties:

(i) for each $y \in D$, $K(\cdot,y) - G(\cdot,y)$ is harmonic in D and continuous on $\bar{D}$
(ii) for each $y \in D$, $x \in \partial D$, $K(x,y) = 0$.

Note: For convenience, we have changed his notation to conform to ours and interchanged the roles of x and y. This interchange makes no difference, since Green's function is symmetric, that is, $K(x,y) = K(y,x)$. (See Folland, page 110.)

It is easy to see that our G_D is equal to the K defined above, since results in the last section imply

(i) $G_D(x,z) - G(x,z) = -\int_{\partial D} \frac{1-|x|^2}{|x-y|^d} G(y,z)\, d\pi(y)$

is harmonic in D,

(ii) if $x_n \to x \in \partial D$,

1.11 Occupation Times for the Sphere

$$\int \frac{1-|x_n|^2}{|x_n-y|^2} G(y,z)\, d\pi(y) \to G(x,z).$$

Now that we have made this connection, it is easy to find the Green's function. We turn to page 123 in Folland (1976) and find that if $y \neq 0$,

(3b) $\qquad G_D(x,y) = G(x,y) - |y|^{2-d} G(x, y/|y|^2).$

To check that this is true, we observe that

(i) since $y/|y|^2 \notin D$, it follows from results in Section 1.8 that the second term is harmonic in D, and
(ii) if $x \in \partial D$,

$$G(x,y) - |y|^{2-d} G(x, y/|y|^2) = \frac{C}{|x-y|^{d-2}} - \frac{1}{|y|^{d-2}} \frac{C}{\left|x - \frac{y}{|y|^2}\right|^{d-2}}$$

$$= \frac{C}{|x-y|^{d-2}} - \frac{C}{|x|y| - y|y|^{-1}|^{d-2}} = 0,$$

since $|x| = 1$ implies

$$\bigl| x|y| - y|y|^{-1} \bigr|^2 = |x|^2 |y|^2 - 2x \cdot y + 1$$
$$= |y|^2 - 2x \cdot y + |x|^2$$
$$= |x-y|^2.$$

At this point we have found the Green's function for the ball in $d \geq 3$, so we turn our attention to $d \leq 2$. In this situation, the first step in our computation fails, so we will begin with the second, that is, we will define

$$H(x,y) = G(x,y) - E_x G(B_\tau, y)$$

and then show that H satisfies (1). Let f be a C^∞ function that has compact support $K \subset D$, and let

$$u(x) = \int H(x,y) f(y)\, dy, \quad u_1(x) = \int G(x,y) f(y)\, dy$$

$$u_2(x) = E_x \int G(B_\tau, y) f(y)\, dy$$

$$= E_x u_1(B_\tau) = \int_{\partial D} k_y(x) u_1(y)\, d\pi(y).$$

A simple but somewhat tedious calculation shows that $\Delta u_1 = -f$ (see John (1982), Section 4.1), and it follows from results in the last section that $\Delta u_2 = 0$, so adding the last two results we see that $\Delta u = -f$ in D. Combining this with the fact that $u(x) = u_1(x_n) - u_2(x_n) = u_1(x_n) - E_{x_n} u_1(B_\tau) \to 0$ as $x_n \to x \in \partial D$ gives us what we need to show $H = G_D$.

(4) Suppose u is C^2 in D and continuous on $\bar{D}$. If $\Delta u = -f$ in D and $u = 0$ on ∂D, then

$$u(x) = E_x \int_0^\tau f(B_t)\,dt.$$

Proof Let $\tau_n = \inf\{t : B_t \notin D(0, 1 - 1/n)\}$. From (5) in Section 1.6,

$$u(B_{t \wedge \tau_n}) - \int_0^{t \wedge \tau_n} \Delta u(B_s)\,ds = u(B_{t \wedge \tau_n}) + \int_0^{t \wedge \tau_n} f(B_s)\,ds$$

is a martingale, and hence

$$u(x) = E_x u(B_{t \wedge \tau_n}) + E_x \int_0^{t \wedge \tau_n} f(B_s)\,ds.$$

Now $E_x \tau < \infty$ and f is bounded, so letting $t \to \infty$ and then $n \to \infty$ and applying the dominated convergence theorem gives

$$u(x) = E_x \int_0^\tau f(B_s)\,ds.$$

From (4) and the remarks above it, we can now conclude that $H = G_D$. The last detail is to find formulas for G_D. In $d = 2$, we have

(5a) $\quad G_D(x, 0) = -\dfrac{1}{\pi} \log|x|.$

(5b) If $y \neq 0$,

$$G_D(x, y) = -\frac{1}{\pi}(\log|x - y| - \log(x|y| - y|y|^{-1})).$$

To get formula (5b), look at the proof of (3b) given above. In $d = 1$, we have

$$G_D(x, y) = G(x, y) - \frac{x+1}{2} G(1, y) - \frac{1-x}{2} G(-1, y)$$

$$= |x - y| - \frac{x+1}{2}(1 - y) - \frac{1-x}{2}(y + 1).$$

and considering the two cases $x < y$ and $x > y$ leads to

(6) $\quad G_D(x, y) = \begin{cases} (1-x)(1+y) & -1 \le y \le x \le 1 \\ (1-y)(1+x) & -1 \le x \le y \le 1. \end{cases}$

The reader should note that if we let $h(x) = G_D(x, y)$, then

(i) $h(-1) = 0$, $h(1) = 0$
(ii) h is linear on $[-1, y]$ and $[y, 1]$, that is, it is harmonic if $x \neq y$
(iii) if $x > y > z$, $h'(x) - h'(z) = -2$, that is, $\frac{1}{2}\Delta h(y) = -\delta_0$

and h is the only function with these properties.

At this point we have, by hook and by crook, found the occupation times and exit distributions. Although the derivations we have given were a little crazy, there was a method to our madness. We were laying the groundwork for explaining the connection

exit distributions ↔ Dirichlet problem

occupation times ↔ Poisson's equation.

In the last two sections we have used solutions to equations on the right-hand side to find the quantities on the left. In Sections 8.5 and 8.6 we will exploit the connection in the other direction, that is, we will run Brownian motion to solve the P.D.E.'s.

Notes on Chapter 1

Robert Brown, an English botanist, was the first to observe that pollen grains in water move continuously and very erratically. The mathematical formulation and study were initiated by Bachelier (1900) and Einstein (1905), who derived the law of the position of the particle and applied this to the determination of molecular diameters.

N. Wiener (1923) was the first to put Brownian motion on a firm mathematical foundation by defining it as a measure on the space of continuous functions, and later with Paley and Zygmund (1933) proved that the paths were nowhere differentiable. At about the same time, Khintchine (1933) proved the law of the iterated logarithm, and the detailed study of the Brownian path was under way. In this connection we must mention the name of P. Lévy, who is responsible for much of our detailed knowledge of the Brownian path. His book (1948) is a classic and still a source of inspiration. See Chung (1976) for a recent exposition of some of Lévy's ideas.

Given the developments above, the reader may find it surprising that the strong Markov property of Brownian motion was first proved by Hunt in 1956. Hunt also noticed the connection between occupation times and Green's functions. At about the same time, Doob (1955b, 1956) noticed the connection between Brownian motion and the heat equation and Dirichlet problem, and the interplay between probability theory and analysis began. We have more to say about this in the following chapters.

2 Stochastic Integration

2.1 Integration w.r.t. Brownian Motion

In this section we will show that, even though with probability 1 $s \to B_s(\omega)$ does not have bounded variation, it is possible to define $\int_0^t H_s \, dB_s$ for processes H_s that are "nonanticipating." The first step is to give a precise description of the collection of integrands. This will require several definitions.

The first and most intuitive concept of being "nonanticipating" is the following:

$H(s, \omega)$ is said to be adapted to $\mathscr{F}_t$, $t \geq 0$, if for each t we have $H_t \in \mathscr{F}_t$.

We encountered this notion in our discussion of the Markov property in Chapter 1. In words, it says that the value at time t can be determined from the information we have at time t. The definition above, while intuitive, is not strong enough. Since there are an uncountable number of times, it does not give us enough control over the behavior of H as a function of (s, ω). We will, therefore, restrict our attention to a smaller class.

Let Λ be the σ-field of subsets of $[0, \infty) \times \Omega$ that is generated by the adapted processes that are right continuous and have left limits.
A process H is said to be optional if $H(s, \omega) \in \Lambda$.

The definition above is very abstract, but for once I want to urge you *not* to think about what it means. The optional σ-field is difficult to describe explicitly, and for the theory we will develop below an explicit description is irrelevant. In all the examples that we will consider below, it is trivial to use the definition above to show that the process under consideration is optional.

The optional processes will be the integrands for our integral with respect to Brownian motion. As in the theory of the Lebesgue integral, we will start with simple integrands and then, little by little, extend to the general case.

2.1 Integration w.r.t. Brownian Motion

$H(s, \omega)$ is said to be a basic optional process if $H(s, \omega) = 1_{[a,b)}(s)C(\omega)$ where $C \in \mathcal{F}_a$. Let $\Lambda_0 =$ the set of basic optional processes.

If $H = 1_{[a,b)}C$, then it is clear that we should define

$$\int H_s \, dB_s = C(\omega)(B(b, \omega) - B(a, \omega))$$

and

$$\int_0^t H_s \, dB_s = \int H_s 1_{[0,t]}(s) \, dB_s.$$

The second formula above defines a process we will denote as $(H \cdot B)_t$, and the first defines a random variable we will call $(H \cdot B)_\infty$. The reason we restrict our attention to optional integrands is so that we have

(1) If $H \in b\Lambda_0 = \{H \in \Lambda_0 : \sup|H(s, \omega)| < \infty\}$, then $(H \cdot B)_t$ is a martingale.

Proof

$$(H \cdot B)_t = \begin{cases} 0 & 0 \leq t \leq a \\ C(B_t - B_a) & a \leq t \leq b \\ C(B_b - B_a) & b \leq t < \infty, \end{cases}$$

so it is clear that $(H \cdot B)_t \in \mathcal{F}_t$ and $E|(H \cdot B)_t| < \infty$. To check the martingale property, it suffices to consider the case $a \leq s < t \leq b$. In this case,

$$E((H \cdot B)_t | \mathcal{F}_s) - (H \cdot B)_s = E((H \cdot B)_t - (H \cdot B)_s | \mathcal{F}_s)$$
$$= E(C(B_t - B_s) | \mathcal{F}_s) = CE(B_t - B_s | \mathcal{F}_s) = 0.$$

The next formula will be important in extending the integral from Λ_0 to larger classes.

(2) If $H, K \in b\Lambda_0$, then

$$E((H \cdot B)_t (K \cdot B)_t) = E \int_0^t H_s K_s \, ds.$$

Proof Replacing H and K by $H 1_{[0,t)}$ and $K 1_{[0,t)}$, it suffices to prove the result when $t = \infty$. Let $H = 1_{[a,b)}C$ and $K = 1_{[c,d)}D$. Since the formula above is linear in H and in K, we can assume without loss that $[a, b)$ and $[c, d)$ are either disjoint or equal.

Case 1: $b \leq c$. In this case, $\int_0^\infty H_s K_s \, ds = 0$, so we need to show that the left-hand side is 0.

$$E((H \cdot B)_\infty (K \cdot B)_\infty | \mathcal{F}_b) = E((H \cdot B)_b (K \cdot B)_d | \mathcal{F}_b)$$
$$= (H \cdot B)_b E((K \cdot B)_d | \mathcal{F}_b)$$
$$= (H \cdot B)_b (K \cdot B)_b = 0,$$

since $(K \cdot B)_t$ is a martingale and $(K \cdot B)_b = 0$.

Case 2: $a = c, b = d$.

$$\begin{aligned} E((H \cdot B)_\infty (K \cdot B)_\infty | \mathscr{F}_a) &= E(CD(B_b - B_a)^2 | \mathscr{F}_a) \\ &= CDE((B_b - B_a)^2 | \mathscr{F}_a) \\ &= CD(b - a) \\ &= \int H_s K_s \, ds, \end{aligned}$$

so taking expected values proves the result.

$H(s, \omega)$ is said to be a simple optional process if H can be written as the sum of a finite number of basic optional processes (multiplying by constants would not enlarge the class). Let $\Lambda_1 =$ the set of all simple optional processes. If $H \in \Lambda_1$ and $H = H^1 + \cdots + H^n$ where the $H^m \in \Lambda_0$, then we let

$$\int H_s \, dB_s = \sum_{m=1}^n \int H_s^m \, dB_s.$$

We will leave it to the reader to prove that this is a good definition (i.e., the integral is independent of how H is written).

Since the sum of a finite number of martingales is a martingale, it is easy to see

(3) If $H \in b\Lambda_1 = \{H \in \Lambda_1 : \sup |H(s, \omega)| < \infty\}$, then $(H \cdot B)_t$ is a martingale.

Since the formula in (2) is linear in H and in K, it generalizes immediately to

(4) If $H, K \in b\Lambda_1$, then

$$E((H \cdot B)_t (K \cdot B)_t) = E \int_0^t H_s K_s \, ds.$$

Taking $H = K$, we get a formula that is the key to our next extension.

(5) If $H \in b\Lambda_1$,

$$E(H \cdot B)_t^2 = E \int_0^t H_s^2 \, ds.$$

Let Λ_2 be the set of all optional processes that have

$$\|H\|_B = (E \int H_s^2 \, ds)^{1/2} < \infty.$$

Let $\mathscr{M}^2$ be the set of all martingales adapted to $\mathscr{F}_t t \geq 0$ that have

$$\|X\|_2 = (\sup_t EX_t^2)^{1/2} < \infty.$$

The next result shows that these are good norms for discussing stochastic integration.

(6) If $H \in b\Lambda_1$, then $\|H \cdot B\|_2 = \|H\|_B$.

2.1 Integration w.r.t. Brownian Motion

Proof Recalling the relevant definitions and using (5) gives

$$\|H\|_B^2 = E\int H_s^2\, ds = \sup_t E\int_0^t H_s^2\, ds$$
$$= \sup_t E(H\cdot B)_t^2 = \|H\cdot B\|_2^2.$$

The definitions above should suggest our strategy for extending the integral from Λ_1 to Λ_2: We will pick a sequence $H^n \in b\Lambda_1$ so that $\|H^n - H\|_B \to 0$, and we will show that $H^n \cdot B$ converges to a limit which is independent of the sequence of approximations chosen. To carry out the first step in this program, we need to show

(7) If $H \in \Lambda_2$, then there is a sequence $H^n \in b\Lambda_1$ so that $\|H^n - H\|_B \to 0$.

Proof Let $G_t^\ell = 2^\ell \int_{t-2^{-\ell}}^t H_s 1_{\{|H_s| \le \ell\}}\, ds$ where $H_s = 0$ for $s < 0$, and let $G_t^{\ell,m} = G^\ell([2^m t]/2^m)$. As $\ell \to \infty$, $\|G^\ell - H\|_B \to 0$, and as $m \to \infty$, $\|G^{\ell,m} - G^\ell\|_B \to 0$, so if we let $H^n = G^{n,m_n}$ and $m_n \to \infty$ fast enough, $\|H^n - H\|_B \to 0$.

To carry out the second step in our program, we need to show

(8) $\mathcal{M}^2$ is complete.

Proof Standard martingale convergence theorems imply that if $X \in \mathcal{M}^2$, then as $t \to \infty$, X_t converges almost surely and in L^2 to a limit X_∞ with $EX_\infty^2 = \sup_t EX_t^2$, and the martingale can be recovered from X_∞ by $X_t = E(X_\infty | \mathcal{F}_t)$. Let $\mathcal{F}_\infty = \sigma(\mathcal{F}_t, t \ge 0)$. Since $X_\infty = \lim X_t \in \mathcal{F}_\infty$, the observation above shows that $X \to X_\infty$ maps $\mathcal{M}^2$ one-to-one into $L^2(\mathcal{F}_\infty)$. On the other hand, if $Y \in L^2(\mathcal{F}_\infty)$, then $Y_t = E(Y|\mathcal{F}_t)$ is a martingale with $Y_t \to Y$ as $t \to \infty$, and Jensen's inequality shows that

$$EY_t^2 = E(E(Y|\mathcal{F}_t)^2) \le E(E(Y^2|\mathcal{F}_t)) = EY^2,$$

so $Y_t \in \mathcal{M}^2$. Combining this with the previous observation shows that $X \to X_\infty$ is an isometry from $\mathcal{M}^2$ onto $L^2(\mathcal{F}_\infty)$ and proves (8).

With the last two results established, we are ready to take limits to define $H \cdot B$ for $H \in \Lambda_2$. Let $H^n \in b\Lambda_1$ be such that $\|H^n - H\|_B \to 0$ as $n \to \infty$. (6) implies $\|H^n - H^m\|_B = \|H^n \cdot B - H^m \cdot B\|_2$, so $H^n \cdot B$ is a Cauchy sequence in $\mathcal{M}^2$, and (8) implies that $H^n \cdot B$ converges to a limit in $\mathcal{M}^2$. Since we have convergence of $H^n \cdot B$ for any sequence of approximations, an easy argument shows that the limit is independent of the sequence chosen, and we can define $H \cdot B$ to be the common value of the limits.

The last step in our definition of the stochastic integral is to make a trivial but useful extension of the class of integrands.

Let Λ_3 be the set of all optional processes that have $\int_0^t H_s^2\, ds < \infty$ a.s. for each t. To define $H \cdot B$ in this generality, let $T_n = \inf\{t : \int_0^t H_s^2\, ds > n\}$ and let $H_s^n = H_s 1_{(s \le T_n)}$. Since $H^n \in \Lambda_2$, we know how to define $H^n \cdot B$. On the other hand, if $m < n$, $H^m \cdot B - H^n \cdot B = (H^m - H^n) \cdot B = 0$ for $t < T_m$, so we can define $H \cdot B$ by setting $(H \cdot B)_s = (H^n \cdot B)_s$ for $s \le T_n$.

Remark: We will show in Section 2.11 that this collection of integrands is essentially unimprovable, that is, if we let

$$T = \inf\left\{t : \int_0^t H_s^2 \, ds = \infty\right\},$$

then on $\left\{\int_0^T H_s^2 \, ds = \infty\right\}$,

$$\limsup_{t \uparrow T} (H \cdot B)_t = \infty$$
$$\liminf_{t \uparrow T} (H \cdot B)_t = -\infty.$$

2.2 Integration w.r.t. Discrete Martingales

Our second step toward the general definition of the stochastic integral is to discuss integration w.r.t discrete time martingales. The definition of the integral in this case is trivial, but by looking at the developments in the right way, we will jump to an important conclusion—if we want our stochastic integrals to be martingales, then the integrands should be "predictable" (a notion we will describe in this section) rather than merely optional.

Let $X_n, n \geq 0$, be a martingale w.r.t. $\mathscr{F}_n$. If $H_n, n \geq 1$, is any process, we can define

$$(H \cdot X)_n = \sum_{m=1}^n H_m(X_m - X_{m-1}).$$

By analogy with results in Section 2.1, you might expect that if $H_n \in \mathscr{F}_n$ and each H_n is bounded, then $(H \cdot X)_n$ is a martingale. A simple example shows that this is false. Let S_n be the symmetric simple random walk, that is, $S_n = \xi_1 + \cdots + \xi_n$ where the ξ_i are independent and have $P(\xi_i = 1) = P(\xi_i = -1) = 1/2$. Let $\mathscr{F}_n = \sigma(\xi_1, \ldots, \xi_n)$. S_n is a martingale w.r.t. $\mathscr{F}_n$, but if we let $H_n = \xi_n$,

$$(H \cdot S)_n = \sum_{m=1}^n \xi_m(S_m - S_{m-1}) = \sum_{m=1}^n \xi_m^2 = n,$$

which is clearly not a martingale.

If we think of ξ_n as the net amount of money we would win for each dollar bet at time n and let H_n be the amount of money we bet at time n, then the "problem" with the last example becomes clear: We should require that $H_n \in \mathscr{F}_{n-1}$ for $n \geq 1$ (and let $\mathscr{F}_0 = \{\emptyset, \Omega\}$), that is, our decision on how much to bet at time n must be based on the previous outcomes $\xi_1, \ldots, \xi_{n-1}$ and not on the outcome we are betting on!

A process H_n that has $H_n \in \mathscr{F}_{n-1}$ for all $n \geq 1$ is said to be predictable since its value at time n can be predicted (with certainty) at time $n-1$. The next result shows that this is the right class of integrands for discrete time martingales.

2.2 Integration w.r.t. Discrete Martingales

(1) Let X_n be a martingale. If H is predictable and each H_n is bounded, then $(H \cdot X)_n$ is a martingale.

Proof The boundedness of the H_n implies $E|(H \cdot X)_n| < \infty$ for each n. With this established, we can compute conditional expectations to conclude

$$E((H \cdot X)_{n+1} | \mathscr{F}_n) = (H \cdot X)_n + E(H_{n+1}(X_{n+1} - X_n) | \mathscr{F}_n)$$
$$= (HX)_n + H_{n+1} E(X_{n+1} - X_n | \mathscr{F}_n) = (H \cdot X)_n,$$

since $H_{n+1} \in \mathscr{F}_n$ and $E(X_{n+1} - X_n | \mathscr{F}_n) = 0$.

The definition and proof given above relied very heavily on the fact that the time set was $\{0, 1, 2, \ldots\}$. The distinction between *predictable* and *optional* becomes very subtle in continuous time, so we will begin by considering a simple example: Let $(\Omega, \mathscr{F}, P)$ be a probability space on which there is defined a random variable T with $P(T \le t) = t$ for $0 \le t \le 1$ and an independent random variable ξ with $P(\xi = 1) = P(\xi = -1) = 1/2$. Let

$$X_t = \begin{cases} 0 & t < T \\ \xi & t \ge T \end{cases}$$

and let $\mathscr{F}_t = \sigma(X_s : s \le t)$. X is a martingale with respect to $\mathscr{F}_t$, but $\int_0^1 X_s dX_s = \xi^2 = 1$, so $Y_t = \int_0^t X_s dX_s$ is not, and hence if we want our stochastic integrals which are martingales we must impose some condition that rules out X.

The problem with the last example is the same as in the first case, and again there is a gambling interpretation that illustrates what is wrong. Consider the game of roulette. After the wheel is spun and the ball is rolled, people can bet at any time before $(<)$ the ball comes to rest but not after $(\ge)$. One way of requiring that our bet be made strictly before T is to require that the amount of money we have bet at time t is left continuous, that is, we cannot react instantaneously to take advantage of a jump in the process we are betting on. Weakening the last requirement we can, by analogy with the optional σ-field, state the following definition.

Let Π be the σ-field of subsets of $[0, \infty) \times \Omega$ that is generated by the left continuous adapted processes. A process H is said to be predictable if $H(s, \omega) \in \Pi$.

The definition of Π, like the definition of Λ, makes it easy to verify that something is predictable, but it does not tell us what sets in Π look like. This time, however, it is easy to describe the σ-field precisely:

(2) $\quad \Pi = \sigma((a, b] \times A : A \in \mathscr{F}_a)$.

Proof Clearly, the right-hand side $\subset \Pi$. Let $H(s, \omega)$ be adapted and left continuous and let $H^n(s, \omega) = H(m2^{-n}, \omega)$ for $m2^{-n} < s \le (m+1)2^{-n}$. Since H is adapted, $H^n(s, \omega) \in \sigma((a, b] \times A : A \in \mathscr{F}_a)$, and since H is left continuous, $H^n(s, \omega) \to H(s, \omega)$ as $n \to \infty$.

Remark: Based on the result above, you might guess that $\Lambda = \sigma([a,b) \times A : A \in \mathscr{F}_a)$, but you would be wrong. I would like to thank Bruce Atkinson for catching this mistake in an earlier version of this chapter.

Exercise 1 Show that if $H(s, \omega) = 1_{(a,b]}(s) 1_A(\omega)$ where $A \in \mathscr{F}_a$, then H is the limit of a sequence of optional processes; therefore, H is optional and $\Pi \subset \Lambda$.

2.3 The Basic Ingredients for Our Stochastic Integral

To define a stochastic integral we need four ingredients:

a probability space $(\Omega, \mathscr{F}, P)$
a filtration $\mathbb{F} = \{\mathscr{F}_t, t \geq 0\}$
a process X_t, $t \geq 0$, that is adapted to $\mathbb{F}$
a class of integrands H_t, $t \geq 0$.

In Section 2.2 we described the class of integrands that we will consider: the predictable processes. In this section, we will describe the assumptions that we will make concerning the process X and the filtration $\mathbb{F}$. This will require a number of definitions and explanations.

(1) X is said to be a local martingale (w.r.t. $\mathbb{F}$) if there are stopping times $T_n \uparrow \infty$ so that $X_{t \wedge T_n}$ is a martingale (w.r.t. $\{\mathscr{F}_{t \wedge T_n} : t \geq 0\}$). The stopping times T_n are said to reduce X.

Remark: In the same way, we can define local submartingale, locally bounded, locally of bounded variation, and so on.

You should think of a local martingale as something that would be a martingale if it had $E|X_t| < \infty$. There are several reasons for working with local martingales rather than with martingales:

(i) It frees us from worrying about integrability. For example, if X_t is a martingale and φ is a convex function, then $\varphi(X_t)$ is always a local submartingale, but we can conclude that $\varphi(X_t)$ is a submartingale only if we know $E|\varphi(X_t)| < \infty$, a fact that may be either difficult to check or false in some applications.
(ii) Often we will deal with processes defined on a random time interval $[0, \tau)$. If $\tau < \infty$, then the concept of martingale is meaningless, but it is trivial to define a local martingale: If there are stopping times $T_n \uparrow \tau$ so that....
(iii) Since most of our theorems will be proved by introducing stopping times T_n to reduce the problem to a question about nice martingales, the proofs are no harder for local martingales defined on a random time interval than for martingales.

Reason (iii) is more than just a feeling. There is a construction that makes

2.3 The Basic Ingredients for Our Stochastic Integral

it almost a theorem. Let X be a local martingale defined on $[0, \tau)$ and let $T_n \uparrow \tau$ be a sequence of stopping times that reduces X. Let

$$\gamma(t) = \begin{cases} t & 0 \leq t \leq T_1 \\ T_1 & T_1 \leq t \leq T_1 + 1 \\ t - 1 & T_1 + 1 \leq t \leq T_2 + 1 \\ T_2 & T_2 + 1 \leq t \leq T_2 + 2 \\ t - 2 & T_2 + 2 \leq t \leq T_3 + 2 \\ \vdots & \vdots \end{cases}$$

In words, $\gamma(t)$ expands $[0, \tau)$ into $[0, \infty)$ by waiting one unit of time each time a T_n is encountered. Since $\gamma(n) \leq T_n$, $X_{\gamma(t)}$, $t \geq 0$, is a martingale (exercise) to which standard theorems can be applied.

In our development of the stochastic integral, we will consider only continuous local martingales. We do this because (i) as we saw in Section 2.2, treating the jumps properly is a delicate matter, (ii) in all our applications the local martingales are continuous, and (iii) the assumption of continuity allows us to considerably simplify many of the main proofs and formulas of the theory.

This assumption is not without its drawbacks (there is no such thing as a free lunch!). At several points below (e.g., in the proof of Girsanov's formula) we will construct martingales by letting $X_t = E(X | \mathcal{F}_t)$. To guarantee that X_t is continuous (which is necessary for our theory to apply), we must assume that *all martingales adapted to* $\mathbb{F}$ *have a continuous version* (i.e., if X_t is a martingale, then there is a Y_t such that $t \to Y_t$ is continuous and $P(X_t = Y_t) = 1$), and this forces us to show (in Section 2.14) that the Brownian filtration has this property. The last result is interesting in its own right and would be in the book in any case, so I don't think this is too great a price to pay for the enormous simplifications that result.

The following is an example of the simplifications mentioned above:

(2) When X has continuous paths, we can always take $T_n = \inf\{t : |X_n| > n\}$ or any other sequence $T_n' \leq T_n$ that has $T_n' \uparrow \infty$ as $n \uparrow \infty$.

In words, every continuous local martingale is locally a uniformly bounded martingale.

Proof Let S_n be a sequence that reduces X. If $s < t$, then applying the optional stopping theorem to $X(r \wedge S_n)$ at times $r = s \wedge T_m'$ and $t \wedge T_m'$ gives

$$E(X(t \wedge T_m' \wedge S_n) | \mathcal{F}(s \wedge T_m' \wedge S_n)) = X(s \wedge T_m' \wedge S_n).$$

As $n \uparrow \infty$, $\mathcal{F}(s \wedge T_m' \wedge S_n) \uparrow \mathcal{F}(s \wedge T_m')$, $X(r \wedge T_m' \wedge S_n) \to X(r \wedge T_m')$ for all $r \geq 0$, and $|X(r \wedge T_m' \wedge S_n)| \leq m$, so it follows from a standard result on convergence of conditional expectations that

$$E(X(t \wedge T_m') | \mathcal{F}(s \wedge T_m')) = X(s \wedge T_m'),$$

proving the desired result.

Note: In the last proof we made a minor mistake that we will repeat several times below: We implicitly assumed that $X_0 = 0$. This mistake is rarely serious, but it does mean that some of the statements that we make are not correct. For example, (2) and the remark after it are not true unless we assume that X_0 is integrable (resp. bounded). Similar "typos" will appear several times below. The reader who is careful enough to detect them should have no trouble correcting them.

In our definition of local martingale in (1), we assumed that $X_{t \wedge T_n}$ is a martingale (w.r.t. $\mathcal{F}_{t \wedge T_n}$, $t \geq 0$). We did this with the proof of (2) in mind. The next exercise shows that we get the same definition if we assume $X_{t \wedge T_n}$ is a martingale (w.r.t. $\mathcal{F}_t$, $t \geq 0$).

Exercise 1 Let S be a stopping time. Then $X_{t \wedge S}$ is a martingale w.r.t. $\mathcal{F}_{t \wedge S}$, $t \geq 0$, if and only if it is a martingale w.r.t. $\mathcal{F}_t$, $t \geq 0$.

2.4 The Variance and Covariance of Continuous Local Martingales

If you go back and look at the proofs in Section 2.1, you will see that in the proofs of (1) and (2) we used only two facts about Brownian motion:

(a) $E(B_t - B_s | \mathcal{F}_s) = 0$
(b) $E((B_t - B_s)^2 | \mathcal{F}_s) = t - s$.

You will see also that after (2) was established, simple considerations of linearity showed that (3) through (5) held, and that after (5) was established, all further considerations used (5) only and no other facts about Brownian motion.

In Section 2.5 we show that the claim in the second part of the paragraph is true, that is, once (5) is suitably generalized, we can repeat the arguments in the last part of Section 2.1 (with some minor modifications) to define the integral w.r.t. a local martingale. This section is devoted to the generalization of (5). The key to doing this is to notice that

$$E((B_t - B_s)^2 | \mathcal{F}_s) = E(B_t^2 | \mathcal{F}_s) - B_s^2,$$

so (b) above says $B_t^2 - t$ is a martingale. This motivates (part of) the following definition.

(1) If X_t is a continuous local martingale, then we define the variance process $\langle X \rangle_t$ to be the unique predictable increasing process A_t that has $A_0 = 0$ and makes $X_t^2 - A_t$ a local martingale.

This result is a special case of

(2) *The Doob-Meyer Decomposition.* If Y_t is a local submartingale, then there is a

2.4 The Variance and Covariance of Continuous Local Martingales

unique predictable increasing process A_t that has $A_0 = 0$ and makes $Y_t - A_t$ a local martingale.

Since the proof of this result is rather technical and the details are irrelevant for later developments, we will content ourselves here with simply trying to give you a feeling for what A is and why predictability is important. The reader can find a nice proof in K. M. Rao (1969), so if you want to keep things strictly self-contained, you should read the heuristic discussion below, put the book down, read Rao's article, and then resume the development of the theory in this section—we will prove everything else.

The first step in understanding (2) is to see why all the conditions are necessary for uniqueness. To do this, we will prove the result for discrete time submartingales, because the construction is trivial in this case. We let $A_0 = 0$ and define for $n \geq 1$

$$A_n = A_{n-1} + E(Y_n | \mathscr{F}_{n-1}) - Y_{n-1}.$$

From the definition, it is immediate that $A_n \in \mathscr{F}_{n-1}$, A_n is increasing (since Y_n is a submartingale), and

$$E(Y_n - A_n | \mathscr{F}_{n-1}) = E(Y_n | \mathscr{F}_{n-1}) - A_n$$
$$= Y_{n-1} - A_{n-1},$$

so A has the desired properties. To see that A is unique, observe that if B is another process with the desired property, then $A_n - B_n$ is a martingale and $A_n - B_n \in \mathscr{F}_{n-1}$. Therefore

$$A_n - B_n = E(A_n - B_n | \mathscr{F}_{n-1}) = A_{n-1} - B_{n-1},$$

and it follows by induction that $A_n - B_n = A_0 - B_0 = 0$.

The key to the uniqueness may be summarized as "any predictable discrete time martingale is constant." The last statement fails miserably if *predictable* is replaced by *optional*, and with a little work (exercise) one can construct a submartingale Y_n for which there are many optional increasing processes with $A_0 = 0$ that make $Y_n - A_n$ a martingale (e.g., $Y_n = \xi_1 + \cdots + \xi_n$ where the ξ_i are independent and have $E\xi_i > 0$).

The last paragraph explains why predictability is needed in discrete time. The same arguments apply in continuous time, that is, the requirement that A_t be predictable and increasing is needed to rule out the possibility of producing another process A'_t by adding a predictable martingale (e.g., Brownian motion) to A_t. The only change is that the uniqueness statement must be formulated more carefully.

(3) Any local martingale that is the difference of two predictable increasing processes is constant.

Remark: This is implicit in the proof of the Doob-Meyer decomposition, but to make it clear that (2) is the only result we are taking for granted, we will show that $(2) \Rightarrow (3)$.

Proof If $X_t = A_t - A'_t$ is a local martingale and A_t, A'_t are predictable and increasing, then $Y_t = A_t$ is a local submartingale that has two Doob-Meyer decompositions, since $Y_t - (A_t - A_0) = A_0$ is a local martingale and $Y_t - (A'_t - A'_0) = X_t - A'_0$ is also. Since the Doob-Meyer decomposition is unique, it follows that $A_t - A_0 = A'_t - A'_0$, that is, $A_t - A'_t = A_0 - A'_0$.

With a little care we can improve (3) to

(4) Any (continuous) local martingale that is predictable and locally of bounded variation is constant.

Remark: The result is true in general, but for simplicity we prove it here only for continuous processes.

Proof In light of (3), all we have to do is show that a predictable process that is locally of bounded variation can be written as a difference of two predictable increasing processes. To do this we open up any real analysis book (e.g., Royden (1968)) and observe that if X_t is optional, continuous, and locally of bounded variation, then the decomposition given there for a function of bounded variation expresses X_t as a difference of two optional continuous increasing processes, proving (4).

The variance process is important for defining and using the stochastic integral. Since we will spend a lot of time considering this process and discussing its properties below, we will drop the subject for the moment and turn to the definition of the covariance of two local martingales.

(5) If X and Y are two local martingales, we let

$$\langle X, Y \rangle_t = \tfrac{1}{4}(\langle X + Y \rangle_t - \langle X - Y \rangle_t).$$

If X and Y are random variables with mean zero,

$$\operatorname{cov}(X, Y) = EXY = \tfrac{1}{4}(E(X + Y)^2 - E(X - Y)^2)$$
$$= \tfrac{1}{4}(\operatorname{var}(X + Y) - \operatorname{var}(X - Y)),$$

so it is natural, I think, to call $\langle X, Y \rangle_t$ the covariance of X and Y. The following result is useful for computing $\langle X, Y \rangle_t$.

(6) $\langle X, Y \rangle_t$ is the unique predictable process A_t that is locally of bounded variation, has $A_0 = 0$, and makes $X_t Y_t - A_t$ a local martingale.

Proof From the definition, it is easy to see that

$$X_t Y_t - \langle X, Y \rangle_t = \tfrac{1}{4}[(X_t + Y_t)^2 - \langle X + Y \rangle_t - (X_t - Y_t)^2 - \langle X - Y \rangle_t]$$

is a local martingale. To prove the converse, observe that if A_t and A'_t are two processes with the desired property, then $A_t - A'_t = (X_t Y_t - A'_t) - (X_t Y_t - A_t)$ is a predictable local martingale that is locally of bounded variation and hence $\equiv 0$.

For some of the developments below, the following result will be important.

Exercise 1 If $S \leq T$ are stopping times and $\langle X \rangle_S = \langle X \rangle_T$, then X is constant on $[S, T]$.

Sketch of Proof It suffices to prove the result when X and $(T - S)1_{(S<\infty)}$ are bounded. In this case, applying Doob's inequality to $Y_t = X_{(S+t) \wedge T} - X_S$ shows
$$E \sup_t Y_t^2 \leq 4E(X_T - X_S)^2 = 4E(X_T^2 - X_S^2) = 0.$$

2.5 Integration w.r.t. Continuous Local Martingales

In this section we will explain how to integrate predictable processes w.r.t. continuous local martingales. To bring out the analogies with the development in Section 2.1, we will start by integrating the simplest class of integrands w.r.t. a continuous martingale and then, little by little, extend to the general case.

$H(s, \omega)$ is said to be a basic predictable process if $H(s, \omega) = 1_{(a,b]}(s) C(\omega)$ where $C \in \mathcal{F}_a$. Let Π_0 = the set of basic predictable processes.

If $H = 1_{(a,b]} C$ and X is a continuous local martingale, then it is clear that we should define
$$\int H_s \, dX_s = C(\omega)(X_b(\omega) - X_a(\omega))$$
and
$$(H \cdot X)_t = \int H_s 1_{[0,t]}(s) \, dX_s.$$

The analogue of (1) of Section 2.1 in this situation is

(1) If X is a martingale and $H \in b\Pi_0 = \{H \in \Pi_0 : \sup |H(s, \omega)| < \infty\}$, then $(H \cdot X)_t$ is a martingale.

Proof
$$(H \cdot X)_t = \begin{cases} 0 & 0 \leq t \leq a \\ C(X_t - X_a) & a \leq t \leq b \\ C(X_b - X_a) & b \leq t < \infty, \end{cases}$$
so it is clear that $(H \cdot X)_t \in \mathcal{F}_t$ and $E|(H \cdot X)_t| < \infty$. To check the martingale property, it suffices to consider the case $a \leq s < t \leq b$. In this case,
$$E((H \cdot X)_t | \mathcal{F}_s) - (H \cdot X)_s = E((H \cdot X)_t - (H \cdot X)_s | \mathcal{F}_s)$$
$$= E(C(X_t - X_s) | \mathcal{F}_s)$$
$$= CE(X_t - X_s | \mathcal{F}_s) = 0.$$

The next formula will be important in extending the integral from Π_0 to larger classes.

(2) If X and Y are bounded martingales and $H, K \in b\Pi_0$, then

$$\langle H \cdot X, K \cdot Y \rangle_t = \int_0^t H_s K_s \, d\langle X, Y \rangle_s$$

and, consequently,

$$E((H \cdot X)_t (K \cdot Y)_t) = E \int_0^t H_s K_s \, d\langle X, Y \rangle_s.$$

Proof Under the assumptions above, $E((H \cdot X)_t (K \cdot Y)_t) = E\langle H \cdot X, K \cdot Y \rangle_t$, so it suffices to prove the first result. Let $H = 1_{(a,b]} C$ and $K = 1_{(c,d]} D$. Since the first formula is linear in H and in K, we can assume without loss that $(a, b]$ and $(c, d]$ are either disjoint or equal.

Case 1: $b \leq c$. In this case, $\int_0^t H_s K_s \, d\langle X, Y \rangle_s = 0$, so we need to show that $\langle H \cdot X, K \cdot Y \rangle_t \equiv 0$, in other words, $(H \cdot X)_t (K \cdot Y)_t$ is a martingale. To prove this we observe that

$$(H \cdot X)_t (K \cdot X)_t = \begin{cases} 0 & 0 \leq t \leq c \\ C(X_b - X_a) D(Y_t - Y_c) & c \leq t \leq d \\ C(X_b - X_a) D(Y_d - Y_c) & d \leq t < \infty, \end{cases}$$

so if we let $J = C(X_b - X_a) D 1_{(c,d]}$, then $(H \cdot X)_t (K \cdot Y)_t = (J \cdot Y)_t =$ a martingale, since $J \in b\Pi_0$.

Case 2: $a = c, b = d$. If $a \leq t < u \leq b$, then we have

$$E((H \cdot X)_u (K \cdot Y)_u - \int_0^u H_s K_s \, d\langle X, Y \rangle_s | \mathscr{F}_t)$$
$$= CDE[(X_u - X_a)(Y_u - Y_a) - \langle X, Y \rangle_u + \langle X, Y \rangle_a | \mathscr{F}_t]$$
$$= CDE[X_u Y_u - \langle X, Y \rangle_u - X_a Y_u - Y_a X_u + X_a Y_a - \langle X, Y \rangle_a | \mathscr{F}_t]$$
$$= CD[X_t Y_t - \langle X, Y \rangle_t - X_a Y_t - Y_a X_t + X_a Y_a - \langle X, Y \rangle_a]$$
$$= (H \cdot X)_t (K \cdot Y)_t - \int_0^t H_s K_s \, d\langle X, Y \rangle_s.$$

With (2) established, then, as we promised in Section 2.4, everything follows as before. $H(s, \omega)$ is said to be a simple predictable process if H can be written as the sum of a finite number of basic predictable processes. If $H \in \Pi_1$ and $H = H^1 + \cdots + H^n$ where the $H^m \in \Pi_0$, then we let

$$\int H_s \, dX_s = \sum_{m=1}^n \int H_s^m \, dX_s$$

(and again we leave it to the reader to show that the right-hand side is independent of how H is written).

2.5 Integration w.r.t. Continuous Local Martingales

Since the sum of a finite number of martingales is a martingale, it is easy to see

(3) If X is a martingale and $H \in b\Pi_1 = \{H \in \Pi_1 : \sup|H(s,\omega)| < \infty\}$, then $(H \cdot X)_t$ is a martingale.

Since the formulas in (2) are linear in H and in K, they generalize immediately to

(4) If X and Y are bounded martingales and $H, K \in b\Pi_1$, then

$$\langle H \cdot X, K \cdot Y \rangle_t = \int_0^t H_s K_s \, d\langle X, Y \rangle_s$$

and, consequently,

$$E((H \cdot X)_t (K \cdot Y)_t) = E \int_0^t H_s K_s \, d\langle X, Y \rangle_s.$$

Taking $H = K$ in the second formula in (4), we get a result that is the key to our next extension.

(5) If X is a bounded martingale and $H \in b\Pi_1$, then

$$E(H \cdot X)_t^2 = E \int_0^t H_s^2 \, d\langle X \rangle_s.$$

Let $\Pi_2(X)$ be the set of all predictable processes that have

$$\|H\|_X = \left(E \int H_s^2 \, d\langle X \rangle_s \right)^{1/2} < \infty.$$

Let $\mathcal{M}^2$ be the set of all martingales adapted to $\mathbb{F}$ that have

$$\|X\|_2 = (\sup_t EX_t^2)^{1/2}.$$

The next result shows that the relationship between these two norms is the same as that between the two corresponding norms in Section 2.1.

(6) If $H \in b\Pi_1$, then $\|H \cdot X\|_2 = \|H\|_X$.

Proof Recalling the relevant definitions and using (5) gives

$$\|H\|_X^2 = E \int H_s^2 \, d\langle X \rangle_s = \sup_t E \int_0^t H_s^2 \, d\langle X \rangle_s$$
$$= \sup_t E(H \cdot X)_t^2 = \|H \cdot X\|_2^2.$$

As the definitions given above should suggest, our strategy for going from Π_1 to Π_2 will be the same as the one used in Section 2.1 to go from Λ_1 to Λ_2. We start with

(7) If $H \in \Pi_2(X)$, then there is a sequence $H^n \in b\Pi_1$ with $\|H^n - H\|_X \to 0$.

Proof Define a measure on the predictable σ-field by setting

$$Q(A \times (s,t]) = E(1_A(\langle X \rangle_t - \langle X \rangle_s)) \quad \text{when} \quad A \in \mathscr{F}_s.$$

Unscrambling the definitions shows that $\Pi_2(X) = L^2(Q)$, and an application of the monotone class theorem proves (7). □

(8) $\mathscr{M}^2$ is complete.

Proof This proof is exactly the same as the proof of (8) in Section 2.1.

With (7) and (8) established, we can define $H \cdot X$ for $H \in \Pi_2$ just as we did in Section 2.1. Let $H^n \in b\Pi_1$ so that $\|H^n - H\|_X \to 0$ and observe that $H^n \cdot X$ converges to a limit in $\mathscr{M}^2$ and that, consequently, the limit is independent of the sequence of approximations chosen.

The extension to $\Pi_3(X) = \{H : \int_0^t H_s^2 d\langle X \rangle_s < \infty \text{ a.s. for all } t \ge 0\}$ is also easy. Let $T_n = \inf\{t : \int_0^t H_s^2 d\langle X \rangle_s > n\}$, let $H_s^n = H_s 1_{(s \le T_n)}$, and observe that if $m < n$, $H^m \cdot X - H^n \cdot X = (H^m - H^n) \cdot X = 0$ for $t < T_m$, we can define $H \cdot X$ by setting $(H \cdot X)_s = (H^n \cdot X)_s$ for $s \le T_n$.

Having extended the definition of the integral to $H \in \Pi_3(X)$, it is natural (and useful) to generalize (3) and (4) to $H \in \Pi_3(X)$. We will prove the extension of (3) now, but (4) will not be proved until Section 2.6.

(9)   If X is a local martingale and $H \in \Pi_3(X)$, then $(H \cdot X)_t$ is a local martingale.

Proof By stopping at $T_n = \inf\{t : \int_0^t H_s^2 d\langle X \rangle_s \text{ or } |X_t| > n\}$, it suffices to show that if X is a bounded martingale and $H \in \Pi_2(X)$, then $(H \cdot X)_t$ is a martingale.

Let H^n be a sequence of elements in $b\Pi_1$ that converges to H in $\Pi_2(X)$. It follows from (3) above that $(H^n \cdot X)_t$ is a martingale. In other words, if $s < t$,

$$E((H^n \cdot X)_t | \mathscr{F}_s) = (H^n \cdot X)_s.$$

To complete the proof, it suffices to show that this equality persists when $n \to \infty$. The first step is to show that $(H^n \cdot X)_s \to (H \cdot X)_s$ in L^2. To do this, we observe that Doob's inequality and the isometry property of $H \to H \cdot X$ imply

$$E(\sup_t |(H^n \cdot X)_t - (H \cdot X)_t|^2) \le 4\|H^n \cdot X - H \cdot X\|_2^2$$
$$= 4\|H^n - H\|_X^2 \to 0.$$

To deal with the conditional expectation, we observe that from Jensen's inequality

$$E[E((H^n \cdot X)_t | \mathscr{F}_s) - E((H \cdot X)_t | \mathscr{F}_s)]^2 \le E((H^n \cdot X)_t - (H \cdot X)_t)^2 \to 0.$$

Therefore $E((H^n \cdot X)_t | \mathscr{F}_s) \to E((H \cdot X)_t | \mathscr{F}_s)$ in L^2, and it follows that

$$E((H \cdot X)_t | \mathscr{F}_s) = (H \cdot X)_s.$$

For several applications that follow (e.g., in Chapters 8 and 9), it is important to make a trivial extension of the integral we have defined above.

S is said to be a continuous semimartingale if S_t can be written as $X_t + A_t$, where X_t is a continuous local martingale and A_t is a continuous adapted process that is locally of bounded variation.

A nice feature of continuous semimartingales that is almost unheard of for their more general counterparts is

(10) Let S_t be a semimartingale. If X_t and A_t are chosen so that $A_0 = 0$, then the decomposition $S_t = X_t + A_t$ is unique.

Proof If $X'_t + A'_t$ is another decomposition, then $A_t - A'_t$ is a continuous local martingale and locally b.v., so $A_t - A'_t$ is constant and hence $\equiv 0$.

Given a unique decomposition, it is easy to extend our definition of the stochastic integral to continuous semimartingales. If $H \in \ell b\Pi =$ the set of locally bounded predictable processes, we can define

$$(H \cdot A)_t(\omega) = \int_0^t H_s(\omega) \, dA_s(\omega)$$

as a Lebesgue-Stieltjes integral (which exists for a.e. ω), and since $\ell b\Pi \subset \Pi_3(X)$, we can define $(H \cdot X)_t$ and

$$(H \cdot S)_t = (H \cdot X)_t + (H \cdot A)_t$$

(since by the uniqueness of the decomposition this is an unambiguous definition.

From the definitions above, it follows immediately that we have

(11) If X is a semimartingale and $H \in \ell b\Pi$, then $(H \cdot X)_t$ is a semimartingale.

2.6 The Kunita-Watanabe Inequality

In this section we will prove a Cauchy-Schwarz inequality due to Kunita and Watanabe and apply this result to extend the formula given in Section 2.5 for the covariance of two stochastic integrals.

(1) If X and Y are local martingales and H and K are two measurable processes, then almost surely

$$\int_0^\infty |H_s K_s| \, |d\langle X, Y\rangle_s| \le \left(\int_0^\infty H_s^2 \, d\langle X\rangle_s\right)^{1/2} \left(\int_0^\infty K_s^2 \, d\langle Y\rangle_s\right)^{1/2},$$

where $|d\langle X, Y\rangle_s|$ stands for dV_s where V_s is the variation of $r \to \langle X, Y\rangle_r$ on $[0, s]$.

Remark: This result is from Meyer (1976). He attributes the proof to P. Priouret. Notice that H and K are not assumed to be predictable. We assume only that $H(s, \omega)$ and $K(s, \omega)$ are measurable with respect to $\mathcal{R} \times \mathcal{F}$ where $\mathcal{R}$ is the Borel subsets of R. The reason we can attain this level of generality is that the notion of martingale does not enter into the proof after the first line.

Proof

Step 1: Observe that if $s \le t$, $\langle X + \lambda Y, X + \lambda Y \rangle_t \ge \langle X + \lambda Y, X + \lambda Y \rangle_s$. If we let $\langle M, N \rangle_s^t = \langle M, N \rangle_t - \langle M, N \rangle_s$, then

$$0 \le \langle X + \lambda Y, X + \lambda Y \rangle_t - \langle X + \lambda Y, X + \lambda Y \rangle_s$$
$$= \langle X, X \rangle_s^t - 2\lambda \langle X, Y \rangle_s^t + \lambda^2 \langle Y, Y \rangle_s^t$$

for all s, t, and λ. Now a quadratic $ax^2 + bx + c$ that is nonnegative at all the rationals and not identically 0 has at most one real root (i.e., $b^2 - 4ac \le 0$), so we have that

$$(\langle X, Y \rangle_s^t)^2 \le \langle X, X \rangle_s^t \langle Y, Y \rangle_s^t.$$

Step 2: Let $0 = t_0 < t_1 < \cdots < t_n$ be an increasing sequence of times, let h_i, k_i, $1 \le i \le n$, be random variables, and define simple measurable processes

$$H(s, \omega) = \sum_{i=1}^n h_i(\omega) 1_{(t_{i-1}, t_i)}(s)$$

$$K(s, \omega) = \sum_{i=1}^n k_i(\omega) 1_{(t_{i-1}, t_i)}(s).$$

From the definition of the integral, the result of Step 1, and the Cauchy-Schwarz inequality, it follows that

$$\left| \int_0^\infty H_s K_s d\langle X, Y \rangle_s \right| \le \sum_{i=1}^n |h_i k_i| \left| \langle X, Y \rangle_{t_{i-1}}^{t_i} \right|$$

$$\le \sum_{i=1}^n |h_i| \left(\langle X, X \rangle_{t_{i-1}}^{t_i} \langle Y, Y \rangle_{t_{i-1}}^{t_i} \right)^{1/2} |k_i|$$

$$\le \left(\sum_{i=1}^n h_i^2 \langle X, X \rangle_{t_{i-1}}^{t_i} \right)^{1/2} \left(\sum_{i=1}^n k_i^2 \langle Y, Y \rangle_{t_{i-1}}^{t_i} \right)^{1/2},$$

proving that for simple measurable processes:

(2) $$\left| \int_0^\infty H_s K_s d\langle X, Y \rangle_s \right| \le \left(\int_0^\infty H_s^2 d\langle X \rangle_s \right)^{1/2} \left(\int_0^\infty K_s^2 d\langle Y \rangle_s \right)^{1/2},$$

which is (1) with the absolute values outside the integral.

Step 3: Let M be a large number and let $T = \inf\{t : \langle X \rangle_t \text{ or } \langle Y \rangle_t > M\}$. By the monotone convergence theorem, it suffices to prove (1) when $H = K = 0$ for $s \ge T$ and $|H_s|, |K_s| \le M$ for $s \le T$. Having restricted our attention to $[0, T]$, $\langle X \rangle$ and $\langle Y \rangle$ are finite measures; so using the bounded convergence theorem, we see that (2) holds for bounded measurable processes. To improve (2) to (1) (and complete the proof), let J_s be a measurable process taking values in $\{-1, 1\}$ such that

2.6 The Kunita-Watanabe Inequality

$$\int_0^t |d\langle X,Y\rangle_s| = \int_0^t J_s \, d\langle X,Y\rangle,$$

and apply (2) to $H_s = |H_s|$ and $K_s = J_s|K_s|$.

With (1) established, we are now ready to generalize (4) of Section 2.5 from $H \in b\Pi_1$ to $H \in \Pi_3(X)$.

(3) If X and Y are local martingales and $H \in \Pi_3(X)$, $K \in \Pi_3(Y)$, then

$$\langle H \cdot X, K \cdot Y \rangle_t = \int_0^t H_s K_s \, d\langle X,Y\rangle_s.$$

Proof What we need to show is that

(∗) $(H \cdot X)_t (K \cdot Y)_t - \int_0^t H_s K_s \, d\langle X,Y\rangle_s$

is a local martingale. By stopping at $T_n = \inf\{t : |X_t|, |Y_t|, \int_0^t |H_s|^2 \, d\langle X\rangle_s$ or $\int_0^t |K_s|^2 \, d\langle Y\rangle_s > n\}$, it suffices to show that if X and Y are bounded, $H \in \Pi_2(X)$, $K \in \Pi_2(Y)$, and Z_t is the quantity defined in (∗), then Z_t is a martingale.

Let H^n and K^n be sequences of elements of $b\Pi_1$ that converge to H and K in $\Pi_2(X)$ and $\Pi_2(Y)$, respectively, and let Z_t^n be the quantity that results when H^n and K^n replace H and K in (∗). By results in Section 2.5, Z_t^n is a martingale, that is, if $s < t$, $E(Z_t^n | \mathscr{F}_s) = Z_s^n$. To complete the proof it suffices to show that this equality persists when $n \to \infty$. The first step is to show $Z_s^n \to Z_s$ in L^1. The triangle inequality implies that

$$E\left|\sup_t |(H^n \cdot X)_t (K^n \cdot Y)_t - (H \cdot X)_t (K \cdot Y)_t|\right|$$
$$\leq E\left|\sup_t |((H^n - H) \cdot X)_t (K^n \cdot Y)_t|\right| + E\left|\sup_t |(H \cdot X)_t ((K^n - K) \cdot Y)_t|\right|.$$

To estimate the first term, we observe that from the inequalities of Cauchy-Schwarz and Doob, it is

$$\leq \{E(\sup_t |((H^n - H) \cdot X)_t|^2) E(\sup_t |(K^n \cdot Y)_t|^2)\}^{1/2}$$
$$\leq 4\|(H^n - H) \cdot X\|_2 \|K^n \cdot Y\|_2 = 4\|H^n - H\|_X \|K^n\|_Y \to 0$$

as $n \to \infty$, since $\|H^n - H\|_X \to 0$ and $\|K^n\|_Y \to \|K\|_Y < \infty$. A similar estimate shows $E|\sup_t |(H \cdot X)_t ((K^n - K) \cdot Y)_t|| \to 0$, so we have shown $(H^n \cdot X)_s (K^n \cdot X)_s \to (H \cdot X)_s (K \cdot Y)_s$ in L^1.

Repeating the last argument and using the Kunita-Watanabe inequality instead of Cauchy-Schwarz, we see that

$$\left|\int_0^t H_s^n K_s^n \, d\langle X,Y\rangle_s - \int_0^t H_s K_s \, d\langle X,Y\rangle_s\right|$$
$$\leq \int_0^t |H_s^n - H_s| |K_s^n| |d\langle X,Y\rangle_s| + \int_0^t |H_s| |K_s^n - K_s| |d\langle X,Y\rangle_s|,$$

and the first term

$$\leq \left(\int_0^t |H_s^n - H_s|^2 d\langle X\rangle_s\right)^{1/2} \left(\int_0^t |K_s^n|^2 d\langle Y\rangle_s\right)^{1/2}.$$

Therefore

$$\int_0^t H_s^n K_s^n d\langle X, Y\rangle_s \to \int_0^t H_s K_s d\langle X, Y\rangle_s$$

in L^1, and it follows that $Z_s^n \to Z_s$ in L^1.

To deal with $E(Z_t^n | \mathscr{F}_s)$, we observe that from Jensen's inequality

$$E|E(Z_t^n | \mathscr{F}_s) - E(Z_t | \mathscr{F}_s)| \leq E|Z_t^n - Z_t| \to 0,$$

so $E(Z_t^n | \mathscr{F}_s) \to E(Z_t | \mathscr{F}_s)$ in L^1, and it follows that $E(Z_t | \mathscr{F}_s) = Z$.

The formula in (3) generalizes readily to sums of stochastic integrals.

(4) If $X = \sum_{i=1}^m H^i \cdot X^i$ and $Y = \sum_{j=1}^m K^j \cdot Y^j$ where each $H^i \in \Pi^3(X^i)$ and each $K^j \in \Pi^3(Y^j)$, then

$$\langle X, Y\rangle_t = \sum_{i,j} \int_0^t H_s^i K_s^j d\langle X^i, Y^j\rangle_s.$$

In the developments that follow there is one application of the Kunita-Watanabe inequality that comes up so often that we have given it a name and left it as an exercise for the reader.

Exercise 1 *The Usual Domination Argument.* If $X, Y \in \mathscr{M}^2$ with $X_0 = Y_0 = 0$, then

$$EX_\infty Y_\infty = E\langle X, Y\rangle_\infty,$$

that is, both expectations exist and are equal.

2.7 Stochastic Differentials, the Associative Law

In some computations, it is useful to write the integral relationship

$$Y_t = \int_0^t K_s dX_s$$

as the formal equation

$$dY_t = K_t dX_t$$

where the dY_t and dX_t are fictitious objects known as "stochastic differentials." A good example is the derivation of the following formula, which (for obvious reasons) we call the associative law:

(∗) $H \cdot (K \cdot X) = (HK) \cdot X.$

2.7 Stochastic Differentials, the Associative Law

Proof Using Stochastic Differentials $d(H \cdot Y)_t = H_t dY_t$. Letting $Y_t = (K \cdot X)_t$ and observing $dY_t = K_t dX_t$ gives $d(H \cdot (K \cdot X))_t = H_t d(K \cdot X)_t = H_t K_t dX_t = d((HK) \cdot X)_t$.

The above proof is not rigorous, but the computation is useful because it tells us what the answer should be. Once we know the answer, it is routine to verify it by checking that it holds for basic predictable processes and then following the extension process we used for defining the integral to conclude that it holds in general.

(1) If $H, K \in \Pi_0$, then (∗) holds.

Proof Let $H = 1_{(a,b]} C$ and $K = 1_{(c,d]} D$. Without loss, we can assume that either (i) $b \leq c$ or (ii) $a = c, b = d$. In case (i), both sides of the equation are $\equiv 0$ and hence equal. In case (ii),

$$(K \cdot X)_t = \begin{cases} 0 & 0 \leq t \leq a \\ D(X_t - X_a) & a \leq t \leq b \\ D(X_b - X_a) & b \leq t < \infty, \end{cases}$$

so

$$(H \cdot (K \cdot X))_t = \begin{cases} 0 & 0 \leq t \leq a \\ CD(X_t - X_a) & a \leq t \leq b \\ CD(X_b - X_a) & b \leq t < \infty, \end{cases}$$

and it follows that $(H \cdot (K \cdot X))_t = ((HK) \cdot X)_t$.

From (1) it follows immediately that if $H, K \in \Pi_1$, then (∗) holds. To extend to more general integrands, we will take limits, and to simplify the argument, we will take the limit for K first and then for H.

(2) If $H \in b\Pi_1$ and $K \in \Pi_2(X)$, then (∗) holds.

Proof Let $K^n \in \Pi_1$ such that $K^n \to K$ in $\Pi_2(X)$. Since H is bounded, $HK^n \to HK$ in $\Pi_2(X)$, and it follows that $HK^n \cdot X \to HK \cdot X$ in $\mathcal{M}^2$. To deal with the left side of (∗), we observe that

$$\|H \cdot (K \cdot X) - H \cdot (K^n \cdot X)\|_2^2 = \|H \cdot ((K - K^n) \cdot X)\|_2^2$$

$$= E \int_0^\infty H_s^2 d\langle (K - K^n) \cdot X \rangle_s$$

$$= E \int_0^\infty H_s^2 (K_s - K_s^n)^2 d\langle X \rangle_s$$

$$\leq C \|K - K^n\|_X^2,$$

so as $n \to \infty$, $H \cdot (K^n \cdot X) \to H \cdot (K \cdot X)$ in $\mathcal{M}^2$ and the result follows.

(3) If $H \in \Pi_2(K \cdot X)$ and $K \in \Pi_2(X)$, then (∗) holds.

Proof Let $H^n \in b\Pi_1$ such that $H^n \to H$ in $\Pi_2(K \cdot X)$. Since $\|H^n \cdot (K \cdot X) - H \cdot (K \cdot X)\|_2 = \|H^n - H\|_{K \cdot X}$, we have $H^n \cdot (K \cdot X) \to H \cdot (K \cdot X)$ in $\mathcal{M}^2$. To deal with the right side of $(*)$, we observe that

$$\|H^n K - HK\|_X^2 = E \int_0^\infty (H_s^n - H_s)^2 K_s^2 \, d\langle X \rangle_s$$
$$= \|H^n - H\|_{K \cdot X}^2 \to 0,$$

so $H^n K \cdot X \to HK \cdot X$ in $\mathcal{M}^2$.

By stopping at $T_n = \inf\{t : \int_0^t K_s^2 \, d\langle X \rangle_s \text{ or } \int_0^t H_s^2 \, d\langle K \cdot X \rangle_s > n\}$ and letting $n \to \infty$, we can extend (3) to

(4) If $H \in \Pi_3(K \cdot X)$ and $K \in \Pi_3(X)$, then $(*)$ holds.

2.8 Change of Variables, Itô's Formula

This section is devoted to a proof of the following useful formula:

(1) If X is a continuous local martingale and f has two continuous derivatives, then with probability 1,

$$f(X_t) - f(X_0) = \int_0^t f'(X_s) \, dX_s + \frac{1}{2} \int_0^t f''(X_s) \, d\langle X \rangle_s.$$

Remark: If A_t is a continuous process that is locally of bounded variation, then

$$f(A_t) - f(A_0) = \int_0^t f'(A_s) \, dA_s.$$

As the reader will see in the proof, the second term comes from the fact that local martingale paths have quadratic variation $\langle X \rangle_t$, and the 1/2 in front of it comes from expanding f in a Taylor series.

Proof By stopping at $T_M = \inf\{t : |X_t| \text{ or } \langle X \rangle_t \geq M\}$, it suffices to prove the result when $|X_t|$ and $\langle X \rangle_t \leq M$. From calculus we know that if $a < b$, there is a $c(a, b) \in [a, b]$ such that

(2) $\quad f(b) - f(a) = (b - a)f'(a) + \frac{1}{2}(b - a)^2 f''(c(a, b))$.

Let t be a fixed positive number. For each $\delta > 0$, define a random partition of $[0, t]$ by

$$t_{i+1} = t \wedge \inf\{s > t_i : s - t_i, |X_s - X_{t_i}|, \text{ or } \langle X \rangle_s - \langle X \rangle_{t_i} > \delta\}.$$

(Note that this partition depends on δ even though we have not recorded the dependence in the notation.) From (2) it follows that

2.8 Change of Variables, Itô's Formula

(3)
$$f(X_t) - f(X_0) = \sum_i f(X_{t_{i+1}}) - f(X_{t_i})$$

$$= \sum_i f'(X_{t_i})(X_{t_{i+1}} - X_{t_i}) + \frac{1}{2}\sum_i g_i(\omega)(X_{t_{i+1}} - X_{t_i})^2$$

where $g_i(\omega) = f''(c(X_{t_i}, X_{t_{i+1}}))$.

Comparing (3) with (1), it becomes clear that we want to show

(4a)
$$\sum_i f'(X_{t_i})(X_{t_{i+1}} - X_{t_i}) \to \int_0^t f'(X_s)\,dX_s$$

(4b)
$$\frac{1}{2}\sum_i g_i(\omega)(X_{t_{i+1}} - X_{t_i})^2 \to \frac{1}{2}\int_0^t f''(X_s)\,d\langle X\rangle_s.$$

The proof of (4a) is easy. If we let $H_s^n = f'(X_{t_i})$ when $s \in [t_i, t_{i+1})$ and $= 0$ otherwise, then

$$\sum_i f'(X_{t_i})(X_{t_{i+1}} - X_{t_i}) = (H^n \cdot X)_t.$$

Let $H_s = f'(X_s)$ when $s < t$ and $= 0$ otherwise. Since f' is uniformly continuous on $[-M, M]$, we see that as $n \to \infty$, $\sup|H_s^n - H_s| \to 0$. Since $\langle X\rangle_\infty \leq M$, it follows that $H^n \to H$ in $\Pi_2(X)$ and hence $H^n \cdot X \to H \cdot X$ in $\mathcal{M}^2$.

To prove (4b), we start by proving the result when $f'' \equiv 1$, that is,

(5) As $\delta \to 0$,
$$\sum_i (X_{t_{i+1}} - X_{t_i})^2 \to \langle X\rangle_t$$

in probability.

Proof Let $\Delta_i = (X_{t_{i+1}} - X_{t_i})^2 - (\langle X\rangle_{t_{i+1}} - \langle X\rangle_{t_i})$ and observe that $E(\Delta_i|\mathcal{F}_{t_i}) = 0$, and if $i < j$, then $\Delta_i \in \mathcal{F}_{t_j}$, so

$$E(\Delta_i \Delta_j) = EE(\Delta_i \Delta_j|\mathcal{F}_{t_j})$$
$$= E(\Delta_i E(\Delta_j|\mathcal{F}_{t_j})) = 0.$$

and it follows that

$$E(\sum_i \Delta_i)^2 = \sum_{ij} E\Delta_i\Delta_j = \sum_i E\Delta_i^2.$$

Using the trivial inequality $(x - y)^2 \leq (2x)^2 + (2y)^2$ now gives

$$\Delta_i^2 \leq 4(X_{t_{i+1}} - X_{t_i})^4 + 4(\langle X\rangle_{t_{i+1}} - \langle X\rangle_{t_i})^2.$$

To estimate the right-hand side, we observe that by the definition of the t_i,

$$\sum_i (\langle X\rangle_{t_{i+1}} - \langle X\rangle_{t_i})^2 \leq \delta \sum_i (\langle X\rangle_{t_{i+1}} - \langle X\rangle_{t_i}) = \delta\langle X\rangle_t$$

$$\sum_i (X_{t_{i+1}} - X_{t_i})^4 \leq \delta^2 \sum_i (X_{t_{i+1}} - X_{t_i})^2.$$

and since martingale increments are orthogonal,

$$E\left(\sum_i (X_{t_{i+1}} - X_{t_i})^2\right) = E\left(\sum_i X_{t_{i+1}} - X_{t_i}\right)^2 = EX_t^2.$$

Combining the estimates above shows

$$\sum_i E\Delta_i^2 \leq 4(\delta M + \delta^2 M^2),$$

and it follows that

$$\sum_i (X_{t_{i+1}} - X_{t_i})^2 \to \langle X \rangle_t \quad \text{in } L^2,$$

proving (5).

With (5) established, we can now prove (4b) as we did (4a). If we let $G_s^n = g_i$ when $s \in [t_i, t_{i+1})$ and $= f''(X_t)$ for $s \geq t$ and let

$$A_s^n = \sum_{t_i \leq s} (X_{t_{i+1}} - X_{t_i})^2,$$

then

$$\sum_i g_i(\omega)(X_{t_{i+1}} - X_{t_i})^2 = \int_0^t G_s^n \, dA_s^n.$$

(5) implies that as $n \to \infty$, A_s^n converges in probability to $\langle X \rangle_s$, and the uniform continuity of f'' implies that $G_s^n \to f''(X_{s \wedge t})$ uniformly in s, so to complete the proof all we have to do is justify taking the limit inside the integral

(6) $$\int_0^t G_s^n \, dA_s^n \to \int_0^t f''(X_s) \, d\langle X \rangle_s.$$

To do this, we observe that by taking subsequences we can suppose that with probability 1, $A_{s \wedge t}^n$ converges weakly to $\langle X \rangle_{s \wedge t}$, in other words, if we fix ω and regard $s \to A_{s \wedge t}^n$ and $s \to \langle X \rangle_{s \wedge t}$ as distribution functions, then the associated measures converge weakly. Having done this, we can fix ω and deduce (6) from the following simple result:

(7) If (i) measures μ_n on $[0, t]$ converge weakly to μ_∞, a finite measure, and (ii) g_n is a sequence of functions with $|g_n| \leq M$ that have the property that whenever $s_n \in [0, t] \to s$ we have $g_n(s_n) \to g(s)$, then as $n \to \infty$

$$\int g_n \, d\mu_n \to \int g \, d\mu_\infty.$$

Proof By letting $\mu_n'(A) = \mu_n(A)/\mu_n([0, t])$, we can assume that all the μ_n are probability measures. A standard construction (see Exercise 1 below) shows that there is a sequence of random variables X_n such that X_n has distribution μ_n and as $n \to \infty$, $X_n \to X_\infty$ a.s. The convergence of g_n to g implies $g_n(X_n) \to g(X_\infty)$, so the result follows from the bounded convergence theorem.

(7) is the last piece in the proof of (1). Tracing back through the proof, we

see that (7) implies (6), which in turn completes the proof of (4b), so adding (4a) and using (3) gives that for each t,

$$f(X_t) - f(X_0) = \int_0^t f'(X_s) dX_s + \frac{1}{2} \int_0^t f''(X_s) d\langle X \rangle_s \quad \text{a.s.}$$

Since each side of the formula is a continuous function of t, it follows that with probability 1 the equality holds for all $t \geq 0$, the statement made in (1).

Exercise 1 Let μ_n be a sequence of probability measures that converges weakly to a limit μ_∞ and let $F_n(x) = \mu_n((-\infty, x])$, $1 \leq n \leq \infty$, be the corresponding sequence of distribution functions. Let U be a random variable with $P(0 < U < u) = u$ for all $u \in (0, 1)$, and let $X_n = F_n^{-1}(U)$ where $F_n^{-1}(y) = \sup\{x : F_n(x) \leq y\}$. Then X_n has distribution μ_n, and $X_n \to X_\infty$ a.s.

Remark: Although (1) is by far the most important formula in this section, (5) is also useful because it says that local martingale paths have quadratic variation $\langle X \rangle_t$. This result has the following useful consequences:

Exercise 2 If $S \leq T$ are stopping times and X is constant on $[S, T]$, then $\langle X \rangle_S = \langle X \rangle_T$.

Exercise 3 If X is a continuous martingale with bounded variation, then $\langle X \rangle \equiv 0$ and hence X is constant.

The reader should note that the proof of the last result is somewhat circular, since the definition of $\langle X \rangle$ relies on the uniqueness of the Doob-Meyer decomposition, which already implies the result above (see (4) in Section 2.4).

2.9 Extension to Functions of Several Semimartingales

In this section, we use the version of Itô's formula that was proved in the last section to prove a much more powerful form of the result. The extension is based on the following simple application of the formula derived in the last section so it is almost an independent proof. If we let $f(x) = x^2$ in (1) in the last section, then we get

(1) $$X_t^2 - X_0^2 = \int_0^t 2X_s dX_s + \langle X \rangle_t,$$

so if $X_0 = 0$,

$$\int_0^t 2X_s dX_s = X_t^2 - \langle X \rangle_t$$

in contrast to $\int_0^t 2s\, ds = t^2$.

From (1), we get the following useful result:

(2)   *Integration by Parts.*
$$X_t Y_t - X_0 Y_0 = \int_0^t Y_s \, dX_s + \int_0^t X_s \, dY_s + \langle X, Y \rangle_t.$$

Proof Applying (1) to $X_t + Y_t$ and $X_t - Y_t$ gives
$$(X_t + Y_t)^2 - (X_0 + Y_0)^2 = \int_0^t 2X_s \, d(X+Y)_s + \int_0^t 2Y_s \, d(X+Y)_s + \langle X+Y \rangle_t$$
$$(X_t - Y_t)^2 - (X_0 - Y_0)^2 = \int_0^t 2X_s \, d(X-Y)_s - \int_0^t 2Y_s \, d(X-Y)_s + \langle X-Y \rangle_t.$$

The linearity of $(H \cdot X)$ in X allows us to break the stochastic integrals into two pieces, using the distributive law:

(3) $H \cdot (X + Y) = (H \cdot X) + (H \cdot Y).$

Multiplying the two equations above by $1/4$ and subtracting gives (2).

If A_s is of bounded variation and Y_s is continuous, then it follows from the theory of Riemann-Stieltjes integration that
$$A_t Y_t - A_0 Y_0 = \int_0^t Y_s \, dA_s + \int_0^t A_s \, dY_s.$$

Adding this to (2), we see that if $R_t = X_t + A_t$ is a continuous semimartingale and Y_t is a continuous local martingale, then
$$R_t Y_t - R_0 Y_0 = \int_0^t Y_s \, dR_s + \int_0^t R_s \, dY_s + \langle X, Y \rangle_t.$$

Repeating the last argument and interchanging the roles of X and Y, we see that if $S_t = Y_t + A'_t$ is another continuous semimartingale, then

(4) $$R_t S_t - R_0 S_0 = \int_0^t S_s \, dR_s + \int_0^t R_s \, dS_s + \langle X, Y \rangle_t.$$

so (2) holds for semimartingales if we define the covariance of two semimartingales to be the covariance of their "martingale parts." (This is unambiguous, since we have a unique decomposition.)

The last observation and induction allow us to prove the following generalization of Itô's formula:

(5)   If $X_t^1, \ldots, X_t^d$ are continuous semimartingales and $f: R^d \to R$ has continuous second-order partial derivatives, then
$$f(X_t) - f(X_0) = \sum_i \int_0^t D_i f(X_s) \, dX_s^i + \frac{1}{2} \sum_{ij} \int_0^t D_{ij} f(X_s) \, d\langle X^i, X^j \rangle_s.$$

Remark: At the end of the proof, we will state a version of this result that requires less differentiability on f, and we will prove the more general result by saying "the same proof works." To prepare for this, the reader should observe that f appears only in the first paragraph of the proof.

2.9 Extension to Functions of Several Semimartingales

Proof By stopping, it suffices to prove the result when $|X_t^i| \leq M$ for all i, t. Since any continuous function on $[-M, M]^n$ can be approximated by polynomials g_n in such a way that g_n, $D_i g_n$, and $D_{ij} g_n$ converge to f, $D_i f$, and $D_{ij} f$ uniformly, it suffices to prove the result when f is a polynomial and, by linearity, when f is a monomial $x^{k_1} x^{k_2} \cdots x^{k_n}$ where $k_1, k_2, \ldots, k_n \in \{1, \ldots, d\}$. (The reader should note that $k_1, \ldots, k_n$ are superscripts, not powers, e.g., our monomial might be $x^1 x^4 x^1 x^1 x^2$.)

If $n = 1$ and $k_1 = k$, then (5) says that

$$X_t^k - X_0^k = \int_0^t 1\, dX_s^k,$$

which is trivially true. To prove the result for a general monomial, we use induction. Let $Y_t = \prod_{m=1}^n X_t^{k(m)}$ be a monomial for which (5) holds, and let $Z_t = X_t^{k(n+1)}$. Applying (4) gives

$$Y_t Z_t = \int_0^t Z_s\, dY_s + \int_0^t Y_s\, dZ_s + \langle Y, Z \rangle_t.$$

Applying (5) to Y gives

$$Y_t = \sum_{i \leq n} \int_0^t \left(\prod_{\substack{m=1 \\ m \neq i}}^n X_s^{k(m)} \right) dX_s^{k(i)} + \frac{1}{2} \sum_{\substack{i,j \leq n \\ i \neq j}} \int_0^t \left(\prod_{\substack{m=1 \\ m \neq i,j}}^n X_s^{k(m)} \right) d\langle X^{k(i)}, X^{k(j)} \rangle_s,$$

so using the associative law,

$$\int Z_t\, dY_t = \sum_{i \leq n} \int_0^t \left(\prod_{\substack{m=1 \\ m \neq i}}^{n+1} X_s^{k(m)} \right) dX_s^{k(i)} + \frac{1}{2} \sum_{\substack{i,j \leq n \\ i \neq j}} \int_0^t \left(\prod_{\substack{m=1 \\ m \neq i,j}}^{n+1} X_s^{k(m)} \right) d\langle X^{k(i)}, X^{k(j)} \rangle_s.$$

By definition,

$$\int Y_t\, dZ_t = \int_0^t \left(\prod_{m=1}^n X_s^{k(m)} \right) dX_s^{k(n+1)}.$$

To evaluate the third term $\langle Y, Z \rangle_t$, we observe that by the formula for the covariance of stochastic integrals,

$$\langle Y, Z \rangle_t = \sum_{i \leq n} \int_0^t \left(\prod_{\substack{m=1 \\ m \neq i}}^n X_s^{k(m)} \right) \cdot 1\, d\langle X^{k(i)}, X^{k(n+1)} \rangle_s.$$

Adding the last three equalities gives

$$Y_t Z_t = \sum_{i \leq n+1} \int_0^t \left(\prod_{\substack{m=1 \\ m \neq i}}^{n+1} X_s^{k(m)} \right) dX_s^{k(i)}$$

$$+ \frac{1}{2} \sum_{\substack{i,j \leq n+1 \\ m \neq i,j}} \int_0^t \left(\prod_{\substack{m=1 \\ m \neq i,j}}^{n+1} X_s^{k(m)} \right) d\langle X^{k(i)}, X^{k(j)} \rangle_s.$$

(Notice that for each i in the sum for $\langle Y, Z \rangle_t$, there are two terms $i = i, j = n+1$, and $i = n+1, j = i$ in the last sum.) The proof is complete.

Remark: For applications to partial differential equations, it is desirable to have a version of (5) that assumes a minimum amount of differentiability. By inspecting the proof given above, the reader can see that we have actually shown the following:

(6) If $X_t^1, \ldots, X_t^d$ are continuous semimartingales and $X_t^{c+1}, \ldots, X_t^d$ are locally b.v., then

$$f(X_t) - f(X_0) = \sum_{i=1}^d \int_0^t D_i f(X_s)\, dX_s^i + \frac{1}{2} \sum_{1 \le i, j \le c} \int_0^t D_{ij} f(X_s)\, d\langle X^i, X^j \rangle_s$$

provided all the derivatives in the formula exist and are continuous.

Proof It suffices to prove the result when f is a polynomial and, by linearity, when f is a monomial ...

2.10 Applications of Itô's Formula

Our first application of Itô's formula shows that it is useful to know that the result holds for semimartingales. Applying Itô's formula with $X_t^1 = X_t$ and $X_t^2 = \langle X \rangle_t$, we obtain

(1) $$f(X_t, \langle X \rangle_t) - f(X_0, 0) = \int_0^t D_1 f(X_s, \langle X \rangle_s)\, dX_s + \int_0^t D_2 f(X_s, \langle X \rangle_s)\, d\langle X \rangle_s$$
$$+ \frac{1}{2} \int_0^t D_{11} f(X_s, \langle X \rangle_s)\, d\langle X \rangle_s.$$

From (1) we see that if $(\frac{1}{2}D_{11} + D_2)f = 0$, then $f(X_t, \langle X \rangle_t)$ is a local martingale. Examples of such functions are $f(x, y) = x$, $x^2 - y$, $x^3 - 3xy$, ... so if we let $X =$ Brownian motion, then we recover two results from Section 1.6: B_t and $B_t^2 - t$ are local martingales, and we also get some new results: $B_t^3 - 3tB_t$, $B_t^4 - 6tB_t^2 + 3t^2$, ... are local martingales. These local martingales are useful for computing expectations for Brownian motion.

Exercise 1 Let $\tau = \inf\{t : |B_t| > a\}$. Then
(i) $E_0 \tau = E_0 B_\tau^2 = a^2$
(ii) $E_0 \tau^2 = E_0(-B_\tau^4 + 6\tau B_\tau^2) = 5a^4$.

If we notice that $f(x, y) = \exp(x - y/2)$ satisfies $(\frac{1}{2}D_{11} + D_2)f = 0$, then we get another useful result.

(2) *The Exponential Formula.* If X is a continuous local martingale, then $\exp(X_t - \frac{1}{2}\langle X \rangle_t)$ is a local martingale.

If we let $Z_t = \exp(X_t - \frac{1}{2}\langle X \rangle_t)$, then (1) says that

2.10 Applications of Itô's Formula

(∗) $Z_t - Z_0 = \int_0^t Z_s \, dX_s,$

or, in stochastic differential notation, that

$dZ_t = Z_t \, dX_s.$

This property gives Z_t the right to be called $\mathscr{E}\mathrm{xp}(X_t)$, the martingale exponential of X_t (the script $\mathscr{E}$ serving to remind us that it is $\exp(X_t - \frac{1}{2}\langle X \rangle_t)$, not $\exp(X_t)$). As in the case of the ordinary differential equation

$f'(x) = f(x)g(x),$

it is possible to prove (under suitable assumptions) that Z is the only solution of (∗). See Doleans-Dade (1970) for details. This martingale will play an important role in Section 2.13 when we discuss Girsanov's transformation. Letting $X_t = \theta B_t$ in (2) gives us a family of (local) martingales $\exp(\theta B_t - \theta^2 t/2)$. These martingales are also useful for computing the distribution of quantities associated with Brownian motion.

Exercise 2 Let $T_a = \inf\{t : B_t = a\}$. Then for $a, \lambda > 0$,

$E_0 \exp(-\lambda T_a) = e^{-a\sqrt{2\lambda}},$

and if you are good at inverting Laplace transforms, you can recover a result we proved in Chapter 1 (see Section 1.5):

$P_0(T_a = s) = (2\pi s^3)^{1/2} a e^{-a^2/2s}.$

Exercise 3 Let $\tau = \inf\{t : |B_t| \geq a\}$ and let $\psi_a(\lambda) = E_0 \exp(-\lambda T_a)$. Applying the strong Markov property at time τ gives

$E_0 \exp(-\lambda T_a) = E_0(\exp(-\lambda \tau); B_\tau = a) + E_0(\exp(-\lambda \tau) \psi_{2a}(\lambda); B_\tau = -a).$

Since τ and B_τ are independent, we have

$\psi_a(\lambda) = \frac{1}{2}(1 + \psi_{2a}(\lambda)) E_0 \exp(-\lambda \tau),$

and solving gives

$E_0 \exp(-\lambda \tau) = 2e^{-a\sqrt{2\lambda}}/(1 + e^{-2a\sqrt{2\lambda}}).$

The next result shows why Brownian motion is relevant to the study of harmonic functions. If B_t is a d-dimensional Brownian motion, then Itô's formula becomes

(3) $f(B_t) - f(B_0) = \sum_i \int_0^t D_i f(B_s) \, dB_s^i + \frac{1}{2} \sum_i \int_0^t D_{ii} f(B_s) \, ds,$

so if $\Delta f = \sum_i D_{ii} f = 0$, then $f(B_t)$ is a local martingale, or more generally,

(4) $f(B_t) - \int_0^t \frac{1}{2} \Delta f(B_s) \, ds$ is a local martingale.

Note: If f and Δf are bounded, then the expression above is bounded and hence a martingale. This proves (4) in Section 1.6, and pays off our debt.

From (4) we immediately get two useful corollaries:

(5) Let $f \in C^2$. If $\Delta f = 0$ in G, $\bar{D}(x,\delta) \subset G$, and $\tau = \inf\{t : B_t \notin \bar{D}(x,\delta)\}$, then

$$f(x) = E_x f(B_\tau) = \int_{\partial D(x,\delta)} f(y)\, d\pi(y).$$

(6) *The Maximum Principle.* Let G be a bounded open set. If f is continuous on $\bar{G}$ and $\Delta f = 0$ in G, then

$$\max_{x \in G} f(x) = \max_{x \in \partial G} f(x).$$

For several developments below it is important to know that there is a converse to (5).

(7) Suppose f is bounded in G and has the averaging property

$$(*)\quad f(x) = \int_{\partial D(x,\delta)} f(y)\, d\pi(y).$$

whenever $\bar{D}(x,\delta) \subset G$, then $f \in C^\infty$ and has $\Delta f = 0$ in G.

Proof We will first show that $f \in C^\infty$. Let $x \in D$ and pick $\delta > 0$ so that $\bar{D}(x, 2\delta) \subset G$. Let φ be a nonnegative, infinitely differentiable function that is not $\equiv 0$ but vanishes on $[\delta^2, \infty)$. It is easy to check that

$$g(y) = \int \psi(|y-x|^2) f(y)\, dy$$

is infinitely differentiable in $D(x,r)$. Changing to polar coordinates and using $(*)$ gives

$$g(y) = \int_{D(0,\delta)} \psi(|z|^2) f(x+z)\, dz$$

$$= C \int_0^\delta dr\, r^{d-1} \psi(r^2) \left(\int_{\partial D(0,r)} f(x+z)\, d\pi(z) \right)$$

$$= C' f(y),$$

so $f \in C^\infty$.

Now that we have shown that $f \in C^\infty$, the rest is easy. Itô's formula implies that

$$f(B_t) - \int_0^t \frac{1}{2} \Delta f(B_s)\, ds \text{ is a local martingale,}$$

so if $\Delta f(x) \neq 0$ at some $x \in G$, then we can pick $\delta > 0$ small enough so that $\bar{D}(x,\delta) \subset G$ and $\Delta f(y) \neq 0$ for all $y \in D(x,\delta)$, and we can apply the optional stopping theorem at time $\tau = \inf\{t : B_t \notin D(x,\delta)\}$ to contradict $(*)$.

For our last application, consider what happens when we take $f(x) = |x|$

2.10 Applications of Itô's Formula

in (3). In this case, $f(B_t) = |B_t|$ is the "radial part" of Brownian motion and is called the (d-dimensional) Bessel process. If $x \neq 0$,

$$D_i|x| = D_i\left(\sum_j x_j^2\right)^{1/2} = \frac{1}{2}\left(\sum_j x_j^2\right)^{-1/2} 2x_i = |x|^{-1}x_i$$

$$D_{ii}|x| = -\frac{1}{4}\left(\sum_j x_j^2\right)^{-3/2}(2x_i)^2 + \left(\sum_j x_j^2\right)^{-1/2},$$

so $\Delta|x| = |x|^{-3}|x|^2 + d|x|^{-1} = (d-1)|x|^{-1}$.

If we let $R_t = |B_t|$ and restrict our attention to $d \geq 2$ to ensure that $T_0 = \inf\{t > 0 : B_t = 0\} = \infty$ a.s., we can apply Itô's formula to Brownian motion stopped at $T_\varepsilon = \inf\{t : |B_t| < \varepsilon \text{ or } \varepsilon^{-1}\}$ and let $\varepsilon \to 0$ to conclude that

(8) $$R_t - R_0 = \sum_i \int_0^t R_s^{-1} B_s^i \, dB_s^i + \frac{1}{2}\int_0^t (d-1) R_s^{-1} \, ds.$$

Replacing t by $t \wedge T_\varepsilon$ in the last equation, taking expectations, and letting $\varepsilon \to 0$ gives

$$E_x R_t = |x| + E_x \frac{1}{2}\int_0^t (d-1) R_s^{-1} \, ds,$$

so

$$\frac{\partial}{\partial t} E_x R_t = E_x \frac{(d-1)}{2} R_t^{-1}.$$

The right-hand side is $(d-1)/2|x|$ at $t = 0$, so we say R_t has "infinitesimal drift" $(d-1)/2|x|$.

Having computed the drift, the next step is to compute the variance process.

(9) $$\langle R \rangle_t \equiv t,$$

or, to use another phrase we will be using later, R_t has "infinitesimal variance" 1 (independent of x).

Proof Let $R_t^i = \int_0^t R_s^{-1} B_s^i \, dB_s^i$.

$$\langle R \rangle_t = \sum_{ij}\langle R^i, R^j \rangle_t = \sum_i \int_0^t (R_s^{-1} B_s^i)^2 \, ds$$

$$= \int_0^t R_s^{-2} R_s^2 \, ds = t.$$

We will gradually explain what the infinitesimal drift and variance mean as the story unfolds below. For the moment, the important thing to understand is how they are computed using Itô's formula.

Exercise 4 Let $S_t = R_t^2$. Show that $S_t - td$ is a martingale and

$$\langle S \rangle_t = \int_0^t 4S_r \, dr,$$

so the infinitesimal drift is d (independent of position) and the infinitesimal variance is four times the value of S (which is $|x|^2$ if the Brownian motion is at x).

Exercise 5 Let $d \geq 2$ and let

$$\varphi(x) = \begin{cases} \log|x| & d = 2 \\ |x|^{2-d} & d \geq 3. \end{cases}$$

Use Itô's formula to show that $\varphi(R_t)$ is a local martingale.

Exercise 6 Suppose

$$X_t - \int_0^t b(X_s)\,ds \text{ is a local martingale}$$

and

$$\langle X \rangle_t = \int_0^t a(X_s)\,ds$$

where $a > 0$ is continuous and b is bounded on compact intervals. Use Itô's formula to show $\varphi(X_t)$ is a local martingale if and only if

$$\tfrac{1}{2}\varphi''(x)a(x) + \varphi'(x)b(x) = 0,$$

so if we normalize φ to have $\varphi(0) = 0$ and $\varphi'(0) = 1$, then

$$\varphi(x) = \int_0^x \exp\left(-\int_0^y \frac{2b(z)}{a(z)}\,dz\right)dy.$$

In Chapter 9 we will show how to construct processes X that satisfy the two conditions above. We will call such an X "a diffusion process with infinitesimal drift $b(x)$ and infinitesimal variance $a(x)$," and we will call φ "the natural scale for this process." One of the reasons for interest in the natural scale is that it allows us to generalize the results in Section 1.7.

Exercise 7 Let X be the process described in Exercise 6 and let $T_a = \inf\{t : X_t = a\}$. Then

$$P_x(T_a < T_b) = \frac{\varphi(b) - \varphi(x)}{\varphi(b) - \varphi(a)}.$$

Exercise 8 Let u and v be harmonic and let $U_t = u(B_t)$, $V_t = v(B_t)$. Use Itô's formula to show

$$\langle U \rangle_t = \int_0^t |\nabla u(B_s)|^2\,ds$$

$$\langle U, V \rangle_t = \int_0^t \nabla u(B_s) \cdot \nabla v(B_s)\,ds.$$

2.11 Change of Time, Lévy's Theorem

Exercise 9 Let $D = \{z : |z| < 1\}$ be the unit ball in R^d and let $\tau = \inf\{t : B_t \notin D\}$. Show that if u and v are harmonic in D, continuous on $\bar{D}$, and have $u(0)v(0) = 0$, then

$$\int_{\partial D} u(y)v(y)\,d\pi(y) = \int_D \nabla u(x) \nabla v(x) G_D(0,x)\,dx$$

where $G_D(0,x)$ is the occupation time density for Brownian motion starting at 0 and killed when it leaves D. Analysts should recognize the last equality as a consequence of Green's theorem.

2.11 Change of Time, Lévy's Theorem

In this section we will prove Lévy's characterization of Brownian motion and use it to show that every continuous local martingale is a time change of Brownian motion.

(1) If X_t is a continuous local martingale with $X_0 = 0$ and $\langle X \rangle_t \equiv t$, then X is a Brownian motion.

Proof It suffices to show for any $s < t$ that $X_t - X_s$ is independent of $\mathcal{F}_s$ and has a normal distribution with mean 0 and variance $t - s$ or, if we introduce $X'_u = X_{u+s} - X_s$ and $\mathcal{F}'_u = \mathcal{F}_{s+u}$, that the last conclusion holds with $s = 0$. By Itô's formula,

$$e^{i\theta X_t} - 1 = i\theta \int_0^t e^{i\theta X_s}\,dX_s - \frac{\theta^2}{2}\int_0^t e^{i\theta X_s}\,ds.$$

Let $A \in \mathcal{F}_0$. The first term on the right is a local martingale, so if we replace t by $t \wedge T_n$ ($T_n = \inf\{t : |X_t| > n\}$), integrate over A, and let $n \to \infty$, we get

$$j(t) - P(A) = 0 - \frac{\theta^2}{2}\int_0^t j(s)\,ds, \text{ where } j(s) = E(e^{i\theta X_s}; A).$$

Since we know *a priori* that $|j(s)| \leq 1$, it follows that j is continuous, j has a continuous derivative ... Differentiating the equation gives

$$j'(t) = \frac{-\theta^2}{2}j(t),$$

which, together with the initial condition $j(0) = P(A)$, shows that $j(t) = P(A)e^{-\theta^2 t/2}$. Since $j(t) = E(e^{i\theta X_t}; A)$, this shows that $E(e^{i\theta X_t}|\mathcal{F}_0) = e^{-\theta^2 t/2}$, that is, X_t is independent of $\mathcal{F}_0$ and has a normal distribution with mean 0 and variance t.

An immediate consequence of (1) is:

(2) Every continuous local martingale with $\langle X \rangle_\infty \equiv \infty$ is a time change of Brownian motion.

Let $\gamma(u) = \inf\{t : \langle X \rangle_t > u\}$ and let $Y_u = X_{\gamma(u)}$. Since $\gamma(u)$, $u \geq 0$, is an increasing family of stopping times, the optional stopping theorem implies that

(3) $Y_u, \mathscr{F}_{\gamma(u)}, u \geq 0$, is a local martingale.

Proof By stopping at $T_n = \inf\{t : |X_t| > n\}$, it suffices to show that if X is a bounded martingale, then Y is a martingale. To do this we observe that if $u_1 < u_2$, then $\gamma(u_1) < \gamma(u_2)$ are stopping times, so we have

$$E(Y(u_2)|\mathscr{F}_{\gamma(u_1)}) = E(X_{\gamma(u_2)}|\mathscr{F}_{\gamma(u_1)})$$
$$= X_{\gamma(u_1)} = Y(u_1),$$

proving (3).

Repeating the argument above shows

(4) $Y_u^2 - u, \mathscr{F}_{\gamma(u)}, u \geq 0$, is a local martingale.

Exercise 1 of Section 2.4 implies that $u \to Y_u$ is continuous, so combining (3) and (4) with Lévy's characterization shows that Y_u is a Brownian motion.

The result above can be easily extended to local martingales with $P(\langle X \rangle_\infty < \infty) > 0$ (and we will see below that this is an important extension). Let $\tau = \langle X \rangle_\infty$. The proofs of (3) and (4) show that Y_u and $Y_u^2 - u$ are local martingales on $[0, \tau)$, and repeating the proof of (1) shows that $Y_u, u < \tau$, is a Brownian motion run for a random amount of time. Since Brownian paths are continuous, it follows immediately from the last observation that

(5) $\lim_{t \uparrow \infty} X_t$ exists on $\{\langle X \rangle_\infty < \infty\}$.

In Chapter 1 we showed that Brownian motion has

(6) $\limsup_{t \to \infty} B_t = \infty$, $\liminf_{t \to \infty} B_t = -\infty$.

Using this observation, we can sharpen (5) to conclude that

(7) The following sets are equal almost surely:

$$C = \{\lim_{t \to \infty} X_t \text{ exists}\}$$
$$B = \{\sup_t |X_t| < \infty\}$$
$$B' = \{\sup_t X_t < \infty\}$$
$$A = \{\langle X \rangle_\infty < \infty\}.$$

Proof Clearly, $C \subset B \subset B'$. (6) implies that $B' \subset A$. (5) shows that $A \subset C$.

Remark: The last result justifies the assertions we made in Sections 2.1 and 2.5 that Λ_3 and $\Pi_3(X)$ are essentially the largest possible classes of integrands.

2.11 Change of Time, Lévy's Theorem

Remark: We return to the result above when we study boundary limits of harmonic functions in Chapter 4. For related results on discrete martingales, see Neveu (1975), Chapter 7. In the discrete case, these results are only true when some regularity assumptions are imposed to control the size of the jumps. One of the joys of working with continuous martingales is that such considerations are unnecessary.

Convergence is not the only property of local martingales that can be studied using (2). Almost any almost-sure property concerning the Brownian path can be translated into a corresponding result for local martingales. This immediately gives us a number of theorems about the behavior of paths of local martingales. We will state only the most famous of these—the law of the iterated logarithm.

(8) Let $L(t) = \sqrt{2t \log \log t}$ for $t \geq e$. Then on $\{\langle X \rangle_\infty = \infty\}$,

$$\limsup_{t \to \infty} X_t/L(\langle X \rangle_t) = 1 \quad \text{a.s.}$$

Proof By time substitution, it suffices to prove the result for Brownian motion, and we have done this in Section 1.3.

Remark: As before, the result is true in the discrete case only if some regularity conditions are imposed on the jumps, and the proofs are more difficult. See Neveu (1975), Chapter 7, for discrete time martingales and Lenglart (1977) and Lepingle (1978) for general continuous time martingales.

Technical Note: In our proofs of (7) and (8), we have swept a little dirt under the rug. If X is a continuous local martingale with $\langle X \rangle_\infty \equiv \infty$ and we define $\gamma(u) = \inf\{t : \langle X \rangle_t > u\}$, $Y_u = X_{\gamma(u)}$, and $\mathscr{G}_u = \mathscr{F}_{\gamma(u)}$, then $\mathscr{G}_t \supset \sigma(Y_s : s \leq t)$, but it may be larger. For example, if $X_t = 0$ for $t \leq 1$ and $= B_t - B_1$ for $t \geq 1$, then $\mathscr{G}_t = \sigma(B_s : s \leq t+1) \neq \sigma(B_s - B_1 : 1 \leq s \leq t+1) = \sigma(Y_s : s \leq t)$. The fact that $\mathscr{G}_t$ is larger than the usual filtration associated with Y is a technical nuisance, but not a real problem. The proofs of (3) and (4) show that Y_u and $Y_u^2 - u$ are local martingales with respect to $\mathscr{G}_u$, so the proof of (1) shows that

$$E(e^{i\theta Y_{s+t}} | \mathscr{G}_s) = e^{-\theta^2 t/2},$$

and Y is what might be called a Brownian motion w.r.t. $\mathscr{G}$. In other words,

(i) $t \to Y_t$ is continuous and adapted to $\mathscr{G}_t$
(ii) for all $s, t \geq 0$, $Y_{t+s} - Y_s$ is independent of $\mathscr{G}_s$ and has a normal distribution with mean 0 and variance t.

It is not hard to show that the results we have proved above for our special presentation are also valid for the more general Brownian motions we have just defined, but that is a big leap to make all at once, so all that we will claim now is the easy-to-verify fact that the results we used in the proofs of (7) and (8) are valid for a general Brownian motion.

2.12 Conformal Invariance in $d \geq 2$, Kelvin's Transformations

In this section we will determine when a vector $(X_t^1, \ldots, X_t^n)$ of continuous local martingales is a time change of a multidimensional Brownian motion. The characterization of Brownian motion given in Section 2.11 generalizes easily, but when there are two or more components, we cannot straighten out all the $\langle X^i, X^j \rangle_t$ with only one time change, and only special vectors of local martingales can be time-changed into Brownian motion. If we consider X_t of the form $f(B_t)$ where $f: R^d \to R^d$, then (modulo trivial modifications) in $d = 2$, f must be analytic, and in $d \geq 3$, $f(x) = Ax$ where A is an orthogonal matrix.

The first step in deriving these results is to prove the following generalization of Lévy's theorem ((1) in the last section)

(1) If $X_t^1, \ldots, X_t^n$ are continuous local martingales with $X_0^i = 0$ and $\langle X^i, X^j \rangle_t = \delta_{ij} t$, then $X_t = (X_t^1, \ldots, X_t^n)$ is an n-dimensional Brownian motion.

Proof If $Y_t = \sum_i \alpha_i X_t^i$ where $\sum_i \alpha_i^2 = 1$, then it follows from the proof of the one-dimensional result that if $s < t$, then

$$E(e^{i\theta(Y_t - Y_s)} | \mathcal{F}_s) = e^{-(t-s)\theta^2/2},$$

or, in terms of the X^i,

$$E(e^{i(\sum_j (\theta \alpha_j)(X_t^j - X_s^j))} | \mathcal{F}_s) = e^{-(t-s)\theta^2/2}.$$

Since any $v \in R^n$ can be written as $\theta \alpha$ where $\sum_i \alpha_i^2 = 1$ and $\theta \in [0, \infty)$, this shows that X_t has independent increments and $X_t - X_s$ has a d-dimensional normal distribution with mean 0 and covariance $(t - s)I$, proving (1).

From (1), it follows immediately that a vector $(X_t^1, \ldots, X_t^n)$ of continuous local martingales is a time change of Brownian motion if and only if there is an increasing process A_t such that $\langle X^i, X^j \rangle_t = \delta_{ij} A_t$. The most important instance of martingales of this type occurs when we compose an analytic function with a complex Brownian motion. To explain this we need some notation.

Let C_t be a complex Brownian motion, that is, $C_t = B_t^1 + iB_t^2$ where B_t^1, B_t^2 are independent Brownian motions. Let f be analytic in C; in other words, if we write $z = x + iy$ and $f = u + iv$, then f satisfies the Cauchy-Riemann equations

$$\frac{\partial u}{\partial x} = \frac{\partial v}{\partial y} \quad \frac{\partial u}{\partial y} = \frac{-\partial v}{\partial x}.$$

Differentiating again gives

$$\frac{\partial^2 u}{\partial x^2} = \frac{\partial^2 v}{\partial x \partial y} \quad \frac{\partial^2 u}{\partial y^2} = \frac{-\partial^2 v}{\partial y \partial x},$$

and since v has partial derivatives of all orders,

2.12 Conformal Invariance in $d \geq 2$, Kelvin's Transformations

$$\Delta u = \frac{\partial^2 u}{\partial x^2} + \frac{\partial^2 u}{\partial y^2} = \frac{\partial^2 v}{\partial x \partial y} - \frac{\partial^2 v}{\partial y \partial x} = 0,$$

that is, u is harmonic. A similar argument shows that v is harmonic, so it follows from results in Section 2.10 that $U_t = u(C_t)$ and $V_t = v(C_t)$ are martingales, and

(2) $$\langle U \rangle_t = \int_0^{t \wedge \tau} |\nabla u(C_s)|^2 \, ds$$

$$\langle V \rangle_t = \int_0^{t \wedge \tau} |\nabla v(C_s)|^2 \, ds$$

where $\nabla u = \left(\frac{\partial u}{\partial x}, \frac{\partial u}{\partial y}\right)$ and $\nabla v = \left(\frac{\partial v}{\partial x}, \frac{\partial v}{\partial y}\right)$.

From the Cauchy-Riemann equations, it follows that $|\nabla u(z)|^2 = |\nabla v(z)|^2$ for all z, so $\langle U \rangle_t = \langle V \rangle_t$, and the time substitution needed to obtain Brownian motion is the same for both coordinates. The last thing to check is that

(3) $$\langle U, V \rangle_t = 0.$$

Proof Suppose $u(C_0) = v(C_0) = 0$. From (3) of Section 2.6,

$$\langle H \cdot X, K \cdot Y \rangle_t = \int_0^t H_s K_s d\langle X, Y \rangle_s.$$

From Itô's formula, if $C_t = B_t^1 + i B_t^2$, then

$$U_t = u(C_t) = \sum_{i=1}^2 \int_0^t D_i u(C_s) \, dB_s^i \equiv U_t^1 + U_t^2,$$

and V_t can be written in a similar way as $V_t^1 + V_t^2$. Now

$$\langle U, V \rangle_t = \sum_{i=1}^2 \sum_{j=1}^2 \langle U^i, V^j \rangle_t$$

$$= \sum_{i=1}^2 \int_0^t D_i u(C_s) D_i v(C_s) \, ds,$$

and from the Cauchy-Riemann equations,

$$(D_1 u)(D_1 v) + (D_2 u)(D_2 v) = -(D_1 u)(D_2 u) + (D_2 u)(D_1 u) = 0,$$

so $\langle U, V \rangle_t \equiv 0$, proving (3).

Combining (1), (2), and (3) above with the results of Section 2.11, we get Lévy's result concerning the conformal invariance of Brownian motion.

(4) If f is analytic and C_t is a complex Brownian motion, then $f(C_t)$ is a time change of a complex Brownian motion. To be precise, if we let

$$\sigma_t = \int_0^t |f'(C_s)|^2 \, ds$$

and

$$\gamma_t = \inf\{s : \sigma_s \geq t\}$$

(which is defined for all t, since the recurrence of C implies $\sigma_\infty \equiv \infty$), then $f(C(\gamma_t))$ is a complex Brownian motion.

Remark: The reader might enjoy looking at Lévy's original proof (see page 270 of Lévy (1948)) for an intuitive but nonrigorous explanation.

Having found that analytic functions map two-dimensional Brownian motion into a time change of itself, it now seems natural to ask what functions do this for n-dimensional Brownian motion. Let $F_t^i = $ the ith component of $f(B_t)$. If $f(B_t)$ is a time change of Brownian motion, we must have $\langle F^i, F^j \rangle_t \equiv \delta_{ij} A_t$, so it follows from Itô's formula that if f_i is the ith component of f, then

(5)
$$\nabla f_i \cdot \nabla f_j = 0 \text{ if } i \neq j$$
$$|\nabla f_i|^2 \text{ is independent of } i.$$

The last two conditions imply that f is conformal (angle-preserving), so a well-known theorem of Liouville (see Spivak (1979), Vol. III, pages 302–310) implies that f must be a composition of mappings from the following list:

(a) translation: $x \to x + y$
(b) multiplication by a constant: $x \to cx$
(c) orthogonal transformation: $x \to Ax$ where A is a matrix with $(Ax, Ay) = (x, y)$ for all x, y
(d) inversion: $J(x) = x/|x|^2$.

We will call anything that can be made from the mappings listed above a *Kelvin transformation*. The first three transformations clearly map Brownian motion into a time change of itself, but if $d \geq 3$, the last transformation does not. The simplest way to prove this is to use two facts from Chapter 1.

First Proof When $d \geq 3$, $|B_t| \to \infty$ as $t \to \infty$, so $J(B_t) \to 0$, something which is impossible for a time change of Brownian motion in $d \geq 3$, since in this case $P(B_t = 0 \text{ for some } t > 0) = 0$ and, as we mentioned above, $|B_t| \to \infty$ as $t \to \infty$.

A second, more tedious, way of proving this is to use Itô's formula to show that the components of $J(B_t)$ are not local martingales.

Second Proof Let $J_i(x) = x_i/|x|^2$. A simple computation shows that

$$D_i J_1 = \begin{cases} x_1 \cdot \dfrac{-2x_i}{|x|^4} & i \neq 1 \\ \dfrac{-2x_1^2}{|x|^4} + \dfrac{1}{|x|^2} & i = 1 \end{cases}$$

$$D_{ii} J_1 = \begin{cases} x_1 \left(\dfrac{-2}{|x|^4} + 2\dfrac{(2x_i)^2}{|x|^6} \right) & i \neq 1 \\ \dfrac{-4x_1}{|x|^4} + 2\dfrac{4x_1^3}{|x|^6} + \dfrac{-2x_1}{|x|^4} & i = 1, \end{cases}$$

2.12 Conformal Invariance in $d \geq 2$, Kelvin's Transformations

so

$$\Delta J_1 = \frac{-6x_1}{|x|^4} + \frac{8x_1^3}{|x|^6} + \sum_{i=2}^{d}\left(\frac{-2x_1}{|x|^4} + \frac{8x_1 x_i^2}{|x|^6}\right)$$

$$= |x|^{-4}\left((-4 - 2d)x_1 + \frac{x_1}{|x|^2}\sum_{i=1}^{d} 8x_i^2\right)$$

$$= \frac{x_1}{|x|^4}(4 - 2d),$$

and we see that J_1 is harmonic if and only if $d = 2$.

Applying Itô's formula now, we see that $J_1(B_t)$ is not a local martingale, and hence $J(B_t)$ cannot be a time change of Brownian motion. This is a shame, because $J(B_t)$ has the right covariances

$$|\nabla J_1|^2 = \frac{4x_1^4}{|x|^8} - \frac{4x_1^2}{|x|^6} + \frac{1}{|x|^4} + \sum_{i=2}^{d} x_1^2 \frac{4x_i^2}{|x|^8}$$

$$= -\frac{4x_1^2}{|x|^6} + \frac{1}{|x|^4} + 4x_1^2 \sum_{i=1}^{d}\frac{x_i^2}{|x|^8} = \frac{1}{|x|^4}$$

and

$$\nabla J_1 \cdot \nabla J_2 = \frac{4x_1^3 x_2}{|x|^8} - \frac{2x_1 x_2}{|x|^6} + \frac{4x_1 x_2^3}{|x|^8} - \frac{2x_1 x_2}{|x|^6} + 4x_1 x_2 \sum_{i=3}^{d}\frac{x_i^2}{|x|^8}$$

$$= 4x_1 x_2 \sum_{i=1}^{d}\frac{x_i^2}{|x|^8} - \frac{4x_1 x_2}{|x|^6} = 0.$$

What kind of process is $J(B_t)$? One clue comes from the first proof given above. $J(B_t) \to 0$ as $t \to \infty$, so if the reader knows something about the results of Chapter 3, he might guess that $J(B_t)$ is "Brownian motion conditioned to converge to 0 as $t \to \infty$" (whatever that means). This guess is correct, but we will not be in a position to explain this until Section 3.4. For the moment, we prove the following result, which is useful for studying the behavior of $J(B_t)$.

(6) If u is harmonic, then $|B_t|^{2-d} u(J(B_t))$ is a local martingale.

Proof Let $f(x) = |x|^{2-d}$ and $g(x) = u(J(x))$. By Itô's formula, it suffices to show that $\Delta(fg) = 0$. Recalling that

$$D_{ii}(fg) = (D_{ii}f)g + 2D_i f D_i g + f D_{ii} g,$$

we take things in three steps.

$$D_i f = \frac{2-d}{2} \frac{2x_i}{|x|^n}$$

$$D_{ii} f = \frac{2-d}{|x|^d} - \frac{(2-d)}{2} \cdot \frac{d}{2} \cdot \frac{(2x_i)^2}{|x|^{d+2}}$$

$$\Delta f = d\frac{2-d}{|x|^d} - (2-d)d\frac{|x|^2}{|x|^{d+2}} = 0 \tag{a}$$

$$D_i g = \sum_{j=1}^{d} D_j u(J(x)) D_i J_j(x)$$

$$= \sum_{j=1}^{d} D_j u(J(x)) \left(\frac{-x_j}{|x|^4}\right)(2x_i) + D_i u(J(x)) \cdot \frac{1}{|x|^2}$$

$$\nabla f \cdot \nabla g = \frac{(2-d)}{|x|^d} \left\{ -\sum_{i=1}^{d}\sum_{j=1}^{d} D_j u(J(x))\frac{2x_i^2 x_j}{|x|^4} + \sum_{i=1}^{d} D_i u(J(x)) \cdot \frac{x_i}{|x|^2} \right\}$$

$$= \frac{(2-d)}{|x|^d} \left\{ -\sum_{j=1}^{d} D_j u(J(x))\frac{x_j}{|x|^2} \right\} \tag{b}$$

$$D_{ii} g = \sum_{j=1}^{d}\sum_{k=1}^{d} D_{jk} u(J(x)) D_i J_j(x) D_i J_k(x) + \sum_{j=1}^{d} D_j u(J(x)) D_{ii} J_j(x).$$

Since $\nabla J_j \cdot \nabla J_k = \delta_{jk}/|x|^4$, it follows that

$$\Delta g = \Delta u(J(x))/|x|^4 + \sum_{j=1}^{d} D_j u(J(x))\Delta J_j(x).$$

If u is harmonic, then $\Delta u = 0$, so multiplying by $f(x) = |x|^{2-d}$ gives

$$f\Delta g = \frac{1}{|x|^{d-2}} \sum_{j=1}^{d} D_j u(J(x)) \frac{x_j}{|x|^4}(4 - 2d), \tag{c}$$

and combining this with (a) and (b) gives

$$\Delta(fg) = (\Delta f)g + 2\nabla f \cdot \nabla g + f(\Delta g) = 0.$$

2.13 Change of Measure, Girsanov's Formula

In this section we will show that if X is a continuous semimartingale with respect to a measure P and Q is a measure equivalent to P, then X is a semimartingale with respect to Q, and we will give explicit formulas for the martingale and the bounded variation processes in its decomposition. We will see in Chapter 8 that this rather esoteric-sounding result is actually quite useful.

Let Q be equivalent to P, let $\alpha = dQ/dP$ (Radon-Nikodym derivative), and let $\alpha_t = E(\alpha|\mathscr{F}_t)$, where this and other nonsubscripted expectations are taken w.r.t. P, and we have chosen versions of the conditional expectations so that $t \to \alpha_t$ is continuous. There is a simple interpretation for α_t.

2.13 Change of Measure, Girsanov's Formula

(1) Let Q_t and P_t be the restrictions of Q and P to $\mathscr{F}_t$. Then

$$\alpha_t = dQ_t/dP_t.$$

Proof Clearly $\alpha_t = E(\alpha|\mathscr{F}_t) \in \mathscr{F}_t$. If $A \in \mathscr{F}_t$, then

$$\int_A \alpha_t\, dP_t = \int_A \alpha_t\, dP,$$

and it follows from the definition of conditional expectation that

$$\int_A \alpha_t\, dP = \int_A \alpha\, dP \equiv Q(A) = Q_t(A),$$

so $\alpha_t = dQ_t/dP_t \ (= dQ_t/dP)$.

The next step in our development is to show

(2) Y_t is a local martingale/Q if and only if $\alpha_t Y_t$ is a local martingale/P.

Proof Let Y_t be a martingale/Q. We want to show that $\alpha_t Y_t$ is a martingale/P, or that for all $s < t$ and $A \in \mathscr{F}_s$,

$$\int_A \alpha_t Y_t\, dP = \int_A \alpha_s Y_s\, dP.$$

To do this, let E_Q denote expectation with respect to Q and observe that

$$E(\alpha_t Y_t 1_A) = E(Y_t 1_A E(\alpha|\mathscr{F}_t))$$
$$= E(E(Y_t 1_A \alpha|\mathscr{F}_t)) = E(Y_t 1_A \alpha)$$
$$= E_Q(Y_t 1_A) = E_Q(Y_s 1_A).$$

Since Y is a martingale/Q, and reversing the steps just executed above and conditioning on $\mathscr{F}_s$ instead of $\mathscr{F}_t$ shows that

$$E_Q(Y_s 1_A) = E(Y_s 1_A \alpha) = E(\alpha_s Y_s 1_A).$$

From the last equation, it follows immediately that if Y_t is a local martingale/Q, then $\alpha_t Y_t$ is a local martingale/P. To prove the converse, observe that (a) if $\beta = dP/dQ$, $\beta_t = E_Q(\beta|\mathscr{F}_t)$, and Z_t is a local martingale/P, then $\beta_t Z_t$ is a local martingale/Q and (b) if P_t and Q_t are the restrictions of P and Q to $\mathscr{F}_t$, then $\beta_t = dP_t/dQ_t$ and $\alpha_t = dQ_t/dP_t$, so $\beta_t = \alpha_t^{-1}$.

With (2) established, we are ready to prove Girsanov's formula.

(3) If X is a local martingale/P and we let $A_t = \int_0^t \alpha_s^{-1} d\langle \alpha, X \rangle_s$, then $X_t - A_t$ is a local martingale/Q.

Proof Although the formula for A looks a little strange, it is easy to see that it must be the right answer. If we suppose that A is locally b.v. and has $A_0 = 0$, integrating by parts gives

(*) $\quad \alpha_t(X_t - A_t) - \alpha_0 X_0 = \int_0^t (X_s - A_s)\, d\alpha_s + \int_0^t \alpha_s\, dX_s - \int_0^t \alpha_s\, dA_s + \langle \alpha, X \rangle_t.$

Remark: At this point, we need the assumption made in Section 2.3 that our filtration only admits continuous martingales, so there is a continuous version of α to which we can apply our integration-by-parts formula.

Since α_s and X_s are local martingales/P, the first two terms on the right in (∗) are local martingales/P, and in view of (2), if we want $X_t - A_t$ to be a local martingale/Q, we need to choose A so that the sum of the third and fourth terms $\equiv 0$, that is,

$$\int_0^t \alpha_s \, dA_s = \langle \alpha, X \rangle_t.$$

From the last equation, it is clear that we want

$$A_t = \int_0^t \alpha_s^{-1} \, d\langle \alpha, X \rangle_s$$

and that this equality will make the bounded variation terms in (∗) cancel each other.

The last detail remaining is to prove that the integral that defines A_t exists. Let $T_n = \inf\{t : \alpha_t \leq n^{-1}\}$. If $t \leq T_n$, then

$$\int_0^t \frac{|d\langle \alpha, X \rangle_s|}{\alpha_s} \leq n \int_0^t |d\langle \alpha, X \rangle_s| < \infty$$

by the Kunita-Watanabe inequality, so A_t is well defined for $t \leq T = \lim_{n \to \infty} T_n$. Since $\alpha_t = E(\alpha | \mathscr{F}_t)$ is uniformly integrable, the optional stopping theorem implies that

$$E(\alpha; T_n < \infty) = E(\alpha_{T_n}; T_n < \infty) \leq 1/n.$$

Letting $n \to \infty$, we see that $\alpha = 0$ a.s. on $\{T < \infty\}$, so $P\{T < \infty\} \leq P\{\alpha = 0\} = 0$ (since Q is equivalent to P), and it follows that the integral that defines A exists.

A typical application of Girsanov's formula is to the following:

Problem Let X_t be a local martingale/P. Given a function b, find a measure Q so that

$$X_t - \int_0^t b(X_s) \, d\langle X \rangle_s$$

is a local martingale/Q.

Solution: From (3), we see that we want

(∗) $$\int_0^t \frac{d\langle \alpha, X \rangle_s}{\alpha_s} = \int_0^t b(X_s) \, d\langle X \rangle_s.$$

If $\alpha_t = 1 + \int_0^t H_s \, dX_s$, then by the formula for the covariance of two stochastic integrals

$$\langle \alpha, X \rangle_t = \int_0^t H_s \, d\langle X \rangle_s.$$

Substituting the last equality into (∗) gives $H_s/\alpha_s = b(X_s)$, or

$$\int_0^t H_s \, dX_s = \int_0^t \alpha_s b(X_s) \, dX_s.$$

The last equality can be written as

$$\alpha_t - 1 = \int_0^t \alpha_s b(X_s) \, dX_s,$$

so it follows from the exponential formula ((2) in Section 2.10) that (∗) is satisfied if and only if

$$\alpha_t = \exp(Y_t - \tfrac{1}{2}\langle Y \rangle_t)$$

where

$$Y_t = \int_0^t b(X_s) \, dX_s.$$

This result will be very useful in Chapter 8.

To get an idea of what α_t and Y_t look like, consider the following simple examples:

Example 1 $b(x) \equiv \mu$ (Brownian motion plus a constant drift):

$$Y_t = \mu(B_t - B_0)$$
$$\alpha_t = \exp\left(\mu(B_t - B_0) - \frac{t}{2}\right).$$

Example 2 $b(x) = -ax$ (the Ornstein-Uhlenbeck process):

$$Y_t = -a \int_0^t B_s \, dB_s = -a(B_t^2 - B_0^2 - t)$$

$$\langle B^2 \rangle_t = \int_0^t |B_s|^2 \, ds$$

$$\alpha_t = \exp\left(-a(B_t^2 - B_0^2 - t) - \frac{a^2}{2}\int_0^t |B_s|^2 \, ds\right).$$

2.14 Martingales Adapted to Brownian Filtrations

Let $\mathbb{B} = \{\mathcal{B}_t, t \geq 0\}$ be the filtration generated by a one-dimensional Brownian motion B_t. In this section we will show (1) all local martingales adapted to $\mathbb{B}$ are continuous and (2) every martingale $X \in \mathcal{M}^2(\mathbb{B})$ can be written as $X_0 +$

$\int_0^t H_s \, dB_s$ where $H \in \Pi_2(B)$. The generalizations of these results to d dimensions will be obvious once the reader has seen the proofs below.

(1) All local martingales adapted to $\mathbb{B}$ are continuous.

Proof Let X_t be a local martingale adapted to $\mathbb{B}$ and let T_n be a sequence of stopping times that reduces X. It suffices to show that for each n, $X_{t \wedge T_n \wedge n}$ is continuous, or in other words, it is enough to show that the result holds for martingales of the form $Y_t = E(Y|\mathcal{B}_t)$ where $Y \in \mathcal{B}_n$. We build up to this level of generality in three steps.

Step 1: Let $Y = f(B_n)$ where f is a bounded continuous function. If $t \geq n$, then $Y_t = f(B_n)$, so $t \to Y_t$ is trivially continuous for $t > n$. If $t < n$, the Markov property of Brownian motion implies that

$$E(Y|\mathcal{B}_t) = h(n - t, B_t)$$

where

$$h(s, x) = \int \frac{1}{\sqrt{2\pi s}} e^{-(y-x)^2/2s} f(y) \, dy.$$

It is easy to see that $h(s, x)$ is a continuous function on $(0, \infty) \times R$, so Y_t is continuous for $t < n$. To check continuity at $t = n$, observe that changing variables $y = x + (z/\sqrt{s})$ gives

$$h(s, x) = \int \frac{1}{\sqrt{2\pi}} e^{-z^2/2} f\left(x + \frac{z}{\sqrt{s}}\right) dz,$$

so the dominated convergence theorem implies that as $t \uparrow n$, $h(n - t, B_t) \to f(B_n)$.

Step 2: Let $Y = f_1(B_{t_1}) f_2(B_{t_2})$ where $t_1 < t_2 \leq n$ and f_1, f_2 are bounded and continuous. If $t \geq t_1$, then

$$Y_t = f_1(B_{t_1}) E(f_2(B_{t_2})|\mathcal{B}_t),$$

so the argument from Step 1 implies that Y_t is continuous on $[t_1, \infty)$. On the other hand, if $t \leq t_1$, then $Y_t = E(Y_{t_1}|\mathcal{B}_t)$ and

$$Y_{t_1} = f_1(B_{t_1}) E(f_2(B_{t_2})|\mathcal{B}_{t_1})$$
$$= f_1(B_{t_1}) g(B_{t_1})$$

where

$$g(x) = \int \frac{1}{\sqrt{2\pi(t_2 - t_1)}} e^{-(y-x)^2/2(t_2 - t_1)} f_2(y) \, dy$$

is a continuous function, so

$$Y_t = E(f_1(B_{t_1}) g(B_{t_1})|\mathcal{F}_t) \quad \text{for} \quad t \leq t_1,$$

2.14 Martingales Adapted to Brownian Filtrations

and it follows from Step 1 that Y_t is continuous on $[0, t_1]$. Repeating the argument above and using induction, it follows that the result holds if $Y = f_1(B_{t_1}) \cdots f_k(B_{t_k})$ where $t_1 < t_2 < \cdots < t_k \leq n$ and $f_1, \ldots, f_k$ are bounded continuous functions.

Step 3: Let $Y \in \mathscr{B}_n$ with $E|Y| < \infty$. It follows from a standard application of the monotone class theorem that for any $\varepsilon > 0$, there is a random variable X^ε of the form considered in Step 2 that has $E|X^\varepsilon - Y| < \varepsilon$. Now

$$|E(X^\varepsilon|\mathscr{B}_t) - E(Y|\mathscr{B}_t)| \leq E(|X^\varepsilon - Y| \,|\, \mathscr{B}_t),$$

and if we let $Z_t = E(|X^\varepsilon - Y| \,|\, \mathscr{B}_t)$, it follows from Doob's inequality that

$$\lambda P(\sup_{t \leq n} Z_t > \lambda) \leq E Z_n = E|X^\varepsilon - Y| < \varepsilon.$$

Now $X^\varepsilon(t) = E(X^\varepsilon|\mathscr{B}_t)$ is continuous, so letting $\varepsilon \to 0$ we see that for a.e. ω, $Y_t(\omega)$ is a uniform limit of continuous functions, so Y_t is continuous.

Remark: I would like to thank Michael Sharpe for telling me about the proof given above.

Having proved (1), we now turn our attention to proving that every martingale $X \in \mathscr{M}^2(\mathbb{B})$ can be written as $X_0 + \int_0^t H_s \, dB_s$ where $H \in \Pi^2(B)$. The first step in doing this is to take a general filtration $\mathbb{F}$ and an $X \in \mathscr{M}^2(\mathbb{F})$ with $X_0 = 0$ and examine $\{H \cdot X : H \in \Pi^2(X)\}$ as a subspace of $\mathscr{M}_0^2 = \{Y \in \mathscr{M}^2 : Y_0 = 0\}$. (Here and below when the filtration is not specified, it is $\mathbb{F}$.)

In Section 2.1, we showed that $X \to X_\infty$ maps $\mathscr{M}^2$ onto $L^2(\mathscr{F}_\infty)$, so $(X, Y) = E X_\infty Y_\infty$ makes $\mathscr{M}^2$ a Hilbert space. In the theory of Hilbert spaces, the following definition applies:

X and Y are orthogonal if $(X, Y) = 0$.

This notion is weaker than the following concept from the theory of martingales:

X and Y are uncorrelated if $\langle X, Y \rangle \equiv 0$, that is, $X_t Y_t$ is a martingale.

The first definition corresponds to $E\langle X, Y \rangle_\infty = 0$, so it is much less restrictive than the second. In some circumstances, however, if Y is orthogonal to all the martingales in a subspace $\mathscr{N}$, then it is also uncorrelated with them.

A subspace $\mathscr{N}$ of $\mathscr{M}^2$ is said to be stable if it is

(a) closed in the $\mathscr{M}^2$ topology
(b) closed under stopping: if $X \in \mathscr{N}$ and T is a stopping time, $X_t^T = X_{T \wedge t} \in \mathscr{N}$.

(3) Let $\mathscr{N}$ be a stable subspace. If $X \in \mathscr{N}$ and $H \in \Pi^2(X)$, then $H \cdot X \in \mathscr{N}$.

Proof The key to the proof is the following:

(4) Let $\mathscr{N}$ be a stable subspace and $\mathscr{N}' = \{Y : (X, Y) = 0 \text{ for all } X \in \mathscr{N}\}$ be the orthogonal complement of $\mathscr{N}$. If $X \in \mathscr{N}$ and $Y \in \mathscr{N}'$, then $X_t Y_t$ is a martingale.

Proof Let $X \in \mathcal{N}$. If T is a stopping time, then $X^T \in \mathcal{N}$, so if $Y \in \mathcal{N}'$, then
$$0 = (X^T, Y) = E(X_T Y_\infty) = E(X_T Y_T),$$
since martingale increments are orthogonal. (3) now follows from

(5) If for all bounded stopping times T we have $EZ_T = EZ_0$, then Z_t is a martingale.

Proof Let $s < t$, $A \in \mathscr{F}_s$, and let $T = s$ on A and $T = t$ on A^c. By hypothesis, $EZ_T = EZ_0 = EZ_t$, so subtracting $EZ_t 1_{A^c}$ from both sides, we find $EZ_s 1_A = EZ_t 1_A$. Since this holds for all $A \in \mathscr{F}_s$, then $Z_s = E(Z_t|\mathscr{F}_s)$, and Z_t is a martingale.

With (4) established, it is easy to prove (3): If $Y \in \mathcal{N}'$, then $\langle X, Y \rangle \equiv 0$, so
$$\langle H \cdot X, Y \rangle = \int_0^\infty H_s d\langle X, Y \rangle_s = 0$$
and $E((H \cdot X)_\infty Y_\infty) = E\langle H \cdot X, Y \rangle_\infty = 0$. This shows that $H \cdot X$ is in the orthogonal complement of $\mathcal{N}'$, that is, $H \cdot X \in \mathcal{N}$. We have proved (3).

Since $\{H \cdot X : H \in \Pi^2(X)\}$ is a stable subspace, it follows from (3) that we have

(6) The smallest stable subspace containing X is $\{H \cdot X : H \in \Pi^2(X)\}$.

Having established some properties of stable subspaces, we are now ready to prove the second result we stated at the beginning of this section.

(2) If $X \in \mathcal{M}^2(\mathbb{B})$ and $X_0 = 0$, then there is an $H \in \Pi^2(B)$ such that
$$X_t = \int_0^t H_s dB_s.$$

Proof Let $\mathcal{N} = \{H \cdot B : H \in \Pi^2(B)\}$. $\mathcal{N}$ is a stable subspace of $\mathcal{M}^2$. Let Y be a martingale in the orthogonal complement $\mathcal{N}'$. We will show that $Y \equiv 0$. The key to doing this is the following formula:

(7) $E(e^{i\theta(B_t - B_s)} Y_t | \mathscr{F}_s) = Y_s e^{-(t-s)\theta^2/2}.$

As we observed in Section 2.11 when we proved a similar formula, it suffices to prove the result when $s = 0$. Itô's formula shows that
$$e^{i\theta B_t} = 1 + i\theta \int_0^t e^{i\theta B_s} dB_s - \frac{\theta^2}{2} \int_0^t e^{i\theta B_s} ds.$$

Let $A \in \mathscr{F}_0$ and multiply both sides by $1_A Y_t$. If we take expectations and let $j(t) = E(e^{i\theta B_t} 1_A Y_t)$, then we get
$$j(t) = E(1_A Y_t) + i\theta E\left(Y_t \int_0^t 1_A e^{i\theta B_s}\right) - \frac{\theta^2}{2} \int_0^t j(s) ds.$$

Since Y_t is a martingale and $A \in \mathscr{F}_0$, $E(1_A Y_t) = E(1_A Y_0)$. To deal with the second term on the right, we observe that $Z_t = \int_0^t 1_A e^{i\theta B_s} dB_s \in \mathcal{N}$, so it follows from (3) that $\langle Y, Z \rangle \equiv 0$ and hence that $E(Y_t Z_t) = 0$. Combining the last two observations with the formula above shows that

$$j(t) = E(1_A Y_0) - \frac{\theta^2}{2} \int_0^t j(s)\,ds.$$

As we argued in Section 2.11, this result implies that

$$j'(t) = \frac{-\theta^2}{2} j(t) \quad j(0) = E(1_A Y_0),$$

so we have $j(t) = e^{-\theta^2 t/2} E(1_A Y_0)$, proving (7).

The next step in the proof is to extend (7) to a statement about an arbitrary finite set of increments of B_t. Let $0 = t_0 < t_1 < \cdots < t_n$ and let $Q_n = \prod_{m=1}^n \exp(i\theta_m(B(t_m) - B(t_{m-1})))$. From the orthogonality of martingale increments and (7), it follows that

$$\begin{aligned} E(Q_n Y_\infty) &= E(Q_n Y(t_n)) \\ &= EE(Q_n Y(t_n) | \mathscr{F}_{t_{n-1}}) \\ &= E(Q_{n-1} E(e^{i\theta_n(B(t_n) - B(t_{n-1}))} Y(t_n) | \mathscr{F}_{t_{n-1}})) \\ &= E(Q_{n-1} Y(t_{n-1})) e^{-(t_n - t_{n-1})\theta_n^2/2}, \end{aligned}$$

so it follows by induction that

(8) $$E(Q_n Y_\infty) = E\left(Y_0 \prod_{m=1}^n e^{-(t_m - t_{m-1})\theta_m^2/2}\right) = 0,$$

since $Y_0 = 0$. Let μ be the signed measure $Y_\infty P$ on $\mathscr{B}_\infty$ where $\mu(A) = \int_A Y_\infty\,dP$. Since $E|Y_\infty| < \infty$, μ has bounded variation. Let ν be the measure on R^n that is the image of μ under the mapping $\omega \to (B_{t_1}(\omega), \ldots, B_{t_n}(\omega) - B_{t_{n-1}}(\omega))$. The measure ν is 0, since (8) shows that its Fourier transform is 0. This fact in turn implies that for any bounded measurable function on R^n,

$$\int_\Omega f(B_{t_1}, \ldots, B_{t_n} - B_{t_{n-1}}) Y_\infty\,dP = \int_{R^n} f\,d\nu = 0$$

and, taking limits, that for any bounded $Z \in \mathscr{B}_\infty$, $E(ZY_\infty) = 0$. Taking $Z = \text{sgn}(Y_\infty)$ ($= 1$ if $Y_\infty > 0$, -1 if $Y_\infty < 0$), we see that $E(|Y_\infty|) = 0$, so $Y_\infty \equiv 0$.

Notes on Chapter 2

Most of Chapter 2 follows Meyer (1976). To be precise, the proofs of the main results in Sections 2.5, 2.6, 2.8, 2.11, 2.13, and 2.14 were obtained by specializing his proofs to the continuous case and translating them from Strasbourg French to southern California English. In writing some of the rest of the material, I have used a number of other sources, primarily McKean (1969), Friedman (1975), and Rogers' article in Williams (1981). The reader can find a beautiful treatment of stochastic integration in Dellacherie (1980). The approach taken there shows that it is "inevitable" that we were led to integrate predictable processes w.r.t. semimartingales.

A Word about the Notes

As the reader has probably already noticed, the note above says nothing about the history of stochastic integration. This situation exists because in my several attempts to write such notes I have found that I have neither the time nor the patience to read all the original papers and to try to figure out who did what and when. Because of this, I have contented myself to merely list the relevant papers in the references and to confine the discussion in the notes to where I got the proofs presented here and where the reader can find more about the subject.

3 Conditioned Brownian Motions

3.1 Warm-Up: Conditioned Random Walks

Let B_t be a d-dimensional Brownian motion, let $H = R^{d-1} \times (0, \infty)$ be the upper half space, and let $\tau = \inf\{t : B_t \notin H\}$ be the exit time from H. In the next two sections, we define Brownian motion conditioned to exit H at 0 and conditioned to never leave H. To prepare the reader for the Brownian motion definitions, this section is devoted to a discussion of the analogous problems for the simple random walk.

To construct the simple random walk in Z^d, let $X_1, X_2, \ldots$ be independent random variables with $P(X_m = v) = 1/2d$ for $v = e_1, -e_1, \ldots, e_d, -e_d$ and define the random walk starting at x by letting $S_0 = x$ and $S_n = S_{n-1} + X_n$ for $n \geq 1$. The simple random walk is the discrete analogue of Brownian motion—it has independent increments and paths that are as continuous as they can be.

Let $N = \inf\{n \geq 0 : S_n^d = 0\}$. The random walk analogue of $(B_{t \wedge \tau} | B_\tau = 0)$ is $(S_{n \wedge N} | S_N = 0)$, but there is a big difference. If the starting point $S_0 = x \in H$, then $P_x(S_N = 0) > 0$, so the process can be defined by elementary conditional probabilities. In order to understand how $(S_{n \wedge N} | S_N = 0)$ behaves, we start by considering what happens at the first step. If $S_0 = x \in H$, then by the definition of conditional probability and the Markov property,

(1) $$P_x(S_1 = y | S_N = 0) = \frac{P_x(S_1 = y, S_N = 0)}{P_x(S_N = 0)} = \frac{P_x(S_1 = y) P_y(S_N = 0)}{P_x(S_N = 0)}.$$

It is easier to see what is going on if we write the right-hand side in more abstract notation. Let $p(x, y)$ be the transition probability for $S_{n \wedge N}$, that is,

$$p(x, y) = \begin{cases} 1/2d & \text{if } x \in H, |x - y| = 1 \\ 1 & \text{if } x \notin H, x = y \\ 0 & \text{otherwise.} \end{cases}$$

Let $h(x) = P_x(S_N = 0)$ and let $q(x, y) = P_x(S_1 = y | S_N = 0)$. By considering the two cases $x \in H$ and $x = 0$, we see that for all x with $h(x) > 0$,

91

(2) $$q(x,y) = h(x)^{-1}p(x,y)h(y).$$

The special relationship between q and p causes some nice cancellation to occur when we consider the joint distribution of the first two steps. If $x, y \in H$, then a simple extension of the computation in (1) shows that

$$\begin{aligned} P_x(S_1 = y, S_2 = z | S_N = 0) &= P_x(S_1 = y, S_2 = z, S_N = 0)/P_x(S_N = 0) \\ &= p(x,y)p(y,z)h(z)/h(x) \\ &= (h(x)^{-1}p(x,y)h(y))(h(y)^{-1}p(y,z)h(z)) \\ &= q(x,y)q(y,z) \end{aligned}$$

(and similar computations show that the result above is true whenever y or $x = 0$), so the joint distribution of the first two steps is the product of the one-step probabilities, a trait indicative of the Markov property.

The reader can easily confirm that the pattern above persists for any finite dimensional distribution and hence that $(S_{n \wedge N} | S_N = 0)$ is a Markov chain with state space $H \cup \{0\}$ and transition probability q given by (2).

The construction above can be generalized considerably. Let $p(x,y)$ be the transition probability for a Markov chain on a countable set S, that is, $p(x,y) \geq 0$ and $\sum_y p(x,y) = 1$. Let $h(x) \geq 0$ be a harmonic function for p, that is,

$$h(x) = \sum_y p(x,y)h(y).$$

Given these two ingredients, we can define a transition probability by setting

$$q(x,y) = h(x)^{-1}p(x,y)h(y)$$

for all x with $h(x) > 0$.

The q defined in the last paragraph is commonly called an h-transform of p. These processes were introduced by Doob in a paper titled "Conditioned Brownian Motion and the Boundary Limits of Harmonic Functions," so the reader should not be surprised that these processes will appear several times below. The first occurrence will be in the next section: by choosing the right harmonic function, we will get an h-transform of Brownian motion with $B_\tau = 0$ a.s. The answer will be obvious when you see it, so you should spend a moment thinking about what the h should be. It is one of the functions in Chapter 1.

In Section 3.3 we will encounter another h-transform: Brownian motion conditioned to never leave H (again you should try to guess the h), so for the rest of this section we will consider the analogous problem for simple random walk. Since the conditioning affects only the last component, it suffices to consider the one-dimensional case. Imitating the proof of (1) in Section 1.7 shows that if $T_x = \inf\{n \geq 0 : S_n = x\}$ and $a \leq x \leq b$, then

(3) $$P_x(T_a < T_b) = \frac{b-x}{b-a} \qquad P_x(T_b < T_a) = \frac{x-a}{b-a}.$$

If we let $a = 0$ and $b \to \infty$, we see that $P_x(T_0 < \infty) = 1$, in other words, the conditioning event has probability 0.

3.1 Warm up: Conditioned Random Walks

To construct $(S_n|T_0 = \infty)$, we take an obvious approximation: we consider $(S_n|T_0 > T_M)$, compute its transition probability $q_M(x, y)$, and let $M \to \infty$. From formula (2) above, we see that if $0 < x < M$ and $|x - y| = 1$, then

$$q_M(x, y) = \frac{1}{2} \cdot \frac{P_y(T_0 > T_M)}{P_x(T_0 > T_M)}$$
$$= \frac{1}{2} \frac{y/M}{x/M} = \frac{y}{2x}.$$

A remarkable aspect of the last formula is that M does not appear on the right-hand side (except in the requirement that $x < M$), so if $x > 0$ and $|x - y| = 1$, then

(4) $$q_\infty(x, y) = \lim_{M \to \infty} q_M(x, y) = \frac{y}{2x}$$

is the transition probability for random walk conditioned to have $T_0 = \infty$.

The reader should observe that q_∞ is itself an h-transform with $h(x) = x$, and the generalization to Brownian motion should then be obvious. The last sentence tells you one of the things I wanted you to guess. For one last hint about the other, observe that in the situations described above, $h = 0$ on the forbidden set and this keeps the h-transform from going there.

Exercise 1 It is interesting to see what happens when we condition an asymmetric random walk, that is, let $P(X_n = 1) = p$, $P(X_n = -1) = q = 1 - p$, and let $S_n = X_1 + \cdots + X_n$ with $X_1, X_2 \ldots$ independent.

(a) Define a function φ by setting $\varphi(0) = 0$ and $\varphi(x) = \varphi(x - 1) + (q/p)^x$, $x \neq 0$, and check that $\varphi(S_n)$ is a martingale.

(b) Repeating the argument for (3) above, we see that

$$P_x(T_b < T_a) = \frac{\varphi(x) - \varphi(a)}{\varphi(b) - \varphi(a)},$$

so if $x > 0$, $P_x(T_0 = \infty) = \varphi(x)/\varphi(\infty)$ where

$$\varphi(\infty) = \sum_{x=1}^{\infty} (q/p)^x \begin{cases} = \infty \text{ if } q \geq p \\ < \infty \text{ if } q < p. \end{cases}$$

(c) If $p > 1/2$, then $P_x(T_0 = \infty) > 0$ and $(S_n|T_0 = \infty)$ (with the elementary definition) is a Markov chain with transition probability

(5) $$q(x, y) = p(x, y) \frac{\varphi(y)}{\varphi(x)} \text{ when } x > 0.$$

As $x \to \infty$, $\varphi(x + 1)/\varphi(x) \to 1$. Therefore $q(x, x + 1) \to p$; in other words, as x gets large, the effect of the conditioning vanishes, and the transition probabilities approach those of the original random walk.

(d) If $p \leq 1/2$, then $P_x(T_0 = \infty) = 0$, but if we define $(S_n|T_0 = \infty)$ as the limit of $(S_n|T_0 > T_M)$, then it is a Markov chain with the transition probability given by (5). This time, as $x \to \infty$, $\varphi(x+1)/\varphi(x) \to q/p$, so $q(x, x+1) \to q$, that is, as x gets large, the conditioned process is again a random walk, but the conditioning changes the probability of $x \to x+1$ from p to $q \geq 1/2$.

3.2 Brownian Motion Conditioned to Exit $H = R^{d-1} \times (0, \infty)$ at 0

Let $H = R^{d-1} \times (0, \infty)$ be the upper half space and let $\tau = \inf\{t : B_t \notin H\}$. In this section, we will define Brownian motion conditioned to have $B_\tau = 0$. As we indicated in the last section, this will be done by introducing a suitable h-transform of the stopped process. To see which h to choose, we begin with some formal calculations. Let $z \in H$ and write $z = (x, y)$ where $x \in R^{d-1}$ and $y > 0$. Let $y' < y$, let $H' = R^{d-1} \times (y', \infty)$, and let $\tau' = \inf\{t : B_t \notin H'\}$. Using the definition of conditional probability, the strong Markov property, and formula (1) from Section 1.9 gives

$$P_z(B(\tau') = (x', y') | B_\tau = 0) = \frac{P_z(B(\tau') = (x', y'), B_\tau = 0)}{P_z(B_\tau = 0)}$$

$$= \frac{P_z(B(\tau') = (x', y')) P_{z'}(B_\tau = 0)}{P_z(B_\tau = 0)}$$

$$= C \frac{y - y'}{|z - z'|^d} \frac{h_0(z')}{h_0(z)},$$

where $C = \Gamma(d/2)/\pi^{d/2}$ and $h_0(z) = y/|z|^d$ is $1/C$ times the probability (density) of exiting H at 0 starting from z, and using a notation that we will use throughout this chapter, we have written $z' = (x', y')$ where $x' \in R^{d-1}$ and $y' > 0$.

The last formula tells us what the distribution of $B(\tau')$ should be under $P_z^0 \equiv P_z(\cdot | B_\tau = 0)$. By extending the last computation, we can compute the distribution of $B(t \wedge \tau')$ under P_z^0. Let $\mathscr{F}' = \mathscr{F}(\tau')$ and $A \in \mathscr{F}'$. Using the definition of conditional probability, two properties of conditional expectation, and, finally, the strong Markov property gives

$$P_z^0(A) = P_z(A, B_\tau = 0)/h_0(z)$$
$$= E_z E_z(1_A 1_{B_\tau = 0} | \mathscr{F}')/h_0(z)$$
$$= E_z(1_A E_z(1_{B_\tau = 0} | \mathscr{F}'))/h_0(z)$$
$$= E_z(1_A h_0(B(\tau')))/h_0(z).$$

The computations above have all been formal, but they have also told us what to guess, so we turn now to the task of defining Brownian motion conditioned to have $B_\tau = 0$. The first task is to say something specific about the probability space on which our conditioned process will be defined. To keep things simple, we will suppose that we have Brownian motion defined on our special probability space $(C, \mathscr{C})$ as a family of measures P_z, $z \in R^d$, that make

3.2 Brownian Motion Conditioned to Exit $H = R^{d-1} \times (0, \infty)$ at 0

the coordinate maps $B_t(\omega) = \omega_t$ a Brownian motion started at z, and we construct a new family of measures P_z^0, $z \in R^d$, that make the coordinate maps a Brownian motion started at z and "conditioned to exit at 0."

Let $H_n = R^{d-1} \times (2^{-n}, \infty)$, let $\tau_n = \inf\{t : B_t \notin H_n\}$, and let $\mathscr{F}_n = \mathscr{F}(\tau_n)$. The first step in our construction is to define P_z^0 on $\mathscr{F}_n$. If $A \in \mathscr{F}_n$, then in light of the formal calculations above, we let

(1) $$P_z^0(A) = E_z(h_0(B_{\tau_n}); A)/h_0(z).$$

The first thing to check is whether these definitions are consistent. Let $m < n$ and $A \in \mathscr{F}_m \subset \mathscr{F}_n$. Since h_0 is harmonic (see Section 1.10, Exercise 1) and bounded in H_n, $h_0(B_{t \wedge \tau_n})$ is a bounded martingale, and it then follows from the optional stopping theorem that

$$h_0(B_{t \wedge \tau_m}) = E_z(h_0(B_{t \wedge \tau_n}) | \mathscr{F}_m).$$

Integrating over $A \in \mathscr{F}_m$ and using the definition of conditional expectation gives

$$E_z(h_0(B_{t \wedge \tau_m}); A) = E_z(h_0(B_{t \wedge \tau_n}); A).$$

With the consistency of our definitions of $(B_{t \wedge \tau_n} | B_\tau = 0)$ established, the next thing to do is to put these processes together to construct $(B_{t \wedge \tau} | B_\tau = 0)$. Let $Y_n(\omega) = B_{\cdot \wedge \tau_n}$ (a random variable taking values in C). (1) specifies the distribution of Y_n, and hence of $(Y_1, Y_2, \ldots, Y_n)$, for any n. In the last paragraph, we showed that these finite-dimensional distributions are consistent, so it follows from the Kolmogorov extension theorem that we can construct on some probability space $(\Omega, \mathscr{F}, P)$ an infinite sequence of random variables with these finite-dimensional distributions. Since the random variables Y_n satisfy $Y_n(t) = Y_{n+1}(t)$, $t \leq \tau_n$, we can define a process on $[0, \tau)$ by letting $Y(t) = Y_n(t)$ for $t \leq \tau_n$. Y is our candidate for $(B_{t \wedge \tau} | B_\tau = 0)$. The first thing to check is that the conditioning has accomplished its goal.

(2) As $t \uparrow \tau$, $Y_t \to 0$ a.s.

Proof Let $G_\delta = \{(x, y) : 0 < y < \delta^{d+1}, |x| > \delta\}$ and let $T_\delta = \inf\{t : Y_t \in G_\delta\}$. Then

$$P(T_\delta < \tau_n) = E_z(h_0(B_{\tau_n}); T_\delta < \tau_n)/h_0(z).$$

Since $h_0(B_{t \wedge \tau_n})$ is a bounded martingale, using the optional stopping theorem at time T_δ shows that

$$P(T_\delta < \tau_n) = E_z(h_0(B(T_\delta)); T_\delta < \tau_n)/h_0(z)$$
$$\leq \sup\{h_0(w) : w \in G_\delta\}/h_0(z)$$
$$\leq \frac{\delta^{d+1}}{\delta^d} \cdot \frac{1}{h_0(z)}.$$

Since the right-hand side is independent of n, this expression gives $P(T_\delta < \tau) \leq \delta/h_0(z)$. Letting $\delta = 2^{-m}$ and summing gives

$P(T_2 - m < \tau$ for infinitely many $m) = 0$.

Combining the last observation with the simple-to-prove fact that $P(B_t^d \to 0$ as $t \uparrow \tau) = 1$ proves the result.

With (2) established, we can extend Y to be a continuous process on $[0, \infty)$ by setting $Y_t = 0, t \geq \tau$, and our construction is almost complete. The last detail is to move the measure back to $(C, \mathscr{C})$. This is no problem, however: $\omega \to Y(\omega)$ maps Ω into C, so we simply let P_z^0 be the image of P under this mapping.

The construction of P_z^0 above can be extended by translation to define measures $P_z^\theta, \theta \in R^{d-1}$, that are "Brownian motion starting at z and conditioned to exit H at $(\theta, 0)$." The next result justifies the description in quotation marks.

(3)   If $A \in \bigcup_n \mathscr{F}_n$, then
$$P_z^\theta(A) = \lim_{\varepsilon \downarrow 0} P_z(A \,|\, |B_\tau - (\theta, 0)| < \varepsilon).$$

Proof   Let $D_\varepsilon = \{\psi : |\varphi - \theta| < \varepsilon\}$ and let
$$h^\varepsilon(z) = P_z(|B_\tau - (\theta, 0)| < \varepsilon)$$
$$= \int_{D_\varepsilon} h_\psi(z) \, d\psi.$$

Since $h_\varepsilon(B_{t \wedge \tau_n})$ is a bounded martingale, we have for each $A \in \mathscr{F}_n$ that
$$P_z(A \,|\, |B_\tau - (\theta, 0)| < \varepsilon) = E_z(h^\varepsilon(B_{\tau_n}); A)/h^\varepsilon(z)$$
$$\to E_z(h_\theta(B_{\tau_n}); A)/h_\theta(z) = P_z^\theta(A)$$

as $\varepsilon \to 0$.

With (3) established, it is easy both to believe and to prove:

(4)   Let $A \in \mathscr{F}_\tau$ and $z \in H$. If we let $g(\theta) = P_z^\theta(A)$, then
$$g(B_\tau) = P_z(A \,|\, B_\tau).$$

Proof   By the bounded convergence theorem, it suffices to prove the result when $A \in \mathscr{F}_n$, $n \geq 0$. To prove the result in this case, we observe that from the definition of conditional expectation and the strong Markov property it follows that if I is a Borel subset of ∂H, then
$$P_z(B_\tau \in I, A) = E_z(P_z(B_\tau \in I | \mathscr{F}_n); A)$$
$$= E_z(P_{B(\tau_n)}(B_\tau \in I); A)$$
$$= E_z(h_I(B(\tau_n)); A)$$

where $h_I(z) =$ the probability of exiting H in I starting from z. Now
$$h_I(z) = \int_I h_\theta(z) \, d\theta,$$

so applying Fubini's theorem (everything is ≥ 0) and recalling the relevant definitions, we find that

$$P_z(B_\tau \in I, A) = \int_I E_z(h_\theta(B_{\tau_n}); A) \, d\theta$$
$$= \int_I h_\theta(z) P_z^\theta(A) \, d\theta$$
$$= \int_I h_\theta(z) g(\theta) \, d\theta$$
$$= E_z(g(B_\tau); B_\tau \in I).$$

Since I is an arbitrary Borel set, this shows that $g(B_\tau)$ is a version of $P_z(A|B_\tau)$ and completes the proof.

Note: The approach we have taken to the definition of P_z^θ follows Appendix II of Brossard (1975) with some minor modifications. This approach is slightly different from the one used in Section 3.1 (see the definition given in (2)), but it is easy to make the connection.

Exercise 1 If f is bounded and $f = 0$ on ∂H, then

$$E_x f(B_t) = \int q_t^\theta(x, y) f(y) \, dy,$$

where

$$q_t^\theta(x, y) = h_\theta(x)^{-1} p_t^H(x, y) h_\theta(y)$$

and p_t^H is the transition probability for Brownian motion killed when it leaves H (see Section 1.9 for a description of this process).

3.3 Other Conditioned Processes in H

As we promised in Section 3.1, in this section we will define Brownian motion conditioned to never leave H. In light of the discussion above, there is not much to say: It is an h-transform with $h(z) = z_d$, and it is defined almost exactly like the processes constructed in the last section. Let $G_n = R^{d-1} \times (0, 2^n)$, let $\tau_n = \inf\{t : B_t \notin G_n\}$, and let $\mathscr{F}_n = \mathscr{F}(\tau_n)$. If $A \in \mathscr{F}_n$, we let

(1) $\qquad P_z^\infty(A) = E_z(h_\infty(B_{\tau_n}); A)/h_\infty(z)$

where $h_\infty(z) = z_d$. As before, it is easy to see that these definitions are consistent, so if we let $Y_n(\omega) = B(\cdot \wedge \tau_n)$, then we can construct $(Y_1, Y_2, \cdots)$ on some probability space with $Y_n(t) = Y_{n+1}(t)$ for $t \leq \tau_n$, and we can define a process on $[0, \lim \tau_n)$ by letting $Y_t = Y_n(t)$ for $t \leq \tau_n$. To see that $\lim \tau_n = \infty$ a.s., we observe that $P_z^\infty(B_{\tau_n} \in \partial H) = 0$ (since $h_\infty = 0$ on ∂H), so the strong Markov property of Brownian motion and an obvious scaling imply that $(\tau_{n+1} - \tau_n)/2^n$, $n = 1, 2, \ldots$, are indpendent and identically distributed, which is more than enough for the desired conclusion.

With two examples of conditioned Brownian motion constructed, it is natural to ask if there are any other interesting examples, or what is the same: Can we describe the set of all nonnegative functions that are harmonic in H? To do this we will start by considering the analogous problem in $D = \{z : |z| < 1\}$.

(2) $u \geq 0$ is harmonic in D if and only if there is a finite measure μ on ∂D such that

$$u(z) = \int k_y(z) \, d\mu(y)$$

where

$$k_y(z) = \frac{1 - |z|^2}{|z - y|^d}.$$

Proof Now $k_y(z)$ is the probability density (with respect to the surface probability measure π) of exiting D at y when we start at z, so as we showed in Section 1.10, each of the functions k_y is harmonic. It is routine to show, using the result on differentiating under the integral sign given in Section 1.10, that

$$\Delta \int k_y(z) \, d\mu(y) = \int \Delta k_y(z) \, d\mu(y) = 0$$

(details are left to the reader), so it follows that all the functions defined above are harmonic.

To prove the converse requires a little more thought, but less tedious computations. For $0 < r < 1$, let μ_r be the measure on ∂D that has density $\varphi(y) = u(ry)$ with respect to π. μ_r has total mass

$$\int u(ry) \, d\pi(y) = u(0).$$

Since all these measures are concentrated on a compact set and have the same total mass, the Helly selection theorem implies that there is a subsequence μ_{r_n} that converges to a limit μ. It follows from the definition of μ_r that

$$\int k_y(z) \, d\mu_r(y) = \int k_y(z) u(ry) \, d\pi(y) = u(rz).$$

Letting $r = r_n$ and $n \to \infty$, we get

$$u(z) = \lim_{n \to \infty} u(zr_n) = \int k_y(z) \, d\mu(y),$$

since μ_{r_n} converges weakly to μ and $y \to k_y(z)$ is continuous.

To translate (2) into a result about H, we will use one of Kelvin's transformations (discussed in Section 2.12) to map D one-to-one onto H. To figure out which transformation we want, we start by calculating what the inversion $J(z) = z/|z|^2$ does to $D(e_d, 1)$, the ball of radius 1 centered at $e_d = (0, \ldots, 0, 1)$.

3.3 Other Conditioned Processes in H

If $z \in \partial D(e_d, 1) = \{z : |z - e_d| = 1\}$, then
$$z_1^2 + \cdots + z_{d-1}^2 + (z_d - 1)^2 = 1.$$

The last component of J is:
$$J_d(z) = z_d/(z_1^2 + \cdots + z_d^2).$$

The first equality above implies that
$$z_1^2 + \cdots + z_{d-1}^2 + z_d^2 - 2z_d = 0,$$

so $J_d(z) = 1/2$ for all $z \in \partial D(e_d, 1)$. A little thought shows that J maps $D(e_d, 1)$ one-to-one onto the half space $H' = \{z : z_d \geq 1/2\}$, so $K(z) = J(z + e_d) - (e_d/2)$ maps D one-to-one onto H.

From (6) in Section 2.12, it follows that there is also a one-to-one correspondence between positive harmonic functions in D and in H, that is, if u is harmonic in D, then $|x|^{2-d} u(K(x))$ is harmonic in H, and, conversely, if u is harmonic in H, then $|x|^{2-d} u(K(x))$ is harmonic in D.

Combining the last observation with (2) gives an integral representation for positive harmonic functions in H. It is easy to see without computation what functions will appear in the decomposition. The $k_y(z)$ with $y \neq -e_d$ get mapped to
$$C_y \frac{z_d}{|z - K(y)|^d},$$

that is, the probability density of exiting D at y gets mapped to a constant (that depends upon y) times the probability density of exiting at $K(y)$, and when $y = -e_d$, k_y gets mapped to Cz_d. To check the second claim, observe that this k_y is mapped to a function harmonic in H that vanishes on ∂H, and that mapping things back to D and using (2) shows that there is (up to a constant multiple) only one such function). Combining the observations above with (2) gives:

(3) If $u \geq 0$ is harmonic in H, then there is a constant C and a measure μ on ∂H such that
$$u(z) = Cz_d + \int_{\partial H} h_\theta(z) \, d\mu(\theta).$$

Remark: The reader should note that the measure μ is not necessarily finite and that this is only an "only if" statement. To assert that such a u is harmonic, we need a condition such as
$$\int_{\partial H} h_\theta(e_d) \, d\mu(\theta) < \infty$$

to assert that $u < \infty$ in H. It turns out that the condition just mentioned is necessary and sufficient for $u \not\equiv \infty$ and harmonic in D, but we will not pursue this here.

(3) gives us the answer to the question asked at the beginning of this section. It shows that all nonnegative functions that are harmonic in H are linear combinations of z_d and the h_θ's, so the processes we have constructed in the last two sections are the only interesting examples of conditioned processes in H. Nontrivial linear combinations being uninteresting, since, for instance, if μ is a probability measure and

$$h(z) = \int h_\theta(z)\,d\mu(\theta),$$

then the corresponding h-transform is just Brownian motion conditioned to have $B_t \stackrel{d}{=} \mu$.

3.4 Inversion in $d \geq 3$, B_t Conditioned to Converge to 0 as $t \to \infty$

Let $J(x) = x/|x|^2$. If B_t is a Brownian motion in $d \geq 3$, then (as we observed in Section 2.12) $J(B_t)$ is not a Brownian motion, since $J(B_t) \to 0$ as $t \to \infty$. What it is, of course, is an h-transform of Brownian motion. To prove this and to discover the h, let u be a harmonic function and let $g(x) = |J(x)|^{2-d}$. Now $J(J(x)) = x$, $g(J(x)) = |x|^{2-d}$, and $|x|^{2-d}u(J(x))$ is harmonic, so

$$E(g(J(B_t))u(J(B_t))|J(B_0) = x) = E(|B_t|^{2-d}u(J(B_t))|B_0 = J(x))$$
$$= |J(x)|^{2-d}u(J(J(x))) = g(x)u(x).$$

Since u is harmonic, we can rewrite the last expression as

$$g(x)u(x) = g(x)\int p_t(x,y)u(y)\,dy$$

where

$$p_t(x,y) = P(B_t = y|B_0 = x) = (2\pi t)^{-d/2}e^{-|x-y|^2/2t}.$$

If we let $\bar{p}_t(x,y) = P(J(B_t) = y|J(B_0) = x)$, then we have shown that

$$\int \bar{p}_t(x,y)g(y)u(y)\,dy = E(g(J(B_t))u(J(B_t))|J(B_0) = x)$$
$$= \int g(x)\frac{p_t(x,y)}{g(y)}g(y)u(y)\,dy$$

for all harmonic functions u. It follows that

$$\bar{p}_t(x,y) = g(x)p_t(x,y)g(y)^{-1},$$

that is, $J(B_t)$ is an h-transform of Brownian motion with $h(x) = 1/g(x) = |x|^{2-d}$, a constant multiple of the potential kernel defined in Section 1.8. After the fact, it should be obvious that this is the right h. It is nonnegative, is harmonic when $x \neq 0$, and converges to 0 as $|x| \to \infty$.

3.4 Inversion in $d \geq 3$, B_t Conditioned to Converge to 0 as $t \to \infty$

Using the last result, we can compute the occupation time density for Brownian motion killed when it leaves $D = \{z; |z| < 1\}$. Let f be bounded and $= 0$ on D^c, and let $\tau = \inf\{t : B_t \notin D\}$. In $d \geq 3$,

$$w(x) = E_x \int_0^\infty f(B_s)\,ds < \infty,$$

and the strong Markov property implies that

$$w(x) = E_x \int_0^\tau f(B_s)\,ds + E_x w(B_\tau).$$

so we have

$$E_x \int_0^\tau f(B_s)\,ds = w(x) - E_x w(B_\tau).$$

To compute the second term on the right, we observe that since $J(x) = x$ on ∂D, it follows that (for $x \neq 0$)

$$E_x w(B_\tau) = E_x w(J(B_\tau))$$
$$= h(J(x))^{-1} E_{J(x)} w(B_\tau) h(B_\tau)$$

where $h(x) = |x|^{2-d}$. Since $h = 1$ on ∂D, it follows that

$$E_x w(B_\tau) = h(J(x))^{-1} E_{J(x)} w(B_\tau)$$
$$= |x|^{2-d} E_{J(x)} \int_\tau^\infty f(B_s)\,ds$$
$$= |x|^{2-d} E_{J(x)} \int_0^\infty f(B_s)\,ds,$$

since $f = 0$ on D^c.

Combining the results of the last two paragraphs shows that (for $x \neq 0$):

$$E_x \int_0^\tau f(B_s)\,ds = w(x) - |x|^{2-d} w(J(x)),$$

so we have

(1) $$E_x \int_0^\tau f(B_s)\,ds = \int G_D(x, y) f(y)\,dy$$

where

$$G_D(x, y) = G(x, y) - |x|^{2-d} G(J(x), y).$$

It is trivial to extend the last formula to $x = 0$. To do this, we observe that as $x \to 0$,

$$|x|^{2-d} G(J(x), y) = c|x|^{2-d}\big||x|^{-2}x - y\big|^{2-d}$$

converges to $C(= \Gamma(d/2)/\pi^{d-2})$. If we set $G_D(0, y) = G(0, y) - C$, then (1) also holds for $d \geq 3$.

3.5 A Zero-One Law for Conditioned Processes

This section is devoted to proving a zero-one law, which is an important tool both for establishing properties of the conditioned processes and for deducing analytical results about boundary limits of harmonic functions from their probabilistic counterparts. Since our only applications will be to the second topic, we will prove the zero-one law only for Brownian motion conditioned to exit H at $(0,0)$. At the end of the proof, however, it will be obvious how to prove the result for other conditioned processes.

To state our result, we need several definitions.

(1) An event $A \in \mathscr{F}_\tau$ is said to be shift invariant if for all stopping times $T < \tau$
$$1_A \circ \theta_T = 1_A \quad P_z \text{ a.s. for all } z \in H$$
where θ_T is the (random) shift operator defined in Section 1.5 by
$$(\theta_T \omega)(t) = \omega(T(\omega) + t).$$

In words, we can determine whether or not a shift invariant event occurs by looking at B_t, $T \le t < \tau$, that is, these events concern the behavior of B_t as $t \uparrow \tau$. The following are typical examples of shift invariant events:

(i) $\{\lim_{t \uparrow \tau} u(B_t) \text{ exists}\}$
(ii) $\{\limsup_{t \uparrow \tau} |u(B_t)| < \infty\}$
(iii) $\{B_t \in A \text{ infinitely often as } t \uparrow \tau\} \equiv$
  $\{\omega: \text{for all } \varepsilon > 0 \text{ there is a } u \in (\tau - \varepsilon, \tau) \text{ with } B_u \in A\}$.

Let $\mathscr{I}$ be the collection of invariant events. Although the examples above suggest that there are a large number of events in $\mathscr{I}$, the next result shows that this impression is wrong. These events are all $\varnothing$ or Ω (give or take a set of measure 0).

(2) If $A \in \mathscr{I}$, then $z \to P_z^\theta(A)$ is constant and the constant is either 0 or 1.

The key to the proof is the following lemma, which is a strong Markov property for P_z^θ:

(3) If $A \in \mathscr{F}_\tau$ and $T < \tau$ is a stopping time, then
$$E_z^\theta(1_A \circ \theta_T | \mathscr{F}_T) = P_{B(T)}^\theta(A).$$

Proof We will prove the result first for $T < \tau_n$. By the bounded convergence theorem, it suffices to prove the result when $A \in \mathscr{F}_m (\equiv \mathscr{F}(\tau_m))$ and $m > n$. If $B \in \mathscr{F}_T$, then by the definition of P_z^θ,
$$E_z^\theta[1_A \circ \theta_T; B] = E_z[h_\theta(B_{\tau_m})(1_A \circ \theta_T); B]/h_\theta(z)$$
$$= E_z[(h_\theta(B_{\tau_m})1_A) \circ \theta_T; B]/h_\theta(z),$$
since $B_{\tau_m}(\omega) = B_{\tau_m}(\theta_T \omega)$. Applying the strong Markov property and using the

3.5 A Zero-One Law for Conditioned Processes

definition of P_z^θ with $z = B(T)$, we find that the above

$$= E_z[E_{B(T)}(h_\theta(B_{\tau_m})1_A); B]/h_\theta(z)$$
$$= E_z[h_\theta(B_T)P_{B(T)}^\theta(A); B]/h_\theta(z).$$

To convert this to the desired form, we observe that the optional stopping theorem implies that if $C \in \mathscr{F}_T$, then $E_z(h_\theta(B_T)1_C) = E_z(h_\theta(B(\tau_n))1_C)$, so the definition of P_z^θ for $C \in \mathscr{F}_T$ may be rewritten as

$$P_z^\theta(C) = E_z(h_\theta(B_T)1_C)/h_\theta(z),$$

and it follows that if $Y \in \mathscr{F}_T$ is bounded,

$$E_z^\theta(Y) = E_z(h_\theta(B_T)Y)/h_\theta(z).$$

Applying this result with $Y = P_{B(T)}^\theta(A)1_B$ and tracing back through the chain of equalities above, we get:

$$E_z^\theta[1_A \circ \theta_T; B] = E_z^\theta[P_{B(T)}^\theta(A); B]$$

for all $B \in \mathscr{F}_T$, which is the desired result in the case $T \leq \tau_n$. To remove this assumption, apply the result above to $T \wedge \tau_n$, let $B_n = B \cap \{T \leq \tau_n\}$, and let $n \to \infty$ to conclude that the last equality holds in general.

Proof of (2) Let $A \in \mathscr{I}$ and let $\varphi(z) = P_z^\theta(A)$. From (3), it follows that if $T < \tau$ is a stopping time, then

$$\varphi(z) = P_z^\theta(A) = P_z^\theta(A \circ \theta_T)$$
$$= E_z^\theta(P_z^\theta(A \circ \theta_T | \mathscr{F}_T))$$
$$= E_z^\theta(P_{B(T)}^\theta(A)) = E_z^\theta(\varphi(B_T))$$
$$= E_z(h_\theta(B_T)\varphi(B_T))/h_\theta(z).$$

Applying this result at $T = \inf\{t : B_t \notin D(z, \delta)\}$, we see that $g(z) = h_\theta(z)\varphi(z)$ has the averaging property

$$g(z) = \int_{\partial D(z, \delta)} g(y) \, d\pi(y),$$

so it follows from (7) in Section 2.10, that g is C^∞ and has $\Delta g = 0$.
  Since $g \geq 0$, it follows from (3) of Section 3.3 that

$$g(z) = Cz_d + \int_{\partial H} h_y(z) \, d\mu(y)$$

and, since $0 \leq \varphi(z) \leq 1$, $g(z) = h_\theta(z)\varphi(z) \leq h_\theta(z)$. Combining the last two observations, we conclude that $\mu(\{\theta\}^c) = 0$ and $C = 0$, so φ is constant. With this result in hand, we can easily complete the proof. If $B \in \mathscr{F}_n (\equiv \mathscr{F}(\tau_n))$, then

$$E_z^\theta(1_A; B) = E_z^\theta(1_A \circ \theta_{\tau_n}; B)$$
$$= E_z^\theta(E_z^\theta(1_A \circ \theta_{\tau_n} | \mathscr{F}_n); B)$$
$$= E_z^\theta(P_{B(\tau_n)}^\theta(A); B).$$

It follows from the definition of φ and the fact that φ is constant that the above

$$= E_z^\theta(\varphi(B_{\tau_n}); B)$$
$$= \varphi(z) P_z^\theta(B)$$
$$= P_z^\theta(A) P_z^\theta(B),$$

and we have shown that

$$P_z^\theta(A \cap B) = P_z^\theta(A) P_z^\theta(B)$$

for all $B \in \mathscr{F}_n$ and hence for all $B \in \mathscr{F}_\tau$. Letting $B = A$ in the last equation gives

$$P_z^\theta(A) = (P_z^\theta(A))^2,$$

so $P_z^\theta(A) = 0$ or 1, and the proof of (2) is complete.

Remark: As we promised at the beginning of this section, it is easy to generalize this result to other conditioned processes, since the keys to the proof above are (a) the strong Markov property of the conditioned process and (b) the fact that if g is harmonic and $0 \leq g(x) \leq h_\theta(x)$, then $g(x) = ch_\theta(x)$. In the terminology of the theory of convex sets, (b) says that h_θ is an extreme point of the cone of nonnegative functions that are harmonic in H, and our general zero-one law may be phrased as follow: If h is an extreme point, then the zero-one law holds for the corresponding h-transform. Looking at the representations given in Section 3.3, we can easily see that in H or D the converse is also true. If h is a nontrivial linear combination, then the asymptotic random variable $\lim_{t \uparrow \tau} B_t$ has a nontrivial distribution.

Note: The proof of the zero-one law in this section, like the definition given in Section 3.2, is from Appendix II of Brossard (1975).

4 Boundary Limits of Harmonic Functions

4.1 Probabilistic Analogues of the Theorems of Privalov and Spencer

Let u be a function that is harmonic in $H = R^{d-1} \times (0, \infty)$. As the section titles in this chapter indicate, we are concerned with the existence of the limit of $u(z)$ when $z \to (\theta, 0) \in \partial H$ in two special ways:

(a) when $z \to (\theta, 0)$ along Brownian paths (i.e., $u(B_t^\theta) \to$ a limit a.s. where B_t^θ is a Brownian motion conditioned to exit H at $(\theta, 0)$)
(b) when $z \to (\theta, 0)$ nontangentially.

To explain notion (b), we need a few definitions.
 For each $a > 0$, let V_a^θ be a cone of height 1 with opening a and peak at θ, that is,

$$V_a^\theta = \{(x, y) \in H : |x - \theta| < ay, y < 1\}.$$

(As before, we write $z \in H$ as (x, y) with $x \in R^{d-1}$ and $y > 0$.)
 Let

$$\mathscr{L}_a = \left\{ \lim_{\theta : n \to \infty} u(z_n) \text{ exists for all sequences } z_n \in V_a^\theta \text{ with } z_n \to (\theta, 0) \right\}.$$

We say that u has a nontangential limit at θ if $\theta \in \mathscr{L} = \bigcap \mathscr{L}_a$.
 From these definitions, it should be clear why the convergence defined above is called *nontangential*. This notion is useful in analysis because it is strong, much better, for instance, than radial convergence ($\lim_{y \to 0} u(\theta, y)$ exists); but it is not too strong, as is unrestricted convergence, which implies that the boundary limit is a continuous function (on the set S where this function is defined, S being given the topology it inherits from R^d).
 Two important problems in the theory of harmonic functions are (a) to

find methods for computing $\mathscr{L}$ and (b) to find conditions which guarantee that $\mathscr{L} = \partial H$. In this chapter we will concern ourselves with two classical results regarding the first problem: nontangential convergence is equivalent (modulo null sets) to nontangential boundedness and to finiteness of the "area function." These are the results I have called the theorems of Privalov and Spencer, even though, as the reader will see below, these designations are slightly inaccurate historically. I will detail who did what, and when, after I describe what has been done.

To state the first result, we need some more notation. Let

$$(N_a u)(\theta) = \sup\{|u(z)| : z \in V_a^\theta\}$$
$$\mathscr{N}_a = \{\theta : (N_a u)(\theta) < \infty\}.$$

We say that u is nontangentially bounded at θ if $\theta \in \mathscr{N} = \bigcap \mathscr{N}_a$.

It is clear that $\mathscr{L} \subset \mathscr{N}$. Privalov's theorem asserts that the opposite inclusion is true modulo a null set, that is,

(1) $\qquad \mathscr{L} = \mathscr{N}$ a.e.

or, to be precise, the symmetric difference $\mathscr{L} \triangle \mathscr{N} = (\mathscr{L} - \mathscr{N}) \cup (\mathscr{N} - \mathscr{L})$ is a null set.

To state Spencer's theorem, we have to define the area function. Let

$$(A_a u)(\theta) = \left(\int_{V_a^\theta} y^{2-d} |\nabla u(z)|^2 \, dz \right)^{1/2}$$

(recall that y is the last coordinate of z). $(A_a u)(\theta)$ is called the area function because if $f = u + iv$ is an analytic function with $\operatorname{Re} f = u$, then $(A_a u)(\theta)$ is the area of the image of V_a^θ under f counting multiplicities (this interpretation fails for $d \geq 3$).

Let

$$\mathscr{A}_a = \{\theta : (A_a u)(\theta) < \infty\}$$
$$\mathscr{A} = \bigcap \mathscr{A}_a.$$

With these definitions in hand, we can state Spencer's theorem (which is actually two theorems).

(2) $\qquad \mathscr{N} = \mathscr{A}$ a.e.

Remark: With the symbols $\mathscr{L}$, $\mathscr{N}$, and $\mathscr{A}$ defined, I can now tell the history of the results given above.

$d = 2$	$\mathscr{L} = \mathscr{N}$		Privalov (1916)
	$\mathscr{N} \subset \mathscr{A}$		Marcinkiewicz and Zygmund (1938)
	$\mathscr{N} \supset \mathscr{A}$		Spencer (1943)
$d \geq 3$	$\mathscr{L} = \mathscr{N}$		Calderon (1950a)
	$\mathscr{N} \subset \mathscr{A}$		Calderon (1950b)
	$\mathscr{N} \supset \mathscr{A}$		Stein (1961)

4.1 Probabilistic Analogues of the Theorems of Privalov and Spencer

From the order in which the results were proved, the reader can see the relative difficulties of the three results.

So much for history. The main task in this chapter is to prove that if u is harmonic in H, then $\mathscr{L} = \mathscr{N} = \mathscr{A}$ a.e. The first step in doing this is to define the analogous probabilistic sets $\mathscr{L}^*$, $\mathscr{N}^*$, and $\mathscr{A}^*$ using Brownian motion and to show that $\mathscr{L}^* = \mathscr{N}^* = \mathscr{A}^*$ a.e.

The generalization of nontangential convergence or boundedness to the Brownian setting is easy. We ask for convergence or boundedness along almost every Brownian path, and to get a subset of R^{d-1} we let

$$\mathscr{L}^* = \left\{\theta : \lim_{t \uparrow \tau} u(B_t) \text{ exists } P^\theta \text{ a.s.}\right\}$$

$$U_\tau^* = \sup_{t < \tau} |u(B_t)|$$

$$\mathscr{N}^* = \{\theta : U_\tau^* < \infty \ P^\theta \text{ a.s.}\}$$

where, as is usual in the theory of Markov processes, P^θ a.s. is a shorthand for "P_z^θ a.s. for all $z \in H$." The zero-one law in Section 3.5 implies that if $A = \{\omega : \lim_{t \uparrow \tau} u(B_t) \text{ exists}\}$ or $A = \{U_\tau^* < \infty\}$, then $z \to P_z^\theta(A)$ is either identically 1 or identically 0, so there is a sharp dichotomy between $\theta \in \mathscr{L}^*$ and $\theta \notin \mathscr{L}^*$.

Finding the Brownian analogue of the area function requires a little more thought. Back in Section 2.10, we saw that if u is harmonic in H and $\tau = \inf\{t : B_t \notin H\}$, then $U_t = u(B_t)$, $t < \tau$, is a local martingale and the variance process of this local martingale is

$$\langle U \rangle_t = \int_0^t |\nabla u(B_s)|^2 \, ds, \quad t < \tau,$$

and in Section 2.11 we saw that

$$\{\lim_{t \uparrow \tau} U_t \text{ exists}\} = \{\langle U \rangle_\tau < \infty\},$$

so combining these ideas suggests the following definition:

$$\mathscr{A}^* = \left\{\theta : \int_0^\tau |\nabla u(B_s)|^2 \, ds < \infty \ P^\theta \text{ a.s.}\right\}.$$

With all the relevant definitions introduced and explained, we can state our probabilistic analogue of the theorems of Privalov and Spencer:

(3) $$\mathscr{L}^* = \mathscr{N}^* = \mathscr{A}^*.$$

The proof of this result is quite short. We start with an apparently weaker statement proved in Section 2.11 and take conditional expectations to prove the desired result. Let

$$\mathscr{L}^\omega = \{\omega : \lim_{t \uparrow \tau} u(B_t) \text{ exists}\}$$

$$\mathscr{N}^\omega = \{\omega : U_\tau^* < \infty\}$$

$$\mathscr{A}^\omega = \{\omega : \langle U \rangle_\tau < \infty\}.$$

In Section 2.11 we showed that

(4) $\quad \mathscr{L}^\omega = \mathscr{N}^\omega = \mathscr{A}^\omega$ a.s.

(Here, as above, a.s. means P_z a.s. for all $z \in H$.)

At first it may appear that (4) has little to do with (3), but a little reflection shows that they are closely related. The fact that $\mathscr{L}^\omega = \mathscr{N}^\omega = \mathscr{A}^\omega$ implies (by the definition of conditional expectation) that

$$P_z(\mathscr{L}^\omega | B_\tau) = P_z(\mathscr{N}^\omega | B_\tau) = P_z(\mathscr{A}^\omega | B_\tau).$$

and using (4) from Section 3.2 now gives that the following three functions of θ are equal a.e.:

$$P_z^\theta(\mathscr{L}^\omega), \; P_z^\theta(\mathscr{N}^\omega), \; P_z^\theta(\mathscr{A}^\omega).$$

To translate this into the desired result, we note that it follows from the definitions of $\mathscr{L}^*$, $\mathscr{N}^*$, and $\mathscr{A}^*$ and the zero-one law that

$$\mathscr{L}^* = \{\theta : z \to P_z^\theta(\mathscr{L}^\omega) \equiv 1\}$$
$$\mathscr{N}^* = \{\theta : z \to P_z^\theta(\mathscr{N}^\omega) \equiv 1\}$$
$$\mathscr{A}^* = \{\theta : z \to P_z^\theta(\mathscr{A}^\omega) \equiv 1\},$$

so $1_{\mathscr{L}*}(\theta) = P_z^\theta(\mathscr{L}^\omega)$, $1_{\mathscr{N}*}(\theta) = P_z^\theta(\mathscr{N}^\omega)$, and $1_{\mathscr{A}*}(\theta) = P_z^\theta(\mathscr{A}^\omega)$ (the last three equalities holding for all z and θ).

Exercise 1 Show that if we let

$$\bar{U}_\tau = \sup_{t < \tau} u(B_t)$$

and

$$\bar{\mathscr{N}}^* = \{\theta : \bar{U}_\tau < \infty \; P^\theta \text{ a.s.}\},$$

then we can add one more set to the chain of inequalities, since $\mathscr{N}^* \subset \bar{\mathscr{N}}^* \subset \mathscr{L}^*$.

Notes: The developments in this section and in most of the rest of this chapter follow Brossard's (1975) *thèse de troisième cycle*. Only the last chapter of his thesis, which contains the material in Section 3.5, has been published, however (see Brossard 1976).

The fact that $\mathscr{L}^\omega = \mathscr{N}^\omega$ is due to Doob (1953), page 382. Doob and others wrote a number of papers concerning the boundary limits of harmonic and analytic functions along Brownian paths (in the "fine topology"). If the reader is interested in this topic, he or she should look at the papers by Doob listed in the references.

4.2 Probability Is Less Stringent Than Analysis

The title of this section refers to the facts that the sets $\mathscr{L}$, $\mathscr{N}$, and $\mathscr{A}$ defined in analysis are always smaller than their probabilistic counterparts $\mathscr{L}^*$, $\mathscr{N}^*$,

4.2 Probability is Less Stringent Than Analysis

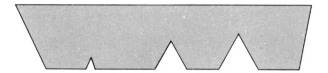

Figure 4.1

and $\mathscr{A}^*$ and that the difference can have positive measure in $d \geq 3$. We will prove the second result in Section 4.4. In this section we will prove that

(1) $\quad \mathscr{N}^* \supset \mathscr{N}_a \supset \mathscr{N}.$

If we combine this statement with the facts that $\mathscr{L}^* = \mathscr{N}^* = \mathscr{A}^*$ and that $\mathscr{L} = \mathscr{N} = \mathscr{A}$ (part of which we will prove in Section 4.5), we see that the probabilistic notions are less strict than their analytical counterparts.

Proof of (1) To simplify drawing pictures, we will prove the result only when $d = 2$. In view of the differences between $d = 2$ and $d \geq 3$ mentioned above, this reduction should worry you a little bit, but, nonetheless, I will leave it to you to check that the same proof with more tedious computations works when $d \geq 3$.

The proof begins with truncation and a standard construction.

Let $M > 0$ and $\mathscr{K} = \{\theta : N_a(\theta) \leq M\}$. If $\theta_n \to \theta$, then $\bigcup_n V_a(\theta_n) \supset V_a(\theta)$, so we have

$$\liminf_{n \to \infty} N_a(\theta_n) \geq N_a(\theta),$$

that is, $\mathscr{K}$ is closed.

Let $\Omega = \bigcup_{\theta \in \mathscr{K}} V_a(\theta)$. Ω is the shaded region in Figure 4.1. For each open interval in $\mathscr{K}^c$, there is a little pyramid in Ω^c. These pyramids combine to give Ω its sawtoothed appearance, or, as other more poetic authors have put it, Ω^c is a little mountain range.

We are interested in Ω because u is bounded by M on Ω. A look at Figure 4.1 suggests the following strategy for showing that $\mathscr{K} \subset \mathscr{N}^*$ a.e. We first show that

(2) If $\sigma = \inf\{t : B_t \notin \Omega\}$ is the exit time from Ω, then for a.e. $\theta \in \mathscr{K}$ there is a $z \in H$ such that $P_z^\theta(\sigma = \tau) > 0$.

If $P_z^\theta(\sigma = \tau) > 0$ and $U_\tau^* = \sup_{t < \tau} |u(B_t)|$, then $P_z^\theta(U_\tau^* < \infty) > 0$, and it follows from the zero-one law (theorem (2) in Section 3.5) that $P_z^\theta(U_\tau^* < \infty) = 1$ for all z, that is, $\theta \in \mathscr{N}^*$.

To prove (1), we need only to prove the innocent-looking fact stated in (2). This result would be easy to prove if $\mathscr{K}$ were always a finite union of intervals—the only exceptional points would be the end points of the intervals—but unfortunately, the set $\mathscr{K}$ can be very ugly. A typical nightmare is $u(z) = \sum_\theta c_\theta h_\theta(z)$ where the sum is over all rational θ, $h_\theta(z)$ is the probability density of exiting H

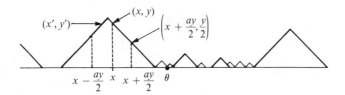

Figure 4.2

at θ, and the $c_\theta > 0$ are chosen so that the sum converges and is harmonic in H. In this case, $\mathcal{K} = \partial H - \bigcup D(\theta, r_\theta)$ where the union is over rational θ and $r_\theta > 0$, so $\mathcal{K}^c$ is dense and it is hard to tell which $\theta \in \mathcal{K}$ will be "good" (that is, have $P_z^\theta(\sigma = \tau) > 0$ for some z). We will eventually derive a criterion (due to Marcinkiewicz) for a $\theta \in \mathcal{K}$ to be good, but the reader would not see the reason for the definition if we gave it now, so we will first see what kind of estimates we can get on $P_z^\theta(\sigma = \tau)$ and then introduce the criterion when it will appear more natural.

Proof of (2) We start by observing that, in order to prove that $P_z^\theta(\sigma = \tau) > 0$ for some z, all we have to do is look at $\partial \Omega$ near θ. Let I, J be intervals centered at θ with radii r and $2r$, respectively. Let

$$\Lambda_1 = \{(x, y) \in \partial \Omega : x \in I, y < 1\}$$
$$\Lambda_2 = \partial \Omega - \Lambda_1$$
$$\sigma_i = \inf\{t : B_t \in \Lambda_i\} \quad i = 1, 2.$$

It is easy to estimate $P_z^\theta(\sigma_2 < \tau)$. By definition,

$$P_z^\theta(\sigma_2 < \tau) = E_z(h_\theta(B(\sigma_2)); \sigma_2 < \tau)/h_\theta(z).$$

Now $C_1 = \sup\{h_\theta(z) : z \in \Lambda_2\} < \infty$, so it follows that

(3) $$P_z^\theta(\sigma_2 < \tau) \leq C_1/h_\theta(z),$$

and if we let $z \to (\theta, 0)$ in such a way that $h_\theta(z) \to \infty$, then $P_z^\theta(\sigma_2 < \tau) \to 0$. This occurs, for instance, if $z \to (\theta, 0)$ in V_b^θ for any $b < \infty$, so to prove (2) we need to estimate $P_z^\theta(\sigma_1 < \tau)$ for $z \in V_b^\theta$. As in the case of σ_2, we start with the equality

$$P_z^\theta(\sigma_1 < \tau) = E_z(h_\theta(B(\sigma_1)); \sigma_1 < \tau)/h_\theta(z),$$

but this time we have to work harder to estimate the right-hand side. We begin by looking at the geometry of $\partial \Omega$ and how it relates to the Poisson kernel.

If $(x, y) \in \Lambda_1$, then $y = d(x, \mathcal{K})/a$ where $d(x, \mathcal{K})$ is the distance from x to the closest point in $\mathcal{K}$. Let $x' \in (x - ay/2, x + ay/2)$ and $y' = d(x', \mathcal{K})/a$. Since $y = d(x, \mathcal{K})/a$, then $x' \in \mathcal{K}^c$ and, consulting the definition of Ω, we see that (x', y') is the point on $\partial \Omega$ that is above x' (see Figure 4.2). As the reader can easily check, we have

$$\frac{ay}{2} < d(x', \mathcal{K}) < \frac{3ay}{2}, \frac{y}{2} < y' < \frac{3y}{2},$$

4.2 Probability is Less Stringent Than Analysis

so

$$\frac{h_\theta(x',y')}{h_\theta(x,y)} = \frac{y'}{y} \cdot \frac{(x-\theta)^2 + y^2}{(x'-\theta)^2 + (y')^2} \geq \frac{1}{2} \cdot \left(\frac{2}{3}\right)^2 = \frac{2}{9}.$$

If we let $g(x') = h_\theta(x',y') \geq \frac{2}{9} h_\theta(x,y)$, we have

$$\int_{x-(ay/2)}^{x+(ay/2)} g(x') h_{x'}(x,y) \, dx' \geq \frac{2}{9} h_\theta(x,y) \int_{x-(ay/2)}^{x+(ay/2)} h_{x'}(x,y) \, dx'$$

and

$$\int_{x-(ay/2)}^{x+(ay/2)} h_{x'}(x,y) \, dx' \geq ay \cdot \frac{y/2}{(ay/2)^2} = \frac{2}{a},$$

so if we let $C_2 = 9a/4$, it follows that if $(x,y) \in \Lambda_1$,

$$h_\theta(x,y) \leq C_2 \int_{x-(ay/2)}^{x+(ay/2)} g(x') h_{x'}(x,y) \, dx'$$

$$\leq C_2 \int_J g(x') h_{x'}(x,y) \, dx'.$$

since $x \in I$, $ay = d(x, \mathcal{K}) \leq r$, and I and J have radii r and $2r$. To estimate $P_z(\sigma_1 < \tau)$, we observe that from the last inequality

$$E_z(h_\theta(B(\sigma_1)); \sigma_1 < \tau) \leq C_2 E_z(E_{B(\sigma_1)}(g(B_\tau); B_\tau \in J))$$
$$\leq C_2 E_z(g(B_\tau); B_\tau \in J)$$

by the strong Markov property, so we have

$$E_z(h_\theta(B(\sigma_1)); \sigma_1 < \tau)/h_\theta(z) \leq C_2 \int_J g(x') h_{x'}(z)/h_\theta(z) \, dx'.$$

Now if $z \in V_b(\theta)$, then

$$\frac{h_{x'}(z)}{h_\theta(z)} = \frac{(x-\theta)^2 + y^2}{(x-x')^2 + y^2} \leq \frac{(x-\theta)^2}{y^2} + 1 \leq b^2 + 1$$

and

$$g(x') = h_\theta(x',y') = \frac{y'}{(x'-\theta)^2 + (y')^2} \leq \frac{d(x',\mathcal{K})/a}{(x'-\theta)^2}.$$

It follows that if $z \in V_b^\theta$, then

(4) $$P_z^\theta(\sigma_1 < \tau) \leq \frac{C_2(b^2+1)}{a} \int_J \frac{d(x',\mathcal{K})}{(x'-\theta)^2} \, dx'.$$

At this point (as we promised earlier), it is clear what we need for a point θ to be good:

$$\varphi_\mathcal{K}(\theta) = \int_{\theta-1}^{\theta+1} \frac{d(x,\mathcal{K})}{(x-\theta)^2} \, d\theta < \infty,$$

for then if J is small, $P_z^\theta(\sigma_1 < \tau) < 1$ for all $z \in V_b(\theta)$. The following lemma due to Marcinkiewicz says that this is a good definition of good.

(5)   For a.e. $\theta \in \mathcal{K}$, $\varphi_{\mathcal{K}}(\theta) < \infty$.

Proof Since $\varphi_{\mathcal{K}}(\theta)$ is determined by $\mathcal{K} \cap (\theta - 1, \theta + 1)$, we can suppose that $\mathcal{K} \subset [-n, n]$. If $\theta \in \mathcal{K}$, then

$$\varphi_{\mathcal{K}}(\theta) \leq \int_{-n-1}^{n+1} \frac{d(x, \mathcal{K})}{(x-\theta)^2} dx,$$

so

$$\int_{\mathcal{K}} \varphi_{\mathcal{K}}(\theta) \, d\theta \leq \int_{\mathcal{K}} \int_{-n-1}^{n+1} \frac{d(x, \mathcal{K})}{(x-\theta)^2} dx \, d\theta$$

$$= \int_{-n-1}^{n+1} \int_{\mathcal{K}} \frac{d(x, \mathcal{K})}{(x-\theta)^2} d\theta \, dx$$

$$= \int_{-n-1}^{n+1} \int_{\mathcal{K}} \frac{d(x, \mathcal{K})}{(x-\theta)^2} 1_{(x \in \mathcal{K}^c)} \, d\theta \, dx,$$

since $d(x, \mathcal{K}) = 0$ if $x \in \mathcal{K}$. Now if $x \in \mathcal{K}^c$ and $\theta \in \mathcal{K}$, then $|x - \theta| \geq d(x, \mathcal{K})$, so we can replace the integral over $\mathcal{K}$ by an integral over $\theta \in [x - d(x, \mathcal{K}), x + d(x, \mathcal{K})]^c$. If we make the replacement and change variables to $y = \theta - x$, it follows that

$$\leq 2 \int_{-n-1}^{n+1} \int_{d(x, \mathcal{K})}^{\infty} \frac{d(x, \mathcal{K})}{y^2} 1_{(x \in \mathcal{K}^c)} \, dy \, dx$$

$$= 2 \int_{-n-1}^{n+1} 1_{(x \in \mathcal{K}^c)} \, dx \leq 4(n+1) < \infty.$$

From the computations above, it follows that

$$\int_{\mathcal{K}} \varphi_{\mathcal{K}}(\theta) \, d\theta < \infty,$$

proving (5).

Having proved (5), we can now easily complete the proof of (2). We have shown that

(3) $P_z^\theta(\sigma_2 < \tau) \leq C_1/h_\theta(z)$

where $C_1 = \sup\{h_\theta(z) : z \in \Lambda_2\}$, and also that if $z \in V_b^\theta$,

(4) $P_z^\theta(\sigma_1 < \tau) \leq \dfrac{C_2(b^2+1)}{a} \int_J \dfrac{d(x', \mathcal{K})}{(x'-\theta)^2} dx'$

where $C_2 = 9a/4$. Therefore, if θ is good (and a.e. θ is good), we can pick r small enough so that $P_z^\theta(\sigma_1 < \tau) < 1/2$ for all $z \in V_b^\theta$ and then $z \in V_\theta^b$ close enough to $(\theta, 0)$ so that $P_z^\theta(\sigma_2 < \tau) < 1/2$.

With (1) finally proved, it is interesting to look back and see what was involved in proving (1). The hard part was to show (2), a purely probabilistic result about exit probabilities for sawtoothed regions, and the rest of the proof consisted of the few lines following the statement of (2), so the result is valid for an arbitrary measurable function u!

Note: The result that $\mathcal{N}^* \supset \mathcal{N}$ is due to Brelot and Doob (1963). The proof given here is again from Brossard (1975).

4.3 Equivalence of Brownian and Nontangential Convergence in $d = 2$

In this section we will show that in two dimensions, $\mathcal{L} = \mathcal{L}^*$, that is, the analytic definition of nontangential convergence agrees a.e. with the probabilistic definition of convergence along the paths of conditioned Brownian motion. The proof of this statement relies heavily on the fact that in two dimensions, Brownian motion can make a loop around a point that cuts it off from the rest of the plane, and hence the maximum principle implies that somewhere along the path the value of u is larger than the value at the point. This technique cannot possibly work in $d \geq 3$, but there is a good reason for this. In Section 4.4 we will see that in $d \geq 3$, we may have $\mathcal{L}^* = \partial H$ and $\mathcal{L} = \emptyset$.

In Section 4.1 we showed that $\mathcal{L}^* = \mathcal{N}^*$, and in Section 4.2 we showed that $\mathcal{N}^* \supset \mathcal{N}_a$. Combining these two facts with the trivial result $\mathcal{N}_a \supset \mathcal{L}_a$, we see that to show $\mathcal{L} = \mathcal{L}^*$ it is enough to show

(1) $\qquad \mathcal{L}_a \supset \mathcal{L}^*,$

for then it follows that $\mathcal{L}^* = \mathcal{N}^* = \mathcal{L}_a = \mathcal{N}_a$ for all $a > 0$ and, hence, that $\mathcal{L}^* = \mathcal{N}^* = \mathcal{L} = \mathcal{N}$.

Remark: From results in Section 4.1, we have that $\mathcal{N}^* = \mathcal{A}^*$, so we can add $\mathcal{A}^*$ to the string of equalities. It is also known (see Stein (1961), or see Brossard (1976) for a probabilistic proof) that in any dimension $\mathcal{N} = \mathcal{A}$, so in two dimensions all six sets are equal.

Proof of (1) As in Section 4.2, the result follows easily from a simple-sounding result about Brownian motion that we will have to struggle to prove. In this case, we will have to work just to state the result.

(2) There is a constant ε (that depends only on a) such that if $(u, 1)$ and $z = (x, y)$ are in V_a^θ and $y \leq 1/3$, then

$$P_{(u,1)}^\theta(B_t, 0 \leq t < \tau, \text{"makes a loop around } z\text{"}) \geq \varepsilon.$$

From the remarks at the beginning of this section, it is probably clear what we

4 Boundary Limits of Harmonic Functions

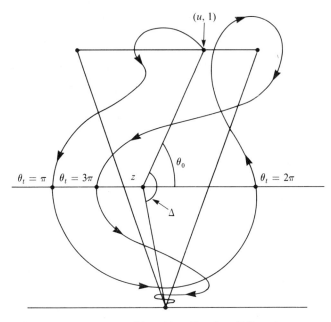

Figure 4.3 $\theta_\tau = 4\pi - (\Delta - \theta_0)$, $W(z) = (\theta_\tau - \theta_0 + \Delta)/2\pi = 2$

mean by "making a loop around z," but to make this description precise and to have a useful criterion for determining when it has occurred, we need to define $w(z)$ = the number of times B_t, $0 \leq t < \tau$, winds around z. Let $(U_t, V_t) = B_t - z$. The probability that $B_t = z$ for some $0 \leq t < \tau$ is zero under $P^\theta_{(u,1)}$ (this is a consequence of (4) in Section 3.2), so there is a unique process θ_t, $0 \leq t < \tau$, with continuous paths such that $\theta_0 \in [0, \pi)$ and

$$\cos \theta_t = V_t/(U_t^2 + V_t^2)^{1/2}.$$

Let $\Delta \in (0, 2\pi)$ be the size of the angle formed by $(u, 1)$, z, and $(\theta, 0)$ when we regard the right-hand side of the arc as the inside. One look at Figure 4.3 shows that $\theta_\tau - \theta_0 + \Delta$ is a multiple of 2π, so $w(z) = (\theta_\tau - \theta_0 + \Delta)/2\pi$ is an integer that counts the number of times B_t, $0 \leq t < \tau$, has wound around z.

With the definitions above, we can state the conclusion of (2) as $P^\theta_{(u,1)}(w(z) \neq 0) \geq \varepsilon$. We will prove this result in a moment, but first, to motivate ourselves for this undertaking, let us observe that (1) follows easily from (2).

Proof of (1) assuming (2) Let $\theta \in \mathscr{L}^*$. It suffices to show the following:

(3) If $z_n \in V_a^\theta$, $z_n \to (\theta, 0)$, and $u(z_n) \to \alpha \in [-\infty, \infty]$, then $u(B_t) \to \alpha$ as $t \uparrow \tau$ P^θ a.s.

Since (3) implies that the number α is independent of the sequence chosen, and hence the limit exists for any sequence converging to $(\theta, 0)$ in V_a^θ.

The first step in the proof of (3) explains our interest in $w(z) \neq 0$. Let G_z be the component of the open set $H - \{B_t : 0 \leq t < \tau\}$ that contains z. If

4.3 Equivalence of Brownian and Nontangential Convergence in $d = 2$

$w(z) \neq 0$, then G_z is bounded and $\partial G_z \subset \{B_t : 0 \leq t < \tau\}$, so it follows from the maximum principle that

$$u(z) \leq \max\{u(B_t) : t \in S_z\}$$

where

$$S_z = \{t \in [0, \tau) : B_t \in \partial G_z\}.$$

Let $A_n = \{w(z_n) \neq 0\}$ and let B be the set of ω that are in infinitely many A_n. B is a shift invariant event (in the sense defined in Section 3.5) and $1_B = \limsup 1_{A_n}$, so Fatou's lemma and (2) imply that for all u,

$$P^\theta_{(u,1)}(B) \geq \limsup P^\theta_{(u,1)}(A_n) \geq \varepsilon,$$

and it follows from our zero-one law ((2) in Section 3.5) that $P^\theta_z(B) = 1$ for all $z \in H$. Combining the last result with the inequality we derived from the maximum principle, we see that (recall $\theta \in \mathscr{L}^*$)

$$P^\theta_z\left(\lim_{t \uparrow \tau} u(B_t) \geq \lim_{n \to \infty} u(z_n)\right) = 1,$$

and repeating the argument upside down gives

$$P^\theta_x\left(\lim_{t \uparrow \tau} u(B_t) \leq \lim_{n \to \infty} u(z_n)\right) = 1,$$

proving that $u(B_t) \to \alpha = \lim_{n \to \infty} u(z_n) P^\theta$ a.s.

To prove (1) now, we need only to prove (2). The idea of the proof is simple: For each u and z, the probability in (2) is positive, so we can use the strong Markov property to prove that the probability is bounded below for $(x, 1/2) \in V^\theta_a$ and then use scaling to improve this to the conclusion in (2).

The first step is to prove:

(4) There is a $\beta > 0$ (that depends only on a) such that if $(u, 1)$ and $z = (x, 1/2)$ are in V^θ_a, then

$$P^\theta_{(u,1)}(w(z) \neq 0) \geq \beta.$$

Proof Let $g_\theta(u, v, x, y) = P^\theta_{(u,v)}(w(x, y) \neq 0)$, let

$$T_y = \inf\{t > 0 : B^2_t < y\},$$

and let $f_\theta(u, v, x, y) = P^\theta_{(u,v)}(B^1(T_y) = x)$. Formula (2) of Section 1.9 and the definition of P^θ_z imply that if $v > y$, then

(5) $$f_\theta(u, v, x, y) = \frac{v - y}{(u - x)^2 + (v - y)^2} \cdot \frac{h_\theta(x, y)}{h_\theta(u, v)},$$

and the strong Markov property implies that if $v > s > y$, then

(6) $$g_\theta(u, v, x, y) \geq \int dr\, f_\theta(u, v, r, s) g_\theta(r, s, x, y).$$

It is easy to see that if $(r, 3/4) \in V_a^\theta$, then $g_\theta(r, 3/4, x, 1/2) > 0$, and there is a constant $\gamma > 0$ such that if $(u, 1), (r, 3/4) \in V_a^\theta$, then $f_\theta(u, 1, r, 3/4) \geq \gamma$. Applying (6) with $v = 1$, $s = 3/4$, and $y = 1/2$ gives (4).

(2) now follows easily. By applying (6) with $v = 1$ and $s = 2y$, doing an obvious scaling, and then using (4), we get

$$g_\theta(u, 1, x, y) \geq \int dr \, f_\theta(u, 1, r, 2y) g_\theta(r, 2y, x, y)$$

$$= \int dr \, f_\theta(u, 1, r, 2y) g_\theta(r/2y, 1, x/2y, 1/2)$$

$$\geq \int_{\theta - ay}^{\theta + ay} dr \, f_\theta(u, 1, r, 2y) \beta.$$

Now if $(u, 1)$ and $(r, 2y) \in V_a^\theta$ and $y \leq 1/3$, it follows from (5) that

$$f_\theta(u, 1, r, 2y) = \frac{1 - 2y}{(u - r)^2 + (1 - 2y)^2} \cdot \frac{2y}{(r - \theta)^2 + 4y^2} \cdot \frac{(u - \theta)^2 + 1}{1}$$

$$\geq \frac{1/3}{1/9} \cdot \frac{2y}{(r - \theta)^2 + 4y^2} \cdot 1.$$

Changing variables to $x = (r - \theta)/ay$, we see that

$$g_\theta(u, 1, x, y) \geq 3 \int_{-1}^{1} \frac{2a}{(a^2 x^2 + 4)} dx,$$

proving (2).

4.4 Burkholder and Gundy's Counterexample ($d = 3$)

In this section we give an example, due to Burkholder and Gundy (1973), of a function u that is harmonic in $H = R^2 \times (0, \infty)$ and that has $\mathcal{N}^* - \mathcal{N}_1 \supset [0, 1]^2$. With a little more work, one can produce an example with $\mathcal{N}^* = \partial H$ and $\mathcal{N}_1 = \emptyset$, but we leave this as an exercise for the idle reader. (Here and throughout the book, $A = B(A \supset B)$ means $A \Delta B \, (A - B)$ is a null set.)

As Burkholder and Gundy say in their paper (I have edited their remarks slightly, since I have modified their counterexample), "Roughly speaking we construct a bed with an infinite number of vertical spines of varying height on the unit square. The function u is defined to be large at the end of each spine and small nearly everywhere else." The spines, placed at locations that are dense in $[0, 1]^2$, are made so small that, with probability 1, Brownian motion will hit only finitely many of them.

The first step in the construction is to introduce the spines.

4.4 Burkholder and Gundy's Counterexample ($d=3$)

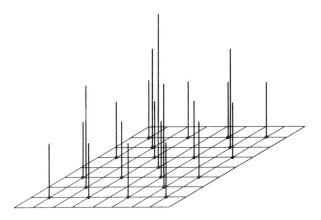

Figure 4.4

$$D_n = \left\{\left(\frac{2j-1}{2^n}, \frac{2k-1}{2^n}, z\right) : 1 \leq j, k \leq 2^{n-1}, 0 \leq z \leq 2^{-(n-1)}\right\}$$

$$E_n = \left\{\left(\frac{2j-1}{2^n}, \frac{2k-1}{2^n}, \frac{1}{2^{n-1}}\right) : 1 \leq j, k \leq 2^{n-1}\right\}.$$

To get a mental picture of D_n, see Figure 4.4. The first thing to show is that there are infinitely many spines in each cone.

(1) If $\theta \in [0,1]^2$, then $V_1(\theta) \cap E_n \neq \emptyset$ for all n.

Proof $V_1(\theta) = \{(x,y) : |\theta - x| < y < 1\}$. The worst case is $\theta = (0,0)$, or more generally, $(2j/2^n, 2k/2^n)$. In this case,

$$|(1/2^n, 1/2^n)| = (2/2^{2n})^{1/2} < 1/2^{n-1},$$

so $(1/2^n, 1/2^n, 1/2^{n-1}) \in V_1(0)$.

If we can construct a harmonic function n with $u(z) \geq n$ for $z \in E_n$, then it will follow from (1) that $N_1 u = \infty$ for all $\theta \in [0,1]^2$. This takes care of $[0,1]^2 \subset \mathcal{N}^c$. The next part of the construction is designed to keep $[0,1]^2 \subset \mathcal{N}^*$. Let

$$A_n = [-b_n, b_n] \times [-b_n, b_n] \times [0, 2b_n]$$
$$B_n = \{z : |z - w| < \varepsilon_n \text{ for some } w \in D_n\}$$
$$S_n = \inf\{t : B_t \notin A_n\}$$
$$T_n = \inf\{t : B_t \in B_n\}.$$

As $b_n \to \infty$, $P_{(0,0,2)}(S_n < \tau) \to 0$, and as $\varepsilon_n \to 0$, $P_{(0,0,2)}(T_n < \infty) \to 0$ (here we use the fact that $d \geq 3$). Therefore we can pick $\varepsilon_n < 1/2^n$ and $b_n > n+2$ so that

$$P_{(0,0,2)}(B_t \in A_n - B_n \text{ for all } 0 \leq t < \tau) \geq 1 - \frac{1}{2^n}.$$

Since $\sum 1/2^n < \infty$, it follows from the last result that with probability 1 there is an N (that will depend on ω) such that $B_t \in A_N - B_N$ for all $0 \le t < \tau$, and hence we can construct u in such a way that it is bounded on each set $A_n - B_n$, it will then follow that $\{\sup_{t<\tau} |u(B_t)| < \infty\}$ has $P_{(0,0,2)}$ probability 1, and from (4) of Section 3.2 that the event has $P^\theta_{(0,0,2)}$ probability 1 for a.e. θ.

At this point our mission is clear: We want to construct a harmonic function that is large on E_n and small on $A_n - B_n$. To do this, we use the following result of J. L. Walsh (1929).

RUNGE'S THEOREM FOR R^d Let K be a compact set in R^d such that $R^d - K$ is connected, and suppose u is harmonic on an open set containing K. Then u can be uniformly approximated by harmonic polynomials on K.

$K_n = (A_n - B_n) \cup E_n$ is compact, and $R^3 - K_n$ is connected. Let U and V be disjoint open sets with $A_n - B_n \subset U$ and $D_n \subset V$. Let $w_n(z)$ equal 0 on U and equal λ_n on V. Then w_n is harmonic on $U \cup V$, and Runge's theorem guarantees that we can find a harmonic polynomial u_n such that

(2)
$$|u_n(z)| < 1/2^n \quad \text{for } z \in A_n - B_n$$
$$|u_n(z) - \lambda_n| < 1/2^n \quad \text{for } z \in E_n.$$

From (2) it follows immediately that we have

(3) $$\sum_{n=1}^\infty |u_n(z)| \text{ converges uniformly on compact subsets of } H.$$

Proof If K is compact, then $K \subset A_n - B_n$ for all n sufficiently large.
Since each u_n is harmonic, that is, if $z \in H$ and $D(z, r) \subset H$, then

(4) $$u_n(z) = \int_{y \in \partial D(x,r)} u_n(y) \, d\pi(y),$$

so it follows from (3) that $u(z) = \sum_{n=1}^\infty u_n(z)$ also satisfies (4) and hence is harmonic in H.

From the construction above, we see that u is bounded on each set $A_n - B_n$, so it remains only to pick the λ_n such that $u(z) \ge n$ for $z \in E_n$. Let $\lambda_1 = 2$. $E_1 = \{(1/2, 1/2, 1)\}$ and $(1/2, 1/2, 1) \in A_n - B_n$ for all $n \ge 2$, so we have (recall $\varepsilon_n < 2^{-n}$)

$$u(1/2, 1/2, 1) = \sum_{n=1}^\infty u_n(1/2, 1/2, 1) \ge 2 - \sum_{n=2}^\infty \left(\frac{1}{2}\right)^n > 1.$$

Suppose now that $\lambda_1, \ldots, \lambda_{n-1}$ have been chosen so that $u(z) \ge m$ for all $z \in E_m$, $m \le n - 1$. If we pick λ_n such that

$$\inf_{z \in E_n} \sum_{m=1}^{n-1} u_m(z) + \lambda_n \ge n + 1,$$

it follows that if $z \in E_n$, then

$$u(z) \ge n + 1 - \sum_{m=n+1}^\infty \left(\frac{1}{2}\right)^m > n.$$

4.5 With a Little Help from Analysis, Probability Works in $d \geq 3$: Brossard's Proof of Calderon's Theorem

In the last section we saw that when $d \geq 3$, $\mathcal{N}^* - \mathcal{N}$ may have positive measure. This result makes morally (if not logically) certain that it is impossible to show that $\mathcal{L} = \mathcal{N}$ by using purely probabilistic methods. In this section, we will show that if we borrow a simple result from analysis, namely,

(1) If u is harmonic and $|u| \leq M$ in H, then u has a nontangential limit at a.e. point of ∂H,

then we can use probability to prove the local version of this result:

(2) If u is harmonic and $N_a u \leq M$ for all $\theta \in S$, then u has a nontangential limit at a.e. point of S.

(2) implies that $\mathcal{N}_a \subset \mathcal{L}$, so when we combine this with the trivial containments $\mathcal{L} \subset \mathcal{L}_a \subset \mathcal{N}_a$, we get Calderon's theorem: $\mathcal{L} = \mathcal{N}$. (1) is a basic fact about harmonic functions and can be found in Chapter 2 of Stein and Weiss (1971), but it is a simple result and the analytical proof has some interesting probabilistic aspects, so I will give the details here. The first step in proving (1) is to prove:

(3) If $|u(x)| \leq M$ for all $x \in H$, then there is a function f with $|f(x)| \leq M$ so that $u(z) = E_z f(B_\tau)$ where $\tau = \inf\{t : B_t \notin H\}$, that is,

$$u(z) = \int h_\theta(z) f(\theta) \, d\theta.$$

Remark: This is Theorem 2.5 of Stein and Weiss (1971). It gives the Poisson integral representation for bounded harmonic functions. Since $u + M \geq 0$ in H, this theorem is, except for the assertion about f, a special case of (3) in Section 3.3.

Notation: Since functions of the form given in (3) will appear many times below, we let

$$\mathcal{P}f(z) = \int h_\theta(z) f(\theta) \, d\theta.$$

Proof of (3) We begin by proving the following semigroup property for $u(\cdot, y)$:

(4) If u is bounded in every $H_\varepsilon = \{(x, y) : y > \varepsilon\}$, then for all $r, s > 0$

$$u(x, r + s) = \int h_\theta(x, r) u(\theta, s) \, d\theta.$$

4 Boundary Limits of Harmonic Functions

Proof Let $\tau = \inf\{t : B_t \notin H_s\}$. Since $u(B_{t \wedge \tau})$ is a bounded martingale,

$$u(x, r+s) = E_{(x, r+s)} u(B_\tau)$$
$$= \int h_\theta(x, r) u(\theta, s) \, d\theta.$$

Remark: The stopping time argument used above is a substitute for (and proof of!) the maximum principle used in the analytic proof (see Stein and Weiss (1971), pages 52–53).

With (4) established, the rest of the proof of (3) is soft analysis. Since $|u| \leq M$ in H, there is a sequence $y_k \downarrow 0$ such that $u(\cdot, y_k)$ converges weakly to a limit f, that is, for all $g \in L^1$

$$\int u(\theta, y_k) g(\theta) \, d\theta \to \int f(\theta) g(\theta) \, d\theta.$$

and so we have in particular that

$$\int h_\theta(x, y) u(\theta, y_k) \, d\theta \to \int h_\theta(x, y) f(\theta) \, d\theta.$$

By (4), the left-hand side is $u(x, y + y_k)$. Since u is continuous in H, $u(x, y + y_k) \to u(x, y)$ as $y_k \downarrow 0$, proving (3).

With the Poisson integral representation established, the proof of (1) is reduced to the task of showing that functions of the form $u = \mathscr{P}f$ have nontangential limits a.e. Since the value of u comes from integrating f with respect to a spherically symmetric kernel, it does not take too much inspiration to guess that the Lebesgue points, that is, the points where

$$\frac{1}{r^n} \int_{|x| < r} |f(x + \theta) - f(\theta)| \, dx \to 0 \quad \text{as } r \to 0,$$

are going to be good. The next result confirms this guess and, since a.e. point is a Lebesgue point, proves (1).

(5) If θ_0 is a Lebesgue point for f, then $u = \mathscr{P}f$ has nontangential limit $f(\theta_0)$ at $(\theta_0, 0)$.

Remark: This result and its proof are classics (see Stein and Weiss (1971), pages 62–63).

Proof Since $h_\theta(x, y) \geq 0$ and $\int h_\theta(x, y) d\theta = 1$,

$$|u(x, y) - f(\theta_0)| = \left| \int h_\theta(x, y) (f(\theta) - f(\theta_0)) \, d\theta \right|$$
$$\leq \int h_\theta(x, y) |f(\theta) - f(\theta_0)| \, d\theta.$$

4.5 Brossard's Proof of Calderon's Theorem

Now if $(x, y) \in V_a(\theta_0)$, then

$$\frac{h_\theta(x,y)}{h_\theta(\theta_0,y)} = \frac{((\theta_0 - \theta)^2 + y^2)^{d/2}}{((x - \theta)^2 + y^2)^{d/2}}$$

$$\leq \begin{cases} (a^2 + 1)^{d/2} & \text{if } |\theta_0 - \theta| \leq ay \\ \left(\dfrac{K^2 a^2 + 1}{(K-1)^2 a^2 + 1}\right)^{d/2} & \text{if } |\theta_0 - \theta| = Kay \text{ where } K \geq 1. \end{cases}$$

The last expression $\to 1$ as $K \to \infty$, so the ratio above is bounded, and it follows that

$$|u(x,y) - f(\theta_0)| \leq C \int h_\theta(\theta_0, y) |f(\theta) - f(\theta_0)| \, d\theta \to 0$$

if θ_0 is a Lebesgue point.

With the proof of (1) completed, we turn now to the main business of this section, namely, the proof of (2). We begin with our usual construction. Let $\mathcal{K} = \{\theta : N_a(\theta) \leq n\}$, let $\Omega = \bigcup_{\theta \in \mathcal{K}} V_a(\theta)$, and let σ be the exit time from Ω. $|u| \leq n$ on Ω, so as $t \uparrow \sigma$, $u(B_t) \to$ a limit that we call $u(B_\sigma)$, and the optional stopping theorem implies that

$$u(z) = E_z u(B_\sigma).$$

Since $\{\sigma = \tau\} \subset \{B_\tau \in \mathcal{K}\}$, we can write

$$u(z) = u_0(z) + u_1(z) + u_2(z)$$

where

$$u_2(z) = E_z(u(B_\sigma) : \sigma < \tau)$$
$$u_0(z) = E_z(u(B_\tau) : B_\tau \in \mathcal{K})$$
$$u_1(z) = -E_z(u(B_\tau) : B_\tau \in \mathcal{K}, \sigma < \tau).$$

Now u_0 is a bounded harmonic function, so by (1) it has nontangential limits a.e., and, furthermore, the limit of u_0 is $f 1_{\mathcal{K}}$. To handle u_1 and u_2, we observe that each is bounded by $nP_z(\sigma < \tau)$, so it suffices to show that for a.e. $\theta \in \mathcal{K}$, $P_z(\sigma < \tau) \to 0$ when $z \to (\theta, 0)$ in $V_b(\theta)$. (This is similar to the result for P_z^θ proved in Section 4.2, but it is simpler.)

To prove this result, we use an idea due to Calderon: it suffices to show that $P_z(\sigma < \tau) \leq CP_z(B_\tau \notin \mathcal{K})$ for the right-hand side is a bounded harmonic function that, by (3) and (5) above, $\to 0$ at a.e. point of $\mathcal{K}^c$. To prove the inequality, observe that if $(x, y) \in H - \Omega$, then either $y \geq 1$ or $y \leq d(x, \mathcal{K})/a$.

If $y \geq 1$ (the trivial case), then

$$P_{(x,y)}(B_\tau \notin \mathcal{K}) \geq P_{(x,y)}(B_\tau \in [-n, n]) \geq \varepsilon_1.$$

If $y \leq d(x, \mathcal{K})/a$, then $D(x, ay) \subset \mathcal{K}^c$ and

$$P_{(x,y)}(B_\tau \notin \mathcal{K}) \leq P_{(x,y)}(B_\tau \in D(x, ay)) \geq \varepsilon_2(a) > 0,$$

so there is an $\varepsilon > 0$ (that depends on a) such that

$$\inf_{z \in H - \Omega} P_z(B_\tau \notin \mathcal{K}) \geq \varepsilon.$$

From the Markov property,

$$\begin{aligned} P_z(B_\tau \notin \mathcal{K}) &= P_z(B_\tau \notin \mathcal{K}, \sigma < \tau) \\ &= E_z(P_{B(\sigma)}(B_\tau \notin \mathcal{K}); \sigma < \tau) \\ &\geq \varepsilon P_z(\sigma < \tau) \end{aligned}$$

by the last inequality, so we have proved Calderon's inequality, and the proof of (2) is complete.

Remarks: Even though probability has borrowed something from analysis to prove (2), I think it has also contributed something. The reader is invited to look at the purely analytical proof of (2) given in Stein and Weiss (1971) on pages 64–66.

It is possible to use probabilistic techniques to prove that $\mathcal{N} \subset \mathcal{A}$ and $\mathcal{A} \subset \mathcal{N}$. Since these proofs require more analytical preliminaries, the reader is referred to Brossard (1976).

Exercise 1 Let $\underline{N}_a u(\theta) = \sup\{u(z) : z \in V_a^\theta\}$

$$\underline{\mathcal{N}}_a = \{\theta : \underline{N}_a(u)(\theta) < \infty\}.$$

Imitate the proof given above and use the integral representation for nonnegative harmonic functions (formula (3) in Section 3.3) in place of (3) above to prove Carleson's (1961) theorem: $\underline{\mathcal{N}}_a \subset \mathcal{L}$.

Remark: The reader should note that Carleson's theorem implies that for Burkholder and Gundy's counterexample, $\underline{\mathcal{N}}_1 \cap [0, 1]^2 = \emptyset$, so although we have done nothing to force u to take on large negative values, it will do so automatically.

5 Complex Brownian Motion and Analytic Functions

5.1 Conformal Invariance, Applications to Brownian Motion

The main purpose of this chapter is to show how Lévy's result concerning the conformal invariance of Brownian motion can be used to study analytic functions. In this section, we will exploit the connection in the other direction: we will use the conformal invariance to derive properties of two-dimensional Brownian motion. We will start with the two path properties proved in Section 1.7.

(1) Let $S_0 = \inf\{t > 0 : B_t = 0\}$. If $x \neq 0$, then

$$P_x(S_0 < \infty) = 0.$$

Remark: As in Section 1.7, it follows from (1) that the result also holds when $x = 0$.

Proof It suffices to prove the result when $x = 1$. To do this, we let C_t be a complex Brownian motion starting at 0, let $F_t = \exp(C_t)$, and observe that Lévy's theorem ((4) in Section 2.12) implies that F_t is a time change of Brownian motion starting at 1, that is, if we let

$$\sigma_t = \int_0^t |\exp(C_s)|^2 \, ds = \int_0^t \exp(2\mathrm{Re}(C_s)) \, ds$$

and

$$\gamma_u = \inf\{t : \sigma_t \geq u\},$$

then (since the recurrence of $\mathrm{Re}\, C_s$ implies $\sigma_t \uparrow \infty$ as $t \uparrow \infty$), $F(\gamma_u), u \geq 0$, is a time change of Brownian motion. Since 0 is not in the range of exp, it follows that

$$P_1(S_0 < \infty) = P(F(\gamma_u) = 0 \text{ for some } u \geq 0) = 0.$$

5 Complex Brownian Motion and Analytic Functions

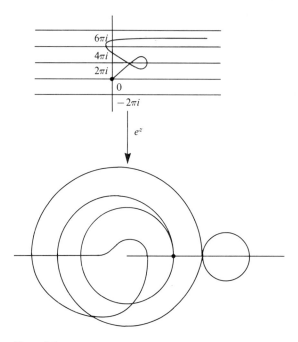

Figure 5.1

(2) Let $S_1 = \inf\{t \geq 0 : |B_t| \leq 1\}$. If $|x| > 1$, then
$$P_x(S_1 < \infty) = 1.$$

Proof $1/z$ is analytic in $C - \{0\}$, so if we let C_t be a complex Brownian motion starting at $1/x$, then Lévy's theorem implies that $F_t = 1/C_t$ is a time change of Brownian motion starting at x, in other words, if we let

$$\sigma_t = \int_0^t |C_s|^{-4} \, ds \qquad ((x^{-1})' = -x^{-2})$$

and

$$\gamma_u = \inf\{t : \sigma_t \geq u\},$$

then $F(\gamma_u)$ is a Brownian motion starting at x. Let $T_1 = \inf\{t \geq 0 : |C_t| > 1\}$. It is trivial to show that $P_{1/x}(T_1 < \infty) = 1$ (see Exercise 1 in Section 1.7 for a more general result). At time $t = T_1$, $|C_t| = 1$, so at time $u = \sigma(T_1)$ (which is $< \infty$, since $|C_t| \neq 0$ for all t), $|F(\gamma_u)| = 1$ and it follows that $P_x(S_1 < \infty) = 1$.

Our next step in investigating two-dimensional Brownian motion is to take another look at the process $F_t = \exp(C_t)$ used in the proof of (1). If we write $C_t = B_t^1 + iB_t^2$, then B_t^1 tells us the size of F_t, that is, $|F_t| = |\exp(C_t)| = \exp(B_t^1)$, and B_t^2 tells us how many times F_t has wound around 0, counting clockwise loops -2π and counterclockwise loops $+2\pi$ (see Figure 5.1). In Section 5.5 we will use this observation to prove Spitzer's result about the winding of Brownian

5.1 Conformal Invariance, Applications to Brownian Motion

motion. At this point, we will content ourselves to make the following simple observation:

(3) Suppose $B_0 \neq 0$ and let θ_t be the (net) number of times $\{B_s : 0 \leq s \leq t\}$ has wound around 0. Then with probability 1,

$$\limsup \theta_t = \infty, \quad \liminf \theta_t = -\infty,$$

and consequently $\theta_t = 0$ i.o.

The exponential function is also useful for computing exit distributions. It maps the strip $G = \{z : |\operatorname{Im} z| < \pi/2\}$ one-to-one onto the half space $H = \{z : \operatorname{Re} z > 0\}$, so Lévy's theorem allows us to compute the exit distributions for G from known formulas for the exit distributions for H. Let C_t be a complex Brownian motion starting at 0. Let $T = \inf\{t : C_t \notin G\}$. Clearly, $T < \infty$ a.s. To compute the distribution of C_T, we observe that if $F_t = \exp(C_t)$ and σ_t, γ_u are the processes defined in the proof of (1), then $B_u = F(\gamma_u)u \geq 0$ is a Brownian motion, and furthermore, if we let $\tau = \inf\{u \geq 0 : B_u \notin H\}$, then $\gamma(\sigma_T) = T$, so

$$\exp(C_T) = F(\gamma(\sigma_T)) = B(\sigma_T) = B_\tau,$$

that is, exp maps the exit distributions from G to corresponding exit distributions from H.

With the last identity established, it is trivial to compute the exit distribution from G.

$$P_0(\operatorname{Im} C_T = \pi/2, \operatorname{Re} C_T \geq a) = P_1(\operatorname{Im} B_\tau \geq e^a)$$
$$= \int_{e^a}^{\infty} \frac{1}{x^2 + 1} \frac{1}{\pi} dx$$

by (2) from Section 1.9. Changing variables $x = \tan y$, $dx = \sec^2 y \, dy$ converts the above to

$$\int_{\tan^{-1}(e^a)}^{\pi/2} \frac{1}{\pi} dy = \frac{1}{2} - \frac{1}{\pi} \tan^{-1}(e^a),$$

and differentiating gives (recall that $(\tan^{-1} x)' = (1 + x^2)^{-1}$)

(4) $$P_0(\operatorname{Im} C_T = \pi/2, \operatorname{Re} C_T = a) = \frac{1}{\pi} \frac{e^a}{e^{2a} + 1}$$
$$= (2\pi \cosh a)^{-1}.$$

The last example is just one of many that can be done with this method. The technique used above can, in principle at least, be used to compute the exit distribution for any simple connected region, since the Riemann mapping theorem implies that any such region (that is not the whole plane) can be mapped one-to-one onto the disk. Carrying out the details of this computation, however, is usually very difficult even in simple examples such as polygons where the Schwarz-Christoffel formula gives an almost explicit formula for the mapping.

Note: Ever since Lévy discovered the conformal invariance, many people have used it to prove results about two-dimensional Brownian motion; see Itô and

McKean (1964), McKean (1969), Davis (1979a, 1979b), and Lyons and McKean (1980). Formulas (1), (2), and (3) each appear several times in these references.

5.2 Nontangential Convergence in D

In the next two sections, we will be concerned with the nontangential limits of analytic functions. For some of the developments, we will need to know that several of the theorems proved in Chapter 4 for the upper half space H are also valid in D. In this section, we will use a combination of proof by analogy and by conformal mapping to prove the results we need. The first thing we have to do is define nontangential convergence in our new setting.

Let $S_\alpha(\theta)$ be the convex hull of the disk $\{|z| \leq \alpha\}$ and the point $e^{i\theta}$. (The S is for Stolz domain.) A function f, analytic or not, is said to have nontangential limit c at $e^{i\theta}$ if for all $\alpha < 1$, $f(z_n) \to c$ whenever $z_n \in S_\alpha(\theta)$ and $z_n \to e^{i\theta}$.

The first step in investigating nontangential convergence in D probabilistically is to define Brownian motion conditioned to exit D at $e^{i\theta}$. It is easy to see that this can be done by imitating the approach used in Section 3.2. We let $\tau_n = \inf\{t : |B_t| = 1 - 1/n\}$, and, if $A \in \mathscr{F}(\tau_n)$, we let

(1) $$P_z^\theta(A) = E_z(k_\theta(B_{\tau_n}); A)/k_\theta(z)$$

where

$$k_\theta(z) = \frac{1 - |z|^2}{|z - e^{i\theta}|^2}$$

is the probability density of exiting D at $e^{i\theta}$ starting from z. Since $k_\theta \geq 0$ is harmonic, repeating the arguments of Section 3.2 shows that we can paste these processes together using Kolmogorov's extension theorem, to get a process Y defined for $t < \tau$. The last problem, then, is to show that $Y_t \to e^{i\theta}$ as $t \to \tau$ by modifying the proof of (2) from Section 3.2. Filling in the details in the sketch above is left as an exercise for the reader.

The first result we want to generalize from H to D is that

(2) $$\mathscr{L}^* = \mathscr{N}^* = \mathscr{L} = \mathscr{N},$$

where these quantities are defined in the obvious way by analogy with the definitions given in Section 4.1. It should not be hard to believe that we could prove this result by patiently working our way through Sections 4.1 to 4.3 and noticing that the proof given there works, with minor modifications, in our new setting. Fortunately, there is another alternative.

Proof $g(z) = (z - 1)/(z + 1)$ maps $H' = \{\operatorname{Re} z > 0\}$ one-to-one onto D. Since g is analytic, if we let $\sigma = \inf\{t : B_t \notin H'\}$, then Lévy's theorem (and a little common sense) shows that $g(B_{t \wedge \sigma})$ is a time change of a Brownian motion starting at $g(B_0)$ and run until it exits D. The last result makes it easy to believe that "g maps B_t conditioned to exit H at y to a time change of B_t conditioned to exit D at $g(y)$," and, in fact, using (3) of Section 3.2, the reader can easily convert this belief into a proof.

With the statement in quotation marks established, the desired result follows immediately, since the last conclusion implies that $g(\mathscr{L}_H^*) = \mathscr{L}_D^*$ and $g(\mathscr{N}_H^*) = \mathscr{N}_D^*$ (where the subscript indicates the domain under consideration), and it follows from the definitions of nontangential convergence and boundedness that $g(\mathscr{L}_H) = \mathscr{L}_D$ and $g(\mathscr{N}_H) = \mathscr{N}_D$.

The argument above can obviously be used to obtain many other results about conditioned processes or harmonic functions in D from facts about the corresponding objects in H. For the arguments in the next two sections, we will also need to generalize results (3) and (5) of Section 4.5 from H to D.

(3) If $|u(z)| \leq M$ for all $z \in D$, then there is a function f with $|f(x)| \leq M$ so that $u(z) = E_z f(B_\tau)$ and, furthermore, u has nontangential limit f at a.e. point of ∂D.

Proof Let g, H', and σ be defined as in the proof of (2). Since g is analytic, a simple computation (or an application of Lévy's theorem) shows that $v(z) = u(g(z))$ is harmonic in $H' = \{\operatorname{Re} z > 0\}$. Since $|v(z)| \leq M$, it follows from (3) and (5) in Section 4.5 that there is a function f_0 defined on $\partial H'$ so that $v(z) = E_z f_0(B_\sigma)$ and v has nontangential limit f_0 at a.e. point of $\partial H'$.

To take these two results back to D, let $f(z) = f_0(g(z))$ for $z \in \partial D - \{1\}$ and let $h(z) = (1 + z)/(1 - z)$ be the inverse of g. Since $h(B_t)$, $t < \sigma$, is a time change of Brownian motion run until it exits D by unscrambling the definitions and applying Lévy's theorem we get $u(z) = E_z f(B_\tau)$. The second conclusion follows from the second paragraph of the proof of (2).

Exercise 1 Let u be a measurable function defined in D and define the two corresponding maximal functions

$$U_\tau^* = \sup\{|u(B_t)| : t < \tau\}$$
$$N_\alpha u(\theta) = \sup\{|u(z)| : z \in S_\alpha(\theta)\}.$$

Use the reasoning we called Calderon's argument in Section 4.5 to show that there is a constant C (whose value depends only on α) such that for all $\lambda > 0$,

$$P_0(U_\tau^* > \lambda) \leq C|\{\theta : N_\alpha u(\theta) > \lambda\}|.$$

A similar result holds in the upper half space:

$$\int dx \, P_{(x,1)}(U_\tau^* > \lambda) \leq C|\{\theta : N_a(\theta) > \lambda\}|.$$

In view of the counterexample in Section 4.4, it is much more difficult to show results in the other direction.

5.3 Boundary Limits of Functions in the Nevanlinna Class N

A function f analytic in $D = \{z : |z| < 1\}$ is said to be in the Nevanlinna class N if

$$\sup_{r<1} \int_0^{2\pi} \log^+ |f(re^{i\theta})| \, d\theta < \infty$$

where $\log^+ x = \max\{\log x, 0\}$. These functions were introduced by F. and R. Nevanlinna (1922), who showed:

(1) An analytic function f is in N if and only if it is a quotient of two bounded analytic functions.

From this representation theorem and the fact that bounded harmonic functions have nontangential limits ((3) in Section 5.2), it follows immediately that

(2) if $f \in N$, then the nontangential limit exists at a.e. point of ∂D.

In this section, we will use Brownian motion to give a proof of (2) that does not rely on the decomposition in (1). The key to our proof is the fact that two-dimensional Brownian motion will not hit the countable set $Z_f = \{z : f(z) = 0\}$, so although it may not be possible to define $\log(f(z))$ as a function analytic in D, it is possible to define $\log(f(B_t))$ so that it is a time change of Brownian motion. Suppose for simplicity (and without loss of generality) that $f(0) = 1$. Since Brownian motion starting from 0 will, with probability 1, not hit Z_f, there is a unique continuous process L_t that has $L_0 = 0$ and $\exp(L_t) = f(B_t)$ for $t < \tau$.

It is easy to see what L_t is for small times. Let $\delta > 0$ be chosen so that $\bar{D}(0, \delta) \cap Z_f = \emptyset$. Since $f(0) = 1$, there is a unique analytic function g that has $g(0) = 0$ and $\exp(g(z)) = f(z)$ for all $z \in \bar{D}(0, \delta)$, so for $t < \tau_\delta = \inf\{t : B_t \notin D(0, \delta)\}$, $L_t = g(B_t)$. The last observation shows that for $t < \tau_\delta$, L_t is an analytic function of Brownian motion. By iterating this result and using the strong Markov property, we get:

(3) L_t, $t < \tau$, is a complex local martingale.

Proof We begin with some notation:

$$A^\varepsilon = \bigcup_{y \in A} \bar{D}(y, \varepsilon)$$
$$T_n^1 = \inf\{t : B_t \in (Z_f \cup D^c)^{1/n}\}$$
$$T_n^2 = \inf\{t : |L_t| > n\}$$
$$T_n = T_n^1 \wedge T_n^2.$$

T_n is the sequence of times that we will use to reduce L. See Figure 5.2 for a picture of $(Z_f \cup D^c)^{1/n}$.

The first thing to check is that $T_n \uparrow \tau$ as $n \to \infty$. Since L_t is continuous, it is immediate that $T_n^2 \uparrow \tau$. To see that $T_n^1 \uparrow \tau$, we observe that $P_0(B_t \in Z_f$ for some $t \geq 0) = 0$, and since the Brownian path is continuous, there is a lower bound on the distance between B_t, $0 \leq t \leq \tau$, and the closed set $Z_f \cap \bar{D}(0, r)$.

5.3 Boundary Limits of Functions in the Nevanlinna Class N

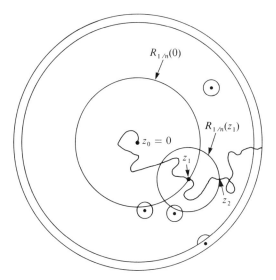

Figure 5.2

To prove that $L(t \wedge T_n)$ is a martingale, we have to further subdivide time. The strategy is simple: For each $z \in D - Z_f$, we will choose a ball $D(z, \delta_z) \subset D - Z_f$, define the associated exit times $\sigma_z = \inf\{t : B_t \notin D(z, \delta_z)\}$, and then exhaust T_n by iterating these exit times. To do this requires a little patience and a lot of notation. Let

$$\delta_\varepsilon(z) = (1 - \tfrac{\varepsilon}{2})\sup\{\delta : D(z, \delta) \subset D - Z_f\}$$
$$R_\varepsilon(z) = D(z, \delta_\varepsilon(z))$$
$$\sigma_\varepsilon(z) = \inf\{t : B_t \notin R_\varepsilon(z)\}$$
$$S_0 = 0, Z_0 = 0$$

and for $m \geq 1$,

$$S_m = \inf\{t > S_{m-1} : B_t \notin R_{1/n}(Z_{m-1})\} \wedge T_n$$
$$Z_m = B(S_m)$$
$$M = \inf\{m : S_m = T_n\}.$$

The next result explains the reason for our choice of $\delta_\varepsilon(z)$.

(4) For each $\varepsilon > 0$, there is a constant $\rho(\varepsilon) > 0$ such that if $z \in D - Z_f$, then

$$P_z(\sigma_\varepsilon(z) \in (Z_f \cup D^c)^\varepsilon) \geq \rho(\varepsilon).$$

Proof Since $\sup\{\delta : D(z, \delta) \subset D - Z_f\} \leq 1$, there is at least one point, say $z_0 \in D^c \cup Z_f$, that is within $\varepsilon/2$ of the boundary of $R_\varepsilon(z)$. Since $D(z_0, \varepsilon) \subset (Z_f \cup D^c)^\varepsilon$, looking at Figure 5.2 and experimenting with a few cases, one finds that

$$\inf_{z \in D - Z_f} P_z(\sigma_z(\varepsilon) \in D(z_0, \varepsilon)) > 0,$$

proving (4).

From (4) and the strong Markov property, it follows immediately that

$$P(M > m) \leq (1 - \rho(1/n))^m,$$

so $P(M < \infty) = 1$. Since $S_M = T_n$, all that remains is to put the pieces together to prove (3). The argument preceding the statement of (3) shows that $L(t \wedge S_1)$ is a martingale. Combining this result with the strong Markov property shows that for any m, $L(t \wedge S_m)$ is a martingale, that is,

$$E(L(t \wedge S_m)|\mathscr{F}_s) = L(s \wedge S_m).$$

Since $|L_t| \leq n$ for $t \leq T_n^2$ and $T_n^2 \geq S_m$, it follows from the bounded convergence theorem that $L(t \wedge T_n)$ is a martingale.

Remark: In Section 6.4, we need to use the fact that if f is an analytic function with $f(0) = 1$ and we define G_t to be the unique continuous process that has $G_0 = 1$ and $G_t^n = f(B_t)$, then G_t, $t < \tau$, is a local martingale. The reader should look back over the last proof now to verify that (a) the proof above (with trivial modifications) proves this fact and (b) in the proof above we never use the fact that $f \in N$.

To use the condition $f \in N$ to conclude something about the convergence of L_t as $t \uparrow \tau$, we have to replace the stopping times T_n by $\tau_r = \inf\{t : |B_t| \geq r\}$. To do this we will discard $\text{Im } L_t$, because we have very little control over its size (and we know that there is a trick for deducing the convergence of $\text{Im } L_t$ from that of $\text{Re } L_t$).

Even $\text{Re } L_t$ is not so nice on $\{z : |z| < r\}$. The function f is bounded, so $\text{Re } L_t = \log|f(B_t)|$ is bounded above, but not below, if f has zeros. To get around this problem and to make the connection with the condition $f \in N$, we truncate below.

(5) If $a > -\infty$ and $U_t^a = (\text{Re } L_t) \vee a$, then $U^a(t \wedge \tau_r)$ is a submartingale.

Proof In the proof of (3) we showed that $L(t \wedge T_n)$ is a martingale, so it follows that $U^a(t \wedge T_n)$ is a submartingale, and the optional stopping theorem gives

$$E(U^a(t \wedge \tau_r \wedge T_n)|\mathscr{F}_s) \geq U^a(s \wedge \tau_r \wedge T_n).$$

Since U_t^a, $t \leq \tau_r$, is uniformly bounded, letting $n \to \infty$ and using the bounded convergence theorem proves (5).

Remark: Continuing to prepare for Section 6.4, we observe that $|G_t| = |F_t|^{1/n}$ is bounded for $t < \tau_r$, so the argument above shows that $G(t \wedge \tau_r)$ is a complex martingale.

Up to this point, we have not used the fact that $f \in N$. The next result ends this generality.

(6) If $f \in N$, then $\lim_{t \uparrow \tau} \text{Re } L_t$ exists a.s.

5.3 Boundary Limits of Functions in the Nevanlinna Class N

Proof Since $f \in N$,

$$EU^a(\tau_r) \leq |a| + \int_0^{2\pi} \log^+ |f(re^{i\theta})| \, d\pi(\theta) \leq C < \infty.$$

Let $T_n = \inf\{t : |B_t| \geq 1 - 1/n\}$ and let $\gamma : [0, \infty) \to [0, \tau)$ be the stretching function defined in Section 2.3. γ is a predictable increasing function that has $\gamma(n) \leq T_n$, so $U^a(\gamma_t)$, $t \geq 0$, is an L^1-bounded submartingale, and it follows that as $t \to \infty$, $U^a(\gamma_t)$ converges to a limit Y^a a.s. If $Y^a > a$, we have $\operatorname{Re} L_t \to Y^a$ as $t \to \tau$. On the other hand, if $Y^a \leq a$ for all $a > -\infty$, then

$$\limsup_{t \uparrow \tau} \operatorname{Re} L_t \leq \inf_a Y^a = -\infty,$$

so the limit exists in this case also, and we have proved (6).

Combining (6) with the results of Section 2.11, we can early conclude that

(7) If $f \in N$, then $\lim_{t \uparrow \tau} \operatorname{Im} L_t$ exists a.s.

Proof Let $U_t = \operatorname{Re} L_t$, $V_t = \operatorname{Im} L_t$. Since L_t is locally an analytic function of B_t, the variance processes $\langle U \rangle_t$ and $\langle V \rangle_t$ are equal. It follows from a result in Section 2.11 that

$$\left\{ \lim_{t \uparrow \tau} U_t \text{ exists} \right\} = \{ \langle U \rangle_\tau < \infty \}$$
$$= \{ \langle V \rangle_\tau < \infty \}$$
$$= \left\{ \lim_{t \uparrow \tau} V_t \text{ exists} \right\}.$$

From (6) and (7), it follows that as $t \uparrow \tau$, $(\operatorname{Re} L_t, \operatorname{Im} L_t)$ converges to a limit and hence so does $f(B_t) = \exp(L_t)$. Combining this result with (2) of the last section, we see that (2) is true.

Remark: From the proof of (7) and the results of Section 2.11, we see that $(\operatorname{Re} L_\tau, \operatorname{Im} L_\tau) \equiv \lim (\operatorname{Re} L_t, \operatorname{Im} L_t)$ is finite a.s. If we let $F_\tau = \exp(L_\tau)$, then $P(|L_\tau| < \infty) = 1$ and $P(F_\tau = 0) = 0$. With a little thought, we can strengthen the last conclusion to:

(8) If $f \in N$ and $f \not\equiv 0$, then $E|\log|F_\tau|| < \infty$.

Proof Simple arguments (left to the reader) show that it is enough to prove the result when $f(0) = 1$. In this case, the arguments above can be applied to conclude that as $r \uparrow 1$, $F(\tau_r) \to F(\tau)$ and

$$\sup_{r<1} E \log^+ |F(\tau_r)| < \infty.$$

so it follows from Fatou's lemma that $E \log^+ |F_\tau| < \infty$. To estimate the negative part, we observe that if T_n is the sequence of stopping times defined in the proof of (3), then $\operatorname{Re} L(t \wedge T_n)$ is a bounded martingale, so $E \operatorname{Re} L(\tau_r \wedge T_n) = 0$. Letting $n \to \infty$, observing that $\operatorname{Re} L(t \wedge \tau_r)$ is bounded above, and using Fatou's lemma (upside down), we see that $E\operatorname{Re} L(\tau_r) \geq 0$, so $E\operatorname{Re} L(\tau_r)^- \leq E\operatorname{Re} L(\tau_r)^+$, and it follows from another application of Fatou's lemma that $E|\log|F_\tau|| < \infty$.

Exercise 1 *Jensen's Formula.*

(a) Show that if $f(z)$ is analytic and free from zeros in $|z| \leq r$, then
$$\log|f(0)| = \int \log|f(re^{i\theta})|\frac{d\theta}{2\pi}.$$

(b) If f has zeros $a_1, \ldots, a_n$ in $|z| < r$ (multiple zeros being repeated according to their multiplicity), then apply (a) to
$$F(z) = f(z)\prod_{i=1}^{n}\frac{r^2 - \bar{a}_i z}{r(z - a_i)},$$
which is free from zeros and has $|F(z)| = |f(z)|$ on $|z| = r$, and conclude that
$$\int \log|f(re^{i\theta})|\frac{d\theta}{2\pi} = \log|F(0)|$$
$$= \log|f(0)| - \sum_{i=1}^{n}\log\left(\frac{r}{|a_i|}\right).$$

What is the probabilistic interpretation of the correction term?

Note: As the date on the Nevanlinnas' paper suggests, these results are classical. For an analytic treatment, see Duren (1970), pages 16–17. (1) is his Theorem 2.1; (2) and (7), when combined, are his Theorem 2.2. The proofs given here, like many of the results in the book, are "new" but can be traced directly to results in the literature. The idea of taking logs along the Brownian path is a natural generalization of a proof in Section 6.4 (where we take nth roots), a tactic that in turn was suggested by a remark in the introduction of Getoor and Sharpe (1972).

5.4 Two Special Properties of Boundary Limits of Analytic Functions

In this section, we will prove two results that show that the possibilities for the boundary behavior of a function analytic in D are much more limited than those of a function harmonic in D. The first result is a very strong uniqueness theorem (that fails miserably for harmonic functions).

(1) If an analytic function f has nontangential limit 0 on a set of positive measure, then $f \equiv 0$.

Remark: This result is due to Privalov (1924). If $f \in N$, (1) is a consequence of (8) in Section 5.3, but the f here is not assumed to satisfy any growth condition.

5.4 Two Special Properties of Boundary Limits of Analytic Functions

Proof If f has nontangential limit 0 on a set of positive measure, then it follows from (2) in Section 5.2 that $P(f(B_t) \to 0$ as $t \uparrow \tau) > 0$. If f is not $\equiv 0$, then f is nonconstant, and it follows from Lévy's theorem that $f(B_t)$, $t < \tau$, is a time change of Brownian motion run for an amount of time

$$\int_0^\tau |f'(B_t)|^2 \, dt > 0.$$

The last conclusion, however, contradicts the first, since if W_t is a two-dimensional Brownian motion, then

$$P_x(W_t = 0 \text{ for some } t > 0) = 0$$

and

$$P_x(W_t \to 0 \text{ as } t \to \infty) = 0.$$

Remark: The proof above is taken from Burkholder (1976). As he says on page 147:

> The advantage of a good probabilistic proof is not that the technical details become easier, although this is sometimes the case, but that the underlying ideas become more transparent. For example, the truth of Privalov's theorem becomes evident once we see that it is a question of whether or not Brownian motion hits a particular point in the complex plane with positive probability.

By using the fact that two-dimensional Brownian motion is recurrent, we can get another result about the boundary behavior of analytic functions that is due to Plessner (1928). As in Chapter 4, it is easy to prove the probabilistic analogue.

(2) With probability 1, either

(i) $\lim_{t \uparrow \tau} f(B_t)$ exists

or

(ii) for all $\varepsilon > 0$, $\{f(B_t) : t \in [\tau - \varepsilon, \tau)\}$ is dense in C.

Proof If we let $U_t = \operatorname{Re} f(B_t)$, then the first event is a.s. equal to $\{\langle U \rangle_\tau < \infty\}$ and the second is a.s. $\{\langle U \rangle_\tau = \infty\}$.

In order to translate this statement into analytical terms, we need a lemma. Let $S_\alpha(\theta)$ be the Stolz domain defined in Section 5.2, and if A is a Borel set, let $S_\alpha(A) = \bigcup_{\theta \in A} S_\alpha(\theta)$.

(3) For each $\alpha < 1$ and Borel set A, as $\varepsilon \to 0$,

$$P(B_\tau \in A \text{ and } B_t \in S_\alpha(A) \text{ for all } t \in [\tau - \varepsilon, \tau)) \to P(B_\tau \in A).$$

Proof Let $h(z) = P_z(B_\tau \in A)$. This result is an immediate consequence of two observations:

(a) A simple generalization of Calderon's argument in Section 4.5 shows that there is an $\varepsilon > 0$ (that depends only on α) such that if $z \notin S_\alpha(A)$, then $h(z) \le 1 - \varepsilon$.

(b) It follows from results in Section 5.2 that $h(B_t) \to 1_A(B_\tau)$ as $t \uparrow \tau$.

With (3) established, we are ready to prove Plessner's theorem.

(4) Except for a set of θ of Lebesgue measure zero, either

(i) f has nontangential limit at $e^{i\theta}$

or

(ii) for all $\alpha > 0$, $f(S_\alpha(\theta))$ is dense in C.

Proof Let A be a set of positive measure such that for all $\theta \in A$, $\lim f(z)$ does not exist as $z \to e^{i\theta}$ in $S_\alpha(\theta)$. If we let P_0^θ be the law of Brownian motion conditioned to exit D at $e^{i\theta}$, it follows from results in Section 5.2 that if $\theta \in A$, then

$$P_0^\theta \left(\lim_{t \uparrow \tau} f(B_t) \text{ exists} \right) = 0,$$

and it follows from (4) in Section 3.2 that

$$P_0 \left(\lim_{t \uparrow \tau} f(B_t) \text{ exists}, B_\tau \in A \right) = 0.$$

Combining the last result with (2) above shows that for all $\varepsilon > 0$,

$$P_0(B_\tau \in A, \{f(B_t) : t \in [\tau - \varepsilon, \tau)\} \text{ is not dense in } C) = 0,$$

which in view of (3) implies that $f(S_\alpha(A))$ is dense in C. Let $A_\alpha = \{\theta : \lim f(z) \text{ does not exist as } z \to e^{i\theta} \text{ within } S_\alpha(\theta)\}$. Since the last conclusion holds for any set $A \subset A_\alpha$ that has positive measure, a simple argument (which is left to the reader) shows that for a.e. $\theta \in A_\alpha$, $S_\alpha(\theta)$ is dense in C.

Note: The main ideas in this section, including the fact that boundary values of analytic functions can be studied via Lévy's theorem, are due to Doob (1961). As usual, however, we are not using the original source. Our proof of Privalov's theorem follows Burkholder (1976). Our proof of Plessner's theorem is from Davis (1979a); specifically, (2), (3), and (4) are his Theorems 4.2, 4.3, and 4.6. For classical proofs of these results, see Theorems 1.9 and 1.10 in Chapter 14 of Zygmund (1959).

5.5 Winding of Brownian Motion in $C - \{0\}$ (Spitzer's Theorem)

Let B_t^1 and B_t^2 be two independent Brownian motions with $B_0^1 = 1$ and $B_0^2 = 0$, and let $C_t = B_t^1 + iB_t^2$ be a complex Brownian motion. Since C_t almost surely never hits 0, we can define the total angle swept out up to time t to be the unique

5.5 Winding of Brownian Motion in $C - \{0\}$ (Spitzer's Theorem)

process θ_t with continuous paths that has $\theta_0 = 0$ and $\sin(\theta_t) = B_t^2/|C_t|$ for all $t > 0$. In words, the process θ_t records the angle and keeps track of the number of times the path has wound around 0, counting clockwise loops -2π and counterclockwise loops $+2\pi$. Spitzer (1958) proved the following limit theorem for θ_t:

(1) As $t \to \infty$,
$$P(2\theta_t/\log t \le y) \to \int_{-\infty}^{y} \frac{dx}{1+x^2} \cdot \frac{1}{\pi}.$$

Spitzer's proof is ingenious but requires a lot of computation (see Itô and McKean (1964), pages 270–271, for a succinct version). In this section, we use the conformal invariance of Brownian motion to give a simple proof of this result. The proof given below is based on Durrett (1982) but incorporates some improvements that I learned from Messulam and Yor (1982) and an earlier work of D. Williams (unpublished).

Let $D_t = A_t + iB_t$ be a complex Brownian motion with $D_0 = 0$, and let $F_t = \exp(D_t)$. By Lévy's theorem, if we let
$$\sigma_t = \int_0^t |\exp(D_s)|^2 \, ds$$

and
$$\gamma_u = \inf\{t : \sigma_t \ge u\},$$

then $C_u = F(\gamma_u)$, $u \ge 0$, is a complex Brownian motion with $C_0 = 1$. The first advantage of constructing C_t as $F(\gamma_t)$ is that we can easily write down the angle process $\theta_t = B(\gamma_t)$. The second, and more crucial, observation is that

(2) $$\sigma_t = \int_0^t |\exp(D_s)|^2 \, ds = \int_0^t \exp(2A_s) \, ds,$$

so γ and B are independent.

Let $S_u = \gamma(e^{2u})$. Time e^{2u} in process C corresponds to time S_u in process D. The main idea of the proof is to show that $\theta(e^{2u}) \equiv B(\gamma(e^{2u})) \equiv B(S_u)$ is approximately $B(T_u)$, where $T_u = \inf\{t : A_t \ge u\}$ (our guess is motivated by (2) above). Once we show this, (1) follows immediately, because a simple scaling gives

(3) $$B(T_u)/u \stackrel{d}{=} B(T_1),$$

and the right-hand side, which is the hitting distribution of $\{(x,y) : x = 1\}$, is (by results in Section 1.9) a Cauchy distribution with parameter 1.

To show that $B(S_u)$ is approximately $B(T_u)$, we start by showing that S_u is approximately T_u. The following estimate is crude, but it is sufficient for proving (1).

(4) If $\varepsilon > 0$, then as $u \to \infty$
$$P(T_{u(1-\varepsilon)} \le S_u \le T_{u(1+\varepsilon)}) \to 1.$$

Proof To get the lower bound, observe that

$$P(S_u \le T_{u(1-\varepsilon)}) = P(\gamma(e^{2u}) \le T_{u(1-\varepsilon)})$$
$$= P(e^{2u} \le \sigma(T_{u(1-\varepsilon)})),$$

and from (2), it follows that

$$\sigma(T_{u(1-\varepsilon)}) \le T_{u(1-\varepsilon)} \exp(2u(1-\varepsilon)),$$

so we have

(5a) $$P(S_u \le T_{u(1-\varepsilon)}) \le P(e^{2u\varepsilon} \le T_{u(1-\varepsilon)}),$$

which $\to 0$ as $u \to \infty$, since $T_{u(1-\varepsilon)} \stackrel{d}{=} u^2 T_{(1-\varepsilon)}$.

To get the upper bound, observe that

$$P(S_u \ge T_{u(1+\varepsilon)}) = P(e^{2u} \ge \sigma(T_{u(1+\varepsilon)}))$$

and let $L_\varepsilon(u)$ be the Lebesgue measure of $\{s \in [0, T_{u(1+\varepsilon)}] : A_s \ge u(1+\varepsilon/2)\}$. From (2), it follows that

$$\sigma(T_{u(1+\varepsilon)}) \ge L_\varepsilon(u) \exp(2u(1+\varepsilon/2)),$$

so we have

(5b) $$P(S_u \ge T_{u(1+\varepsilon)}) \le P(e^{-\varepsilon u} \ge L_\varepsilon(u) \ge 0),$$

which $\to 0$ as $u \to \infty$ since $L_\varepsilon(u) \stackrel{d}{=} u L_\varepsilon(1)$.

With (4) established, completing the proof of (1) is routine. If we let

$$\Delta_\varepsilon(u) = \sup\{|B(t) - B(T_u)|/u : t \in [T_{u(1-\varepsilon)}, T_{u(1+\varepsilon)}]\}$$

and recall that $\theta(e^{2u}) = B(S_u)$, then (4) implies that for fixed $\varepsilon > 0$,

$$P\left(\left|\frac{B(T_u)}{u} - \frac{\theta(e^{2u})}{u}\right| \le \Delta_\varepsilon(u)\right) \to 1$$

as $u \to \infty$. Now the distribution of $\Delta_\varepsilon(u)$ is independent of u and tends to 0 as $\varepsilon \to 0$, and (3) says that $B(T_u)/u \stackrel{d}{=} B(T_1)$, so it follows from a routine computation that

$$\frac{\theta(e^{2u})}{u} \Rightarrow B(T_1) \quad \text{as } u \to \infty$$

(where $\Rightarrow$ stands for convergence in distribution).

With a little effort, (1) can be improved to a limit theorem that describes when the winding occurs. Using the notation introduced in the proof of (1), we can state this result as the following:

(6) As $u \to \infty$, the finite-dimensional distributions of $\theta(e^{2tu})/u$, $t \ge 0$, converge to those of $B(T_t)$, $t \ge 0$ (a Cauchy process with parameter 1).

To understand what (6) says, look at Figure 5.3 and observe that if $u = 1 - \varepsilon$ and ε is small, then $B(T_{1-\varepsilon})$ is almost $B(T_1)$. Therefore, if $N = e^{2u}$, then most of

5.5 Winding of Brownian Motion in $C - \{0\}$ (Spitzer's Theorem)

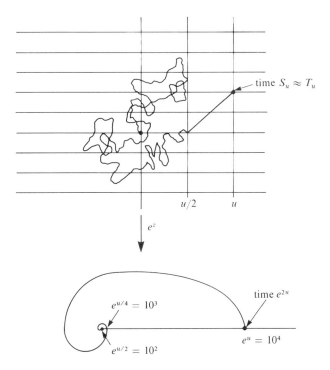

Figure 5.3

the winding has occurred by time $e^{2u(1-\varepsilon)} = N^{1-\varepsilon} = o(N)$. To see when in $[0, N^{1-\varepsilon}]$ most of this winding occurred, observe that if $t_1 < t_2 < \cdots < t_{n(\delta)}$ are the times of the jumps $> \delta$ in magnitude, the winding at time N is almost a sum of contributions from times proportional to $N^{t_1}, N^{t_2}, \ldots, N^{t_{n(\delta)}}$.

Technical Remark: Some probabilists may have expected (6) to contain a claim of weak convergence as a sequence of random elements of D (the space of functions that are right continuous and have left limits—see Billingsley (1968) for details). We have refrained from doing this because the claim is false: C = the set of continuous functions that is closed in the usual (Skorokhod J_1) topology, so we cannot have weak convergence.

Refinement 1: Rate of Convergence In his unpublished paper, Williams said, "The reader will find it a useful exercise to see how good a 'rate of convergence' ... he can get by refining our method." We will give a partial solution below and leave the rest of the details to the reader. Looking back at the proof, we see that the difference between $\theta(e^{2u})/u$ and $B(T_u)/u \stackrel{d}{=} B(T_1)$ is determined by the size of $S_u - T_u$, which in turn was controlled by the estimates

(5a) $\quad P(S_u \le T_{u(1-\varepsilon)}) \le P(e^{2u\varepsilon} \le T_{u(1-\varepsilon)})$

(5b) $\quad P(S_u \ge T_{u(1+\varepsilon)}) \le P(e^{-\varepsilon u} \ge L_\varepsilon(u) \ge 0)$.

To estimate the right-hand side of (5a), we observe that

$$P(e^{2u\varepsilon} \le T_{u(1-\varepsilon)}) = P\left(\frac{e^{2u\varepsilon}}{u^2(1-\varepsilon)^2} \le T_1\right),$$

so if we let $\varepsilon(u) = (\log u)^{1+\delta}/u$ where $\delta > 0$, then

(7a) $$P(S_u \le T_{u-(\log u)^{1+\delta}}) \le P\left(T_1 \ge \frac{u^{2+2\delta}}{u^2}\right) \to 0$$

as $u \to \infty$.

To improve (5b), we observe that, by considering T_{u+K}, we see that

$$\text{meas}\{s \in [0, T_{u+2K}] : A_s \ge u + K\} \stackrel{d}{=} \text{meas}\{s \in [0, T_K] : A_s \ge 0\},$$

so if we let $M_a = \text{meas}\{s \in [0, T_a] : A_s \ge 0\}$ and let $K(u) \to \infty$, then repeating the argument for (5b) shows

(7b) $$P(S_u \ge T_{u+K(u)}) \le P(e^{-K(u)} \ge M_{K(u)}) \to 0$$

as $u \to \infty$.

The last two results show that $S_u - T_u = 0(\log u)$ and hence that

$$\frac{\theta(e^{2u})}{u} - \frac{B(T_u)}{u} = 0\left(\frac{(\log u)^{1/2}}{u}\right),$$

or in terms of $t = e^{2u}$, the error is smaller than $C(\log \log t)^{1/2}/\log t$. This estimate is not sharp (since (7a) is not), but it is not far off.

Refinement 2: Joint Distribution of the Winding about Two Points It is an interesting and difficult problem to determine the limiting behavior of $(2/\log t)(\theta_t^a, \theta_t^b)$ where θ_t^a and θ_t^b denote the winding about a and $b \ne a$. After partial results of Messulam and Yor (1982) (see Theorem 3.2 of their paper), this problem has recently been solved by Jim Pitman and Marc Yor. The final result is one of those tantalizing theorems that are easy to discover but difficult to prove, so I will only give you a hint of what the result is and let you figure out the precise statement for yourself.

(8) If $a \ne b$, then as $t \to \infty$

$$\frac{2}{\log t}(\theta_t^a, \theta_t^b) \Rightarrow (X + Z, Y + Z).$$

The limit is independent of a and b. Z represents the contribution of big windings around $\{a, b\}$, and X and Y are the contribution of little windings around $\{a\}$ and $\{b\}$, respectively.

Note: I would like to thank Marc Yor for several useful conversations concerning the material in this section, and Jim Pitman for pointing out an error in the original version of the technical note. Formulating a good weak convergence theorem in this context is tricky; you need to take a very careful look at Figure 5.3.

5.6 Tangling of Brownian Motion in $C - \{-1, 1\}$ (Picard's Theorem)

If f is a nonconstant function that is analytic in the entire complex plane and B_t is a complex Brownian motion, then Lévy's theorem tells us that $f(B_t)$ is a time change of Brownian motion. Since two-dimensional Brownian motion is recurrent (see Section 1.7), it is immediate that if f is a nonconstant entire function, then the range $f(C)$ is dense in C. Picard's little theorem asserts that more is true.

(1) If f is a nonconstant entire function, then the range of f omits at most one complex number.

In this section, we give Davis's (1975) proof of Picard's little theorem. The idea of the proof is simple: Without loss of generality, we can assume that $\{-1, 1\} \not\subset f(C)$ and $f(0) = 0$. Let B_t be a Brownian motion starting at 0. If we can find times $t_0 < t_1$ such that $B_{t_0} = B_{t_1}$ but $f(B_t)$, $t_0 \leq t \leq t_1$ is not homotopic to a constant in $C - \{-1, 1\}$, then we will have a contradiction that proves Picard's theorem, because B_t, $t_0 \leq t \leq t_1$, can be continuously shrunk to 0 and $f(B_t)$, $t_0 \leq t \leq t_1$, cannot.

In order to carry out the plan described in the last paragraph, it is convenient to make a minor modification in what we are trying to prove. Identify the points of $\{z : |z| \leq .1\}$, call this set $\hat{0}$, and let $\hat{C}$ be $C - \{-1, 1\}$ with the points of $\hat{0}$ identified. Let $\tilde{0}$ be the component of $f^{-1}(\hat{0})$ containing 0, and let $\tilde{C}$ be C with the points of $\tilde{0}$ identified. To prove (1), we will prove:

(2) There is a time σ with $P_0(\sigma < \infty) = 1$ such that $B_\sigma \in \tilde{0}$ and $f(B_t)$, $0 \leq t \leq \sigma$, is not homotopic to $\hat{0}$ in $\hat{C}$.

Remark: This result has to be formulated carefully to be true. If $f(z) = e^z - 1$, then $-1 \notin f(C)$, and Spitzer's theorem shows that there is an infinite sequence of times $\sigma_n \uparrow \infty$ such that $f(B_{\sigma_n}) \in \hat{0}$ and $f(B_t)$, $0 \leq t \leq \sigma_n$, is not homotopic to 0 in $\hat{C}$, but this can never happen when $B(\sigma_n) \in \tilde{0}$.

To prove (2), it suffices to show that

(3) There is a time τ with $P_0(\tau < \infty) = 1$ such that for all $T \geq \tau$, $f(B_t)$, $0 \leq t \leq T$, is not homotopic to $\hat{0}$ in $\hat{C}$.

For then we can let $\sigma = \inf\{t > \tau : C_t \in \tilde{0}\}$. There are then a number of ways of proving (3). Davis gave one proof (1974) and later suggested another one ((1979a), page 920). The proof given below is our version of his second proof. Most of the effort is spent introducing a lot of notation for keeping track of the tangling around the two points. We start by introducing six sets (see Figure 5.4),

$$A_0 = \{x + iy : -1 < x < 1, y = 0\}$$
$$A_1 = \{x + iy : x < -1, y = 0\}$$
$$A_2 = \{x + iy : x > 1, y = 0\}$$

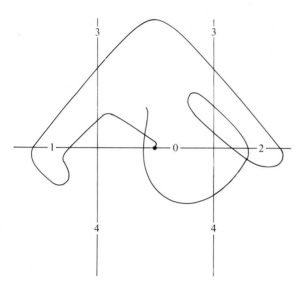

Figure 5.4

$$A_3 = \{x + iy : |x| = 1, y > 0\}$$
$$A_4 = \{x + iy : |x| = 1, y < 0\}$$
$$A = \bigcup_{i=0}^{4} A_i,$$

and an associated sequence of stopping times. Let B_t be a complex Brownian motion with $B_0 = 0$. Let $T_0 = 0$ and, for $n \geq 0$, let $T_{n+1} = \inf\{t > T_n : f(B_t) \subset A - A_{i_n}\}$ where the sequence of indices i_n is defined by $f(B_{T_n}) \in A_{i_n}$ for $n \geq 0$ (recall that we have assumed $f(0) = 0$, so $i_0 = 0$).

It is clear that if we are given the sequence of indices, then we can compute the homotopy class of the path, but this sequence contains a lot of irrelevant information, so we define a reduced sequence inductively by the following procedure:

(a) At time 0, we write 0.
(b) At time T_n, if there is only one number in the string, we add i_n to the end of the string.
(c) At time T_n, if there are $k > 1$ numbers in the string and the new number i_n is the same as the next to last, we erase the last number. Otherwise we add i_n to the end of the string.

Convention (c) allows us to delete irrelevant loops from the sequence of numbers. (To check your understanding of the procedure, you should try the example drawn in Figure 5.4. Its reduced sequence is 03240.) Besides cutting down on the length of the sequence, the erasing also drastically reduces the number of possibilities for the sequence. The reader can easily check that between two successive occurrences of 0, the sequence must be one of four types:

5.6 Tangling of Brownian Motion in $C - \{-1, 1\}$ (Picard's Theorem)

(i) 03240, 0324130, 032413240, ...
(ii) 03140, 0314230, 031423140, ...
(iii) 04230, 0423140, 042314230, ...
(iv) 04130, 0413240, 041324130, ...

From the observations above, it is clear that we can compute the homotopy class of the path from the reduced sequence and that to prove the theorem it suffices to show that the length of the reduced sequence $\to \infty$ almost surely. The fact that the reduced sequence $\to \infty$ can be made obvious by the observation that if the last two digits of the string are 03 or 04, then the probability we will add a number is greater than 1/2, while in any other case the probability we will add a number equals 1/2.

To make the last argument precise, we would need to obtain an estimate on how frequently $i_n = 0$. Rather than trying to do this, we will adopt a slightly different approach to avoid this difficulty. Let $N_0 = 0$, and for $k \geq 1$ let $N_k = \inf\{n > N_{k-1} : i_n = 0$ and the reduced sequence corresponding to $i_{N_{k-1}} \ldots i_n$ is not $0\}$. At time $R_k = T_{N_k}$, the number of zeros in the reduced string will either be the same or differ by one from the number of zeros at time R_{k-1}. To describe the conditions that result in the three possible outcomes, we need some definitions.

Any of the sequences that appear in the lists (i), (ii), (iii), and (iv) above is called a zero-block.

A zero-block b_2 is said to be the inverse of the zero-block b_1 if b_2 is obtained by reading b_1 backwards, that is, if $b_1 = 0324130$, then $b_2 = 0314230$.

The reason for the name *inverse* is explained by the next result. Let a_k be the last zero-block in the reduced string at time R_k, and let b_k be the zero-block that is obtained by reducing $i_{N_{k-1}}, \ldots, i_{N_k}$. If $\xi_k = $ the number of zeros at time R_k minus the number of zeros at time R_{k-1}, then

$$\xi_k = \begin{cases} -1 & \text{if } b_k \text{ is the inverse of } a_k \\ +1 & \text{if the first nonzero digit of } b_k \text{ (which must be 3 or 4) is} \\ & \text{different from the last nonzero digit of } a_k \\ 0 & \text{otherwise.} \end{cases}$$

At this point we have almost, but not quite, avoided the difficulty referred to above. It follows from the strong Markov property and symmetry that

$$P_0(\xi_{k+1} = 1 | \mathscr{F}(R_k)) = \tfrac{1}{2}$$
$$P_0(\xi_{k+1} = -1 | \mathscr{F}(R_k)) = h(a_k, f(B_{R_k})),$$

where h is a function that has $h(a, x) < 1/2$ for all a, x. It is somewhat unfortunate that $\sup h(a, x) = 1/2$ (contrary to what we claimed in an earlier version of this proof), but this difficulty is easy to remedy.

The culprits are the zero-blocks of length five: 03240, 03140, 04230, and 04130. In this first or second case we have

$$\lim_{x \uparrow 1} h(a, x) = 1/2,$$

since the next winding will, with a probability approaching 1, be around 1, and symmetry implies that the two possibilities have equal probability. A similar argument shows that in the third or fourth case

$$\lim_{x \downarrow -1} h(a, x) = 1/2,$$

but the other zero-blocks, luckily, are our friends. It is easy to check that if a is not one of the four zero-blocks mentioned above, then

$$\sup_{x \in (-1, 1)} h(a, x) = C(a) < 1/2$$

because each such a involves winding about both -1 and 1, and hence has

$$\lim_{x \uparrow 1} h(a, x) = \lim_{x \downarrow -1} h(a, x) = 0.$$

Combining this with the observation that if a_n is a sequence of zero-blocks with length $\to \infty$ and $x_n \in (-1, 1)$ then $h(a_n, x_n) \to 0$, it follows that if we exclude the four bad zero-blocks then

$$\sup_{a, x} h(a, x) < 1/2.$$

To complete the proof now, it suffices (as we will show below) to ignore the bad zeros and count only the good ones. Let $N_0' = 0$ and for $k \geq 1$ let $N_k' = \inf\{n > N_{k-1}' : i_n = 0$ and the reduced sequence which corresponds to $L_{N_{k-1}'} \ldots L_n$ contains a zero-block that has length greater than $5\}$. If we call a zero that is at the right end of a zero-block of length greater than 5 a "good zero," then arguments given above show that at time $R_k' = T_{N_k'}$ the number of good zeros in the reduced string will be the same or differ by one from the number of zeros at time R_{k-1}', and, furthermore, if we let a_k', b_k', and ξ_k' be defined in the obvious way then we can repeat the arguments above with "primes" attached.

Having done this we come at last to a point where (at last!?) we can use the strong Markov property and symmetry to conclude that

$$P_0(\xi_{k+1}' = 1 | \mathscr{F}(R_k')) = 1/2$$
$$P_0(\xi_{k+1}' = -1 | \mathscr{F}(R_k')) = h'(a_k', f(B_{R_k'}))$$

where h' is a function that has $h'(a', x) < 1/2$ for all x and a'. The last observation makes it clear that $S_k = \xi_1' + \cdots + \xi_k' \to \infty$ a.s. To prove this, let η_k be a sequence of random variables that have $\eta_k = \xi_k'$ on $\{\xi_k' \neq 0\}$ and are defined on $\{\xi_k' = 0\}$ in such a way that

$$P_0(\eta_k = -1 | \mathscr{F}(R_{k-1})) = q$$
$$P_0(\eta_k = 0 | \mathscr{F}(R_{k-1})) = \tfrac{1}{2} - q.$$

If we let $S_k' = \eta_1 + \cdots + \eta_k$, then S_k' is a sum of independent random variables, so

$$\frac{S_k'}{k} \to \frac{1}{2} - q > 0 \text{ a.s.}$$

and it follows that $S_k \geq S_k' \to \infty$ a.s.

5.6 Tangling of Brownian Motion in $C - \{-1, 1\}$ (Picard's Theorem)

Problem 1 Having shown that the length of the reduced string that describes the homotopy class $\to \infty$ a.s., it is natural to inquire about the rate of growth. If L_t denotes the length at time t, then it is trivial that

$$|L_t| \geq 4 \min\left(\frac{|\theta_t^{-1}|}{2\pi}, \frac{|\theta_t^1|}{2\pi}\right),$$

where θ_t^a is the winding around a defined in Section 5.5. This suggests (to me at least) that $|L_t|$ should grow faster than $\log t$. What is the right rate of growth? Someone who understands Jacobi's modular function well can probably solve this problem by looking carefully at how Brownian motion exits H (see McKean (1969)). Once one proves Picard's little theorem using Brownian motion, it is natural to try to prove his "great theorem" as well. This theorem states that if f has an essential singularity at $z = a$, then in each neighborhood of a, f assumes each complex number (with one possible exception) an infinite number of times. See Conway (1978), page 300, for the relevant definitions and an analytic proof. Davis (1979a) has given a probabilistic proof, but since his proof still requires a fair amount of complex analysis, we refer the reader to pages 928–930 of his paper for the details.

A second type of extension of the results above is to ask if Brownian motion in $C - \{-1, 1\}$ gets tangled in other senses as well. McKean (1969) claimed on page 112 that "the plane Brownian path undoes itself i.o. as $t \uparrow \infty$ from the point of view of homology with integral coefficients." This is incorrect. Lyons and McKean (1980) "corrects this error and takes the matter further."

Note: I would like to thank Jim Pitman for pointing out an error in the original version of the proof given above. He is not to be blamed, however, if my "first aid" fails to save the victim.

6 Hardy Spaces and Related Spaces of Martingales

6.1 Definition of H^p, an Important Example

Let $D = \{z : |z| < 1\}$ be the unit disk in the complex plane. A function u that is harmonic in D is said to be in h^p, $0 < p < \infty$, if

$$d_p(u) = \sup_{r<1} \int_0^{2\pi} |u(re^{i\theta})|^p \, d\pi(\theta) < \infty.$$

Extending this notion, we say an analytic function f is in H^p if

$$d_p(f) = \sup_{r<1} \int_0^{2\pi} |f(re^{i\theta})|^p \, d\pi(\theta) < \infty$$

(here and in what follows, $d\pi(\theta) = d\theta/2\pi$).

The class H^p has been studied extensively since the time of Hardy (1915). In this chapter, we will describe some of the results that have been obtained. To prepare for these developments, we devote this section to the consideration of one example that will appear at several points in the development of our story:

$$\rho(z) = (1+z)/(1-z).$$

ρ sends $-1 \to 0$, $0 \to 1$, $1 \to \infty$, and a little thought shows that ρ maps D one-to-one onto the half plane $\operatorname{Re} z > 0$. To find $u = \operatorname{Re} \rho$ and $v = \operatorname{Im} \rho$, we write

$$\rho(z) = \frac{1+z}{1-z} \cdot \frac{1-\bar{z}}{1-\bar{z}} = \frac{1+z-\bar{z}-|z|^2}{|1-z|^2},$$

so

$$u(z) = \frac{1-|z|^2}{|1-z|^2} \quad v(z) = \frac{2\operatorname{Im} z}{|1-z|^2}.$$

6.1 Definition of H^p, an Important Example

We have seen $u(z)$ many times above. It is the probability density (w.r.t. π) that a Brownian motion starting at z will exit D at 1. From this interpretation of u and the strong Markov property it follows that

$$\int_0^{2\pi} u(re^{i\theta})\,d\pi(\theta) = u(0) = 1$$

for all $r < 1$ and also that if $\theta \neq 0$, then as $r \to 1$, $u(re^{i\theta}) \to 0$. Combining the last two observations shows that if $p > 1$,

$$\int_0^{2\pi} |u(re^{i\theta})|^p\,d\pi(\theta) \to \infty$$

as $r \to 1$ (if not, then $u \to 0$ in L^1), so $u \in h^1$ but $u \notin h^p$ for any $p > 1$.

At this point, we might hope that $\rho \in H^1$. In Section 6.5, we will see that the limiting behavior of u precludes this, but for the moment we do not know this, so we will show that $\rho \notin H^1$ by computing

$$\int_0^{2\pi} |v(re^{i\theta})|\,d\pi(\theta) = 2\int_0^{\pi} \frac{2r\sin\theta}{1 - 2r\cos\theta + r^2}\frac{d\theta}{2\pi}$$

$$= \frac{1}{\pi}\log(1 - 2r\cos\theta + r^2)\Big|_0^{\pi}$$

$$= \frac{2}{\pi}\log\left(\frac{1+r}{1-r}\right) \to \infty \quad \text{as } r \to 1.$$

It doesn't miss by much since

$$\lim_{r\uparrow 1} \frac{2r\sin\theta}{1 + r^2 - 2r\cos\theta} = \frac{\sin\theta}{1 - \cos\theta} \quad \theta \neq 0,$$

and as $\theta \to 0$,

$$\frac{\sin\theta}{1 - \cos\theta} = \frac{\theta + O(\theta^3)}{\frac{\theta^2}{2} + O(\theta^4)} \sim \frac{2}{\theta}.$$

so we have

$$\lim_{r\uparrow 1} v(re^{i\theta}) \in L^p \quad \text{for all } p < 1.$$

As our experience with u indicates, this result is not, in general, enough to conclude that $v \in h^p$ (in that case $\lim_{r\uparrow 1} u(re^{i\theta}) = 0 \in L^\infty$). This time, however, nothing bad happens. For $r \geq 0$, $1 + r^2 \geq 2r \geq 2r\cos\theta$, so

$$\frac{|2r\sin\theta|}{1 + r^2 - 2r\cos\theta} \leq \frac{|2r\sin\theta|}{2r - 2r\cos\theta} = \frac{|\sin\theta|}{1 - \cos\theta}.$$

and if $p < 1$, it follows that

$$d_p(v) \leq \int_0^{2\pi} \left(\frac{|\sin\theta|}{1 - \cos\theta}\right)^p d\pi(\theta) < \infty,$$

that is, $v \in h^p$.

To sum up, what we have found is that

$u \in h^p$ for $p \le 1$
$v \in h^p$ for $p < 1$,

so

$\rho \in H^p$ for $p < 1$.

The list above shows that we may have $\operatorname{Re} f \in h^1$ but $\operatorname{Im} f \notin h^1$. In Section 6.7, we will show that this example is the worst behavior we can encounter when $\operatorname{Re} f \in h^1$, and that this cannot happen if 1 is replaced by $p \in (1, \infty)$, that is, if $1 < p < \infty$, $\operatorname{Re} f \in h^p$ implies $\operatorname{Im} f \in h^p$. The last result is not true for

$$h^\infty = \left\{ z : \sup_{z \in D} |u(z)| < \infty \right\}.$$

A counterexample is $i \log \rho(z)$. Since ρ maps D one-to-one onto $\operatorname{Re} z > 0$, $\log \rho$ maps D one-to-one onto the strip $\{z : -\pi/2 < \operatorname{Im} z < \pi/2\}$, and we see that $\operatorname{Re}(i \log \rho) \in h^\infty$, but $\operatorname{Im}(i \log \rho) \notin h^\infty$.

The last example is just one of many that we can produce by modifying ρ. In Section 6.7, when we study conjugation (i.e., the map $\operatorname{Re} f \to \operatorname{Im} f$), the functions

$$\rho(z)^\beta, \quad \frac{2}{\pi} \log \rho(z), \quad \frac{\rho(z) + \rho(-z)}{2} = \frac{2z}{1 - z^2}$$

will be used to show that certain inequalities are sharp (experts are invited to guess the inequalities). At first, the ubiquitous appearance of ρ may seem mysterious. In Section 6.5, we show that it is in some sense inevitable: If $f \in H^p$, $p \ge 1$, then there is a function $\varphi \in L^p(\partial D, \pi)$ such that

$$f(z) = \int \rho(z e^{-i\theta}) \varphi(e^{i\theta}) \, d\pi(\theta),$$

so any $f \in H^p$, $p \ge 1$, is a linear combination of copies of ρ composed with rotations of D. (Here and in what follows, the integration is over any convenient interval of length 2π if the limits are not explicitly indicated.)

6.2 First Definition of $\mathcal{M}^p$, Differences Between $p > 1$ and $p = 1$

In this section, we will define the spaces $\mathcal{M}^p$, $p > 0$, that are the martingale analogues of the Hardy spaces introduced in Section 6.1. The definition is designed so that if $f \in H^p$ and B_t is a Brownian motion starting at 0, then $\operatorname{Re} f(B_t)$, $t < \tau = \inf\{t : B_t \notin D\}$, will be an element of $\mathcal{M}^p$, but we will not be able to show this until the end of Section 6.4. For the moment, our goal is to define $\mathcal{M}^p$ and use martingale theory to prove some simple results that point out some differences between $\mathcal{M}^1$ and $\mathcal{M}^p$, $p > 1$.

6.2 First Definition of $\mathcal{M}^p$, Differences Between $p > 1$ and $p = 1$

Let B_t be a Brownian motion starting at 0 (defined for convenience on our special probability space $(C, \mathscr{C})$) and let $\mathbb{F}$ be the filtration generated by B_t. Let $X^* = \sup_t |X_t|$ and let $\mathcal{M}^p = \{X : X \text{ is a local martingale w.r.t. } \mathbb{F} \text{ and } E|X^*|^p < \infty\}$. If $p \geq 1$, then $E|X^*| < \infty$, so X is uniformly integrable, and a standard martingale convergence theorem implies that

(i) as $t \to \infty$, $X_t \to X_\infty$ a.s.
(ii) $X_t = E(X_\infty | \mathscr{F}_t)$.

It is clear that $E|X_\infty|^p \leq E|X^*|^p$. When $p > 1$, there is a converse inequality.

(1) If $X \in \mathcal{M}^p$ with $p > 1$, then

$$E|X_\infty|^p \leq E|X^*|^p \leq \left(\frac{p}{p-1}\right)^p E|X_\infty|^p.$$

Remark: This is a standard result from martingale theory, but aspects of the proof are important for what follows, so we will prove it here. The key to the proof is the following result (Doob's inequality), which for later purposes we will prove as an equality.

(2) Let X_t be a continuous martingale with $X_0 = c$ and let $\bar{X}_t = \sup_{s \leq t} X_s$. Then for $\lambda > c$,

$$\lambda P(\bar{X}_t > \lambda) = E(X_t; \bar{X}_t > \lambda).$$

Proof Let $T = \inf\{s : X_s > \lambda\} \wedge t$. Since T is a bounded stopping time, it follows from the optional stopping theorem that

$$E(X_t | \mathscr{F}_T) = X_T.$$

Since X_t has continuous paths and $X_0 = c < \lambda$, we have $X_T = \lambda$ on $\{T < t\} = \{\bar{X}_t > \lambda\}$. Combining this result with the definition of conditional expectation, we have

$$\lambda P(\bar{X}_t > \lambda) = E(X_T; \bar{X}_t > \lambda)$$
$$= E(X_t; \bar{X}_t > \lambda).$$

Remark: The same proof (with a few minor changes) shows the following:

(2)′ If X is a continuous submartingale, then

$$\lambda P(\bar{X}_t > \lambda) \leq E(X_t; \bar{X}_t > \lambda)$$

without any restriction on X_0 or λ, since in this case

$$E(X_t | \mathscr{F}_T) \geq X_T$$

and

$$X_T \geq \lambda \quad \text{on} \quad \{T < t\} = \{\bar{X}_t > \lambda\}.$$

To obtain (1) from (2), we integrate and use Fubini's theorem (everything is nonnegative) as follows:

$$\begin{aligned}E\bar{X}_t^p &= \int_0^\infty p\lambda^{p-1} P(\bar{X}_t > \lambda)\, d\lambda \\ &\le \int_0^\infty p\lambda^{p-1}\left(\lambda^{-1}\int_{\{\bar{X}_t > \lambda\}} X_t^+\, dP\right) d\lambda \\ &= \int_\Omega X_t^+ \left(\int_0^{\bar{X}_t} p\lambda^{p-2}\, d\lambda\right) dP \\ &= \frac{p}{p-1}\int_\Omega X_t^+ \bar{X}_t^{p-1}\, dP.\end{aligned}$$

If we let $q = p/(p-1)$ be the exponent conjugate to p and apply Hölder's inequality, we see that the above

$$\le q(E|X_t^+|^p)^{1/p}(E|\bar{X}_t|^p)^{1/q}.$$

At this point, we would like to divide both sides of the inequality above by $(E|\bar{X}_t|^p)^{1/q}$ to get

(*) $\qquad (E\bar{X}_t^p)^{1/p} \le q(E|X_t^+|^p)^{1/p}.$

Unfortunately, the laws of arithmetic do not allow us to divide by something that may be ∞. To remedy this difficulty, we observe that $P(\bar{X}_t \wedge n > \lambda) \le P(\bar{X}_t > \lambda)$, so repeating the proof above shows that

$$(E(\bar{X}_t \wedge n)^p)^{1/p} \le q(E|X_t^+|^p)^{1/p},$$

and letting $n \to \infty$ proves (*).

The last step in proving (1) is to let $t \to \infty$ and use the monotone and dominated convergence theorems to conclude that if we let $\bar{X} = \sup_t X_t$, then we have

(1') $\qquad E|\bar{X}|^p \le \left(\dfrac{p}{p-1}\right)^p E|X_\infty^+|^p,$

a one-sided result from which (1) follows immediately.

Let $\|X\|_p = (E|X^*|^p)^{1/p}$. It is easy to see that if $p \ge 1$, $\|\cdot\|_p$ defines a norm on $\mathcal{M}^p$. From formula (1) and the discussion preceding it, we see that $X \to X_\infty$ maps $\mathcal{M}^p$ one-to-one into a subspace of $L^p(\mathcal{F}_\infty)$, where $\mathcal{F}_\infty = \sigma(B_t : t \ge 0) = \mathcal{C}$. Since we only allow continuous martingales in $\mathcal{M}^p$, the image of $\mathcal{M}^p$ under this map is, in general, $\ne L^p(\mathcal{F}_\infty)$, however, we also only consider Brownian filtrations, so it follows from (1) and results in Section 2.14 that if $p > 1$, $X \to X_\infty$ maps $\mathcal{M}^p$ one-to-one onto $L^p(\mathcal{F}_\infty)$, and furthermore that $\mathcal{M}^p$ and $L^p(\mathcal{F}_\infty)$ are equivalent as metric spaces. When $p = 1$, the last two claims are false. Rather than giving an explicit counterexample, I will prove the following result, which I learned from Gundy (1980a). The theorem originally appeared in Gundy (1969).

(3) $\qquad$ If $X \ge 0$ is a continuous martingale with $X_0 = 1$, then

6.2 First Definition of $\mathcal{M}^p$, Differences Between $p > 1$ and $p = 1$

$$EX^* \geq 1 + EX_\infty \log^+ X_\infty,$$

where $X_\infty = \lim X_t$ (which exists, since $X \geq 0$).

Proof Let $T_\lambda = \inf\{t : X_t > \lambda\}$. If $\lambda > 1$, then $X(T_\lambda) = \lambda$ on $\{T_\lambda < \infty\} = \{X^* > \lambda\}$, so it follows from (2) that

$$\lambda P(T_\lambda < t) = E(X_t 1_{(T_\lambda < t)}).$$

Now as $t \to \infty$, $X_t \to X_\infty$ and $1_{(T_\lambda < t)} \to 1_{(T_\lambda < \infty)}$ a.s., so it follows from Fatou's lemma that

$$\lambda P(T_\lambda < \infty) \geq E(X_\infty 1_{(T_\lambda < \infty)}) \geq E(X_\infty 1_{(X_\infty > \lambda)}).$$

We therefore have

$$EX^* = 1 + \int_1^\infty P(T_\lambda < \infty) \, d\lambda$$
$$\geq 1 + \int_1^\infty \lambda^{-1} E(X_\infty 1_{(X_\infty > \lambda)}) \, d\lambda$$
$$= 1 + E\left(X_\infty \int_1^\infty \lambda^{-1} 1_{(X_\infty > \lambda)} \, d\lambda\right)$$
$$= 1 + E(X_\infty (\log X_\infty)^+).$$

Let M^1 be the image of $\mathcal{M}^1$ under the mapping $X \to X_\infty$. The last result shows that if Y is a nonnegative random variable with $EY = 1$ and $EY(\log Y)^+ = \infty$, then $Y \in L^1 - M^1$. The next result shows that these are the only nonnegative random variables in $L^1 - M^1$.

(4) If $X \geq 0$ is a continuous submartingale and $X_t^* = \sup\{|X_s| : s \leq t\}$, then

$$EX_t^* \leq 2(1 + EX_t \log^+ X_t).$$

Proof To simplify the computation, we will first transform the estimate obtained in (2).

(5) Under the hypotheses of (2) or (2)', if $\lambda > 0$, then

$$\lambda P(\bar{X}_t > 2\lambda) \leq E(X_t; X_t > \lambda).$$

Proof By assumption, we have

$$2\lambda P(\bar{X}_t > 2\lambda) \leq E(X_t; \bar{X}_t > 2\lambda)$$
$$\leq E(X_t; X_t > \lambda) + E(X_t; X_t \leq \lambda, \bar{X}_t > 2\lambda)$$
$$\leq E(X_t; X_t > \lambda) + \lambda P(\bar{X}_t > 2\lambda),$$

and subtracting $\lambda P(\bar{X}_t > 2\lambda)$ from both sides proves (5).

With (5) established, it is routine to prove (4). Since $X_t^* = \bar{X}_t$, applying (5) gives

$$E(X_t^*/2) = \int_0^\infty P(X_t^* > 2\lambda)\,d\lambda$$

$$\leq 1 + \int_1^\infty \lambda^{-1} \int_{\{X_t > \lambda\}} X_t\,dP\,d\lambda$$

$$= 1 + \int_\Omega \int_1^{X_t \vee 1} \lambda^{-1} X_t\,d\lambda\,dP$$

$$= 1 + EX_t(\log X_t)^+.$$

Remark: I learned the trick above from one of my students (R. Banuelos), but as several people who have read preliminary versions of this text have pointed out, the idea has appeared in the literature many times before. To fully appreciate the simplifications that follow from using (5) rather than (2), the reader is invited to integrate (2) to prove that

$$EX_t^* \leq \frac{e}{e-1}(1 + EX_t \log^+ X_t).$$

This is Problem 7 on page 355 in Chung (1974). A solution can be found on page 317 of Doob (1953).

If we let $Y_t = E(Y|\mathcal{F}_t)$ and apply (5) to $X_t = |Y_t|$, we get

$$EY_t^* \leq 2(1 + E|Y_t|(\log|Y_t|)^+).$$

Now $(x \log x)' = 1 + \log x$, which is increasing, so $\varphi(x) = (1 + x \log x) \vee 1$ is convex and increasing, and it follows that the right-hand side is

$$E\varphi(|E(Y|\mathcal{F}_t)|) \leq E\varphi(E(|Y||\mathcal{F}_t))$$
$$\leq E(E(\varphi(|Y|)\,\mathcal{F}_t)) = E\varphi(|Y|),$$

so letting $t \to \infty$ gives

(6) $$EY^* \leq 2(1 + E|Y|\log^+|Y|)$$

and justifies the remark we made above (4).

As we mentioned then, (3) shows that the mapping $X \to X_\infty$ sends $\mathcal{M}^1$ into a proper subspace of L^1. Since every $X \in L^1(\mathcal{F}_\infty)$ gives rise to a martingale $X_t = E(X|\mathcal{F}_t)$ with $X_\infty = X$, the following definition seems natural: Let $\mathcal{L}^1 = \{$martingales X that can be written as $X_t = E(X|\mathcal{F}_t)$ with $X \in L^1(\mathcal{F}_\infty)\} = $ the set of uniformly integrable martingales.

Another space of martingales that is important in the theory of martingales is $\mathcal{K}^1 = \{$martingales X such that $\sup_t E|X_t| < \infty\} = L^1$ bounded martingales. It is clear that $\mathcal{K}^1 \supset \mathcal{L}^1$. The canonical example of a martingale in $\mathcal{K}^1 - \mathcal{L}^1$ is $X_t = B_{t \wedge T}$, where B_t is a Brownian motion starting at 1 and $T = \inf\{t > 0 : B_t = 0\}$. The optional stopping theorem implies that $EX_t = 1$, so $E|X_t| = EX_t = 1$ and $X \in \mathcal{K}^1$. On the other hand, $X_t \to 0$ a.s., so $X_t \neq E(X_\infty|\mathcal{F}_t)$ and $X \notin \mathcal{L}^1$.

6.2 First Definition of $\mathcal{M}^p$, Differences Between $p > 1$ and $p = 1$

Note: As the story unfolds, the reader will see that there is a close connection between this example and $\text{Re}\,\rho$, where $\rho(z) = (1 + z)/(1 - z)$ is the function we studied in Section 6.1.

If we wanted to, we could define $\mathcal{K}^p$ and $\mathcal{L}^p$ for $p > 1$ in the obvious ways,

$$\mathcal{K}^p = \{\text{martingales } X \text{ with } \sup E|X_t|^p < \infty\}$$
$$\mathcal{L}^p = \{X_t = E(X|\mathcal{F}_t) \text{ with } X \in L^p\},$$

but there is no reason to do this. Standard results from martingale theory imply that $\mathcal{K}^p = \mathcal{L}^p = \mathcal{M}^p$ for $p > 1$. Because of this fact, we will not consider $\mathcal{K}^p$ and $\mathcal{L}^p$, $p > 1$, below, and we will simplify our notation by defining

$$\|X\|_{\mathcal{K}} = \sup_t E|X_t|$$
$$\|X\|_{\mathcal{L}} = E|X_\infty| \quad \text{if} \quad X_t = E(X_\infty|\mathcal{F}_t).$$

Another thing that might occur to you is to define

$$\mathcal{K}\log\mathcal{K} = \{\text{martingales } X \text{ with } \sup_t E|X_t|\log^+|X_t| < \infty\}.$$

The proof of (6) given above shows that $\mathcal{K}\log\mathcal{K} \subset \mathcal{M}^1$, but it is easy to show that the inclusion is strict. Let $Z \geq 0$ be a random variable with $EZ < \infty$ and $E|Z|\log^+|Z| = \infty$. If we let B_t be a Brownian motion that is independent of Z, let $T = \inf\{t: B_t \notin [-Z, Z]\}$, and let $X_t = B_{t \wedge T}$, then $X^* = Z$, so $X \in \mathcal{M}^1$, but

$$\lim_{t \to \infty} E|X_t|\log^+|X_t| = E|Z|\log^+|Z| = \infty,$$

so $X \notin \mathcal{K}\log\mathcal{K}$. Generalizing this example, we see that $\mathcal{M}^1$ cannot be characterized as $\{\text{martingales } X \text{ with } \sup E\psi(X_t) < \infty\}$ for any ψ.

Two formulas sum up this section:

$$\mathcal{K}^1 \supset \mathcal{L}^1 \supset \mathcal{M}^1 \supset \mathcal{K}\log\mathcal{K}$$
$$\mathcal{K}^p = \mathcal{L}^p = \mathcal{M}^p \quad \text{for} \quad p > 1.$$

Exercise 1 Let

$$\mathcal{L}^\infty = \{X: X_t = E(X|\mathcal{F}_t), X \in L^\infty\}$$
$$\mathcal{M}^\infty = \{\text{martingales } X: X^* \in L^\infty\}.$$

Show that $\mathcal{L}^\infty = \mathcal{M}^\infty$.

Remark: It is interesting to note that although the proof of (1) is very simple, the constant is the best possible. For a proof of this assertion and a lot more, see Dubins and Gilat (1978). They conjecture that there is no martingale that is $\not\equiv 0$ and has

$$E|X^*|^p = \left(\frac{p}{p-1}\right)^p E|X_\infty|^p.$$

Pitman (1979) has proved this for the case $p = 2$ and has shown that, although the inequality is sharp, it can be improved to

$$E\left(\sup_t X_t^+ - \inf_t X_t^-\right)^2 \leq 4EX_\infty^2.$$

6.3 A Second Definition of $\mathcal{M}^p$

In this section, we will introduce a second equivalent norm for the $\mathcal{M}^p$ spaces. This norm is the analogue of the area function in analysis and, as we shall see below, it is much easier to work with in many circumstances. In Section 2.1, we associated with each continuous local martingale X_t, $t \geq 0$, a predictable increasing process $\langle X \rangle_t$, with $\langle X \rangle_0 \equiv 0$, that makes $X_t^2 - \langle X \rangle_t$ a local martingale. Since $t \to \langle X \rangle_t$ is increasing, $\lim_{t \to \infty} \langle X \rangle_t$ exists. If we let $\langle X \rangle_\infty$ denote the limit, and if $p \geq 1$, we can define a norm by letting

$$\langle\!\langle X \rangle\!\rangle_p = \|\langle X \rangle_\infty^{1/2}\|_p,$$

where $\|Y\|_p$ denotes the L^p norm $(E|Y|^p)^{1/p}$ (the $1/2$ is necessary because $\langle cX \rangle_t = c^2 \langle X \rangle_t$).

One rationale for this definition can be found in Section 2.11. If we let $\gamma_t = \sup\{s : \langle X \rangle_s \leq t\}$, then $X(\gamma_t)$ is a Brownian motion run for an amount of time $\langle X \rangle_\infty$, so the Brownian scaling relationship $B_t \stackrel{d}{=} t^{1/2} B_1$ suggests that X^* and $\langle X \rangle_\infty^{1/2}$ should be about the same size. It is a delicate problem to formulate a result which makes this precise. If $X_t = B_{t \wedge 1}$, then $\langle X \rangle_\infty \equiv 1$, but since $P(X^* > \lambda) > 0$ for all λ, there can be no inequality of the form $P(X^* > \lambda) \leq AP(\langle X \rangle_\infty^{1/2} > \lambda/B)$. The other direction is just as bad. If $T = \inf\{t : |B_t| = 1\}$ and $X_t = B_{t \wedge T}$, then $X^* \equiv 1$, but $P(\langle X \rangle_\infty^{1/2} > \lambda) > 0$ for all λ. To get around these problems, we change the left-hand side of the desired inequalities.

(1) Let $\beta > 1$ and $\delta > 0$. If τ is a stopping time for Brownian motion, then

(1a) $P(B_\tau^* > \beta\lambda, \tau^{1/2} \leq \delta\lambda) \leq \dfrac{\delta^2}{(\beta - 1)^2} P(B_\tau^* > \lambda)$

(1b) $P(\tau^{1/2} > \beta\lambda, B_\tau^* \leq \delta\lambda) \leq \dfrac{\delta^2}{\beta^2 - 1} P(\tau^{1/2} > \lambda).$

Remark: As we will see in the proof of (2), the key point here is that for some $\beta < \infty$, we have an inequality of the form $P(f > \beta\lambda, g \leq \delta\lambda) \leq C_\delta P(f > \lambda)$, where $C_\delta \to 0$ as $\delta \to 0$.

Proof It is enough to prove the result for τ bounded, for if the result holds for $\tau \wedge n$ for all $n \geq 1$, it also holds for τ.

Let $S_1 = \inf\{t : |B(\tau \wedge t)| > \lambda\}$.
Let $S_2 = \inf\{t : |B(\tau \wedge t)| > \beta\lambda\}$.
Let $T \equiv (\delta\lambda)^2$.

6.3 A Second Definition of $\mathcal{M}^p$

It is easy to check that if $B_t^* = \sup_{s \leq t} |B_s|$, then

$$P(B_\tau^* > \beta\lambda, \tau^{1/2} \leq \delta\lambda)$$
$$\leq P(|B(\tau \wedge S_2 \wedge T) - B(\tau \wedge S_1 \wedge T)| > (\beta - 1)\lambda)$$
$$\leq (\beta - 1)^{-2}\lambda^{-2} E(B(\tau \wedge S_2 \wedge T) - B(\tau \wedge S_1 \wedge T))^2.$$

Now if $R_1 \leq R_2$, are both bounded stopping times, we have

$$E(B(R_2) - B(R_1))^2 = E(B(R_2)^2 - B(R_1)^2) = E(R_2 - R_1),$$

so the above

$$= (\beta - 1)^{-2}\lambda^{-2} E((\tau \wedge S_2 \wedge T) - (\tau \wedge S_1 \wedge T))$$
$$\leq (\beta - 1)^{-2}\lambda^{-2} (\delta\lambda)^2 P(S_1 < \infty)$$
$$= (\beta - 1)^{-2}\delta^2 P(B_\tau^* > \lambda),$$

proving the first inequality.

To prove the other inequality

Let $S_1 = \inf\{t : (\tau \wedge t)^{1/2} > \lambda\}$.
Let $S_2 = \inf\{t : (\tau \wedge t)^{1/2} > \beta\lambda\}$.
Let $T = \inf\{t : |B(\tau \wedge t)| > \delta\lambda\}$.

Again, it is easy to check that

$$P(\tau^{1/2} > \beta\lambda, B_\tau^* \leq \delta\lambda) \leq P(\tau \wedge S_2 \wedge T - \tau \wedge S_1 \wedge T \geq \beta^2\lambda^2 - \lambda^2)$$
$$\leq (\beta^2 - 1)^{-1}\lambda^{-2} E(\tau \wedge S_2 \wedge T - \tau \wedge S_1 \wedge T),$$

and using the stopping time result mentioned in the first part of the proof, it follows that the last expression above

$$= (\beta^2 - 1)^{-1}\lambda^{-2} E(B(\tau \wedge S_2 \wedge T)^2 - B(\tau \wedge S_1 \wedge T)^2)$$
$$\leq (\beta^2 - 1)^{-1}\lambda^{-2} (\delta\lambda)^2 P(S_1 < \infty)$$
$$= (\beta^2 - 1)^{-1}\delta^2 P(\tau^{1/2} > \lambda),$$

proving the second inequality.

Remark: The inequalities above are called "good λ" inequalities, although the reason for the name is obscured by our formulation (which is from Burkholder (1973)). The name "good λ" comes from the fact that early versions of this and similar inequalities (see Theorems 3.1 and 4.1 in Burkholder and Gundy (1970), or page 148 in Burkholder, Gundy, and Silverstein (1971)) were formulated as $P(f > \lambda) \leq C_{\beta,K} P(g > \lambda)$ for all λ that satisfy $P(g > \lambda) \leq KP(g > \beta\lambda), \beta, K > 1$.

The next result shows why we are interested in good λ inequalities. First, we need a definition. A function φ is said to be moderately increasing if φ is a nondecreasing function with $\varphi(0) = 0$ and if there is a constant K such that $\varphi(2\lambda) \leq K\varphi(\lambda)$ for all $\lambda > 0$.

Examples

(i) $\varphi(x) = x^p, 0 < p < \infty, K = 2^p$
(ii) $\varphi(x) = x + x \log^+ x, K < 2 + 2\log 2$.

Proof for (ii) Since $\log^+ xy \le \log^+ x + \log^+ y$,
$$\frac{2x + 2x\log^+ 2x}{x + x\log^+ x} \le 2 + \frac{2\log 2}{1 + \log^+ x}.$$
The bound now follows by considering two cases, $x \le 1$ and $x \ge 1$.

(2) If φ is a moderately increasing function, then there are constants $c, C \in (0, \infty)$ (that depend only on the growth rate K) such that
$$cE\varphi(\tau^{1/2}) \le E\varphi(B^*_\tau) \le CE\varphi(\tau^{1/2}).$$

Remark: We will only use the result for $\varphi(x) = x^p$, but it is nice to know that the only property of x^p we need for the proof is that $\varphi(2\lambda) \le K\varphi(\lambda)$ for all $\lambda > 0$.

Proof To prove the result, it suffices to show:

(3) If $X, Y \ge 0$ satisfy
$$P(X > 2\lambda, Y \le \delta\lambda) \le \delta^2 P(X > \lambda) \quad \text{for all } \delta \ge 0$$
and φ is a moderately increasing function, then there is a constant C (that depends only on the growth rate K) such that
$$E\varphi(X) \le CE\varphi(Y).$$

Proof It is enough to prove the result for bounded φ, for if the result holds for $\varphi \wedge n$ for all $n \ge 1$, it also holds for φ. φ is the distribution function, a measure on $[0, \infty)$ that has
$$\varphi(h) = \int_0^h d\varphi(\lambda) = \int_0^\infty 1_{(h > \lambda)} \, d\varphi(\lambda).$$
If Z is a nonnegative random variable, taking expectations and using Fubini's theorem gives
$$E\varphi(Z) = E \int_0^\infty 1_{(Z > \lambda)} \, d\varphi(\lambda) = \int_0^\infty P(Z > \lambda) \, d\varphi(\lambda).$$
From our assumption it follows that
$$P(X > 2\lambda) = P(X > 2\lambda, Y \le \delta\lambda) + P(Y > \delta\lambda)$$
$$\le \delta^2 P(X > \lambda) + P(Y > \delta\lambda),$$
or integrating $d\varphi(\lambda)$,
$$E\varphi(2^{-1}X) = \int_0^\infty P(2^{-1}X < \lambda) \, d\varphi(\lambda)$$
$$\le \delta^2 E\varphi(X) + E\varphi(\delta^{-1}Y).$$

Pick $\delta^2 K < 1$ and then pick $N \ge 0$ so that $2^N > \delta^{-1}$. From the growth condition and the monotonicity of φ, it follows that

$$E\varphi(\delta^{-1}Y) \leq K^N E\varphi(Y).$$

Combining this with the last inequality and using the growth condition gives

$$E\varphi(X) \leq KE\varphi(2^{-1}X)$$
$$\leq K\delta^2 E\varphi(X) + K^{N+1} E\varphi(Y).$$

Solving for $E\varphi(X)$ now gives

$$E\varphi(X) \leq \frac{K^{N+1}}{1 - K\delta^2} E\varphi(Y),$$

proving (3) and hence (2).

Applying (2) to the case $\varphi(x) = x^p$ and recalling the results of Section 2.11, we get the following inequality that will be useful in studying the spaces $\mathcal{M}^p$.

(4) There are constants c, $C \in (0, \infty)$ (that depend only on p) so that for all $0 \leq t \leq \infty$,

$$cE\langle X\rangle_t^{p/2} \leq E(X_t^*)^p \leq CE\langle X\rangle_t^{p/2}.$$

As the reader might expect, the constants in (4) are not very good. If $p = 2$, then $EX_\infty^2 = E\langle X\rangle_\infty$, so it follows from (1) of Section 6.2 that

$$E\langle X\rangle_\infty = EX_\infty^2 \leq E|X^*|^2 \leq 4EX_\infty^2 = 4E\langle X\rangle_\infty,$$

but in the proof above, $K = 4$, so if we take $\delta = 1/3$ and $N = 2$, then

$$C = \frac{K^{N+1}}{1 - K\delta^2} = 115.2.$$

Remark: There are other ways of proving (4) directly; see Getoor and Sharpe (1972) for an interesting proof using stochastic integration.

6.4 Equivalence of H^p to a Subspace of $\mathcal{M}^p$

Let B_t be a complex Brownian motion starting at 0 and let $\tau = \inf\{t : |B_t| = 1\}$. In this section, we will show that the mapping $f \to \operatorname{Re} f(B_t)$, $t < \tau$, maps $H_0^p = \{f \in H^p : f(0) = 0\}$ one-to-one into $\mathcal{M}_\tau^p = \{X : X$ is a local martingale on $[0, \tau)$ and $X_\tau^* = \sup_{t < \tau} |X_t| \in L^p\}$, and that furthermore, the H^p norm of f is equivalent to the $\mathcal{M}_\tau^p$ norm of its image, that is, if we let $u = \operatorname{Re} f$, $U_t = u(B_t)$, $t < \tau$, and $U_\tau^* = \sup_{t < \tau} |U_t|$, then:

(1) There are constants c, $C \in (0, \infty)$ (that depend only on p) such that for all $f \in H_0^p$,

$$cE|U_\tau^*|^p \leq d_p(f) \leq CE|U_\tau^*|^p.$$

The first step in proving (1) is to prove the result with U_τ^* replaced with $F_\tau^* = \sup_{t < \tau} |F_t|$, where $F_t = f(B_t)$, $t < \tau$.

(2) There is a constant $C \in (0, \infty)$ (that depends only on p) such that for all $f \in H^p$,
$$d_p(f) \le E|F_\tau^*|^p \le C d_p(f).$$

Remark: It is curious that the best constant we can obtain from the proof given below is $C = e$ (independent of p).

Proof If $r < 1$ and $\tau_r = \inf\{t : |B_t| = r\}$, then $|F(\tau_r)| \le F_\tau^*$, so we have
$$\int_0^{2\pi} |f(re^{i\theta})|^p \, d\pi(\theta) = E|F(\tau_r)|^p \le E(F_\tau^*)^p.$$

Taking the supremum over $r < 1$ gives $d_p(f) \le E(F_\tau^*)^p$.

To prove the other inequality takes some work. We start with the trivial case, $p > 1$. Since f is analytic in D and bounded on $D(0, r) = \{z : |z| < r\}$, it follows from results in Chapter 2 that $f(B(t \wedge \tau_r))$ is a complex martingale and, hence, that $|f(B(t \wedge \tau_r))|$ is a submartingale. Noting that $|f(B(\tau_r))| \in L^p$, $p > 1$, and applying Doob's inequality, we get
$$E|F^*(\tau_r)|^p \le \left(\frac{p}{p-1}\right)^p E|f(B(\tau_r))|^p.$$

Letting $r \uparrow 1$ and using the monotone convergence theorem gives
$$E|F_\tau^*|^p \le \left(\frac{p}{p-1}\right)^p d_p(f).$$

proving (1) in the case $p > 1$ with a constant slightly larger than the one advertised in the remark, since
$$\frac{d}{dx} \log\left(\frac{x}{x-1}\right)^x = \frac{d}{dx}(x \log x - x \log(x-1))$$
$$= \log x - \log(x-1) - (x-1)^{-1}$$
$$= \int_{x-1}^x y^{-1} - (x-1)^{-1} \, dy < 0.$$

To prove the result when $0 < p \le 1$, we will have to be more devious. First consider $1/2 < p \le 1$ and let us now suppose $f(0) = 1$. Since $f(B_{t \wedge \tau})$ is a time change of a complex Brownian motion starting at 1, $P(f(B_t) = 0$ for some $0 \le t < \tau) = 0$, and we can define a pathwise square root G_t by requiring G to be continuous and to have $G_0 = 1$, and $G_t^2 = F_t$ for all $0 \le t < \tau$. As we mentioned in Section 5.3,

(3) G_t, $t < \tau$, is a local martingale

and furthermore

(4) If $r < 1$, $G(t \wedge \tau_r)$ is a martingale.

(See the remarks after the proofs of (3) and (5) in Section 5.3.) With (4) proved, the rest is easy. Since

6.4 Equivalence of H^p to a Subspace of $\mathcal{M}^p$

$$E|G(\tau_r)|^{2p} = \int_0^{2\pi} |f(re^{i\theta})|^p \, dm(\theta) \leq d_p(f)$$

and $2p > 1$, it follows from Doob's inequality that

$$E|G^*(\tau_r)|^{2p} \leq \left(\frac{2p}{2p-1}\right)^{2p} E|G(\tau_r)|^{2p}.$$

Taking a supremum over r and applying the monotone convergence theorem gives

$$E|F_\tau^*|^p = E|G^*(\tau)|^{2p} \leq \left(\frac{2p}{2p-1}\right)^{2p} d_p(f).$$

This proves the result when $f(0) = 1$. If $f(0) \neq 0$, applying the result above to $g(x) = f(x)/f(0)$ proves the inequality for f. To extend the result to f with $f(0) = 0$, consider $g_\varepsilon(x) = f(x) + \varepsilon$ and let $\varepsilon \to 0$.

The argument above can be extended to $p > 1/n$ by taking nth roots, with the result that

$$E|F_\tau^*|^p \leq \left(\frac{np}{np-1}\right)^{np} d_p(f).$$

Letting $n \to \infty$, the last inequality gives

$$E|F_\tau^*|^p \leq e d_p(f),$$

proving (2) with $C = e$.

The result in (2) shows that the H^p norm of f and the $\mathcal{M}_\tau^p$ norm of $f(B_t)$, $t < \tau$, are equivalent. As we mentioned above, we want to go one step further and consider H^p as a space of real local martingales by using the mapping $f \to \text{Re} f(B_t)$, $t < \tau$. Since every harmonic function u has a conjugate harmonic function $\bar{u}$ defined by the requirements that $\bar{u}(0) = 0$ and $f = u + i\bar{u}$ is analytic in D, the mapping is one-to-one on $H_0^p = \{f \in H^p : f(0) = 0\}$. The next result shows that if we restrict our attention to this subspace, then the H^p norm of f and the $\mathcal{M}_\tau^p$ norms of $\text{Re} f(B_t)$, $t < \tau$, are equivalent (i.e., (1) holds).

(5) There is a constant $C \in (0, \infty)$ (that depends only on p) such that for all $f \in H_0^p$,

$$E|U_\tau^*|^p \leq E|F_\tau^*|^p \leq CE|U_\tau^*|^p.$$

Proof Since $|a| \leq (a^2 + b^2)^{1/2}$ for all a, b, we have $|u(z)| \leq |f(z)|$ for all z, proving the inequality on the left. To prove the other inequality, we observe that if we let $\bar{U}_t = \bar{u}(B_t)$, $t < \tau$, then it follows from results in Sections 2.11 and 6.4 that

$$E(\bar{U}_\tau^*)^p \leq CE\langle \bar{U} \rangle_\tau^{p/2}$$
$$\langle \bar{U} \rangle_\tau^{p/2} = \langle U \rangle_\tau^{p/2}$$
$$E\langle U \rangle_\tau^{p/2} \leq CE(U_\tau^*)^p$$

(here, as before and ever shall be, the value of C is unimportant and will change from line to line). It is trivial that

$$E(F_\tau^*)^p \le 2^p(E(U_\tau^*)^p + E(\bar{U}_\tau^*)^p).$$

Combining this with the other three estimates proves (5).

Given the results above and in Chapter 4, it is now easy to prove the "maximal function characterization of H^p" due to Burkholder, Gundy, and Silverstein (1971). Let $S_\alpha(\theta)$ be the convex hull of the disk $\{|z| \le \alpha\}$ and the point $e^{i\theta}$ and let

$$(N_\alpha u)(\theta) = \sup\{|u(z)| : z \in S_\alpha(\theta)\}$$

be the nontangential maximal function. A simple generalization of the argument given at the end of Section 4.5 shows that

$$P_0\left(\sup_{0 \le s < \tau} u(B_s) > \lambda\right) \le C|\{\theta : N_\alpha u(\theta) > \lambda\}|,$$

where C depends only on α, and integrating gives

$$E_0(U_\tau^*)^p \le C \int_0^{2\pi} |N_\alpha u|^p(\theta)\, d\pi(\theta)$$

(a result that holds for any measurable u).

To prove the other inequality, we use (2) from Section 4.3 (which generalizes easily from H to D). This result implies that there is an $\varepsilon > 0$ such that if $z \in S_\alpha(\theta)$, then

$$P_0^\theta(B_t, 0 \le t < \tau, \text{ makes a loop around } z) \ge \varepsilon,$$

so it follows from the maximum principle that

$$P_0^\theta(U_\tau^* > N_\alpha u(\theta)/2) \ge \varepsilon,$$

and hence that

$$E_0(U_\tau^*)^p \ge \varepsilon \int_0^{2\pi} |N_\alpha u|^p(\theta)\, d\pi(\theta).$$

Combining this with the other inequality and (1) above gives the maximal function characterization of H^p

$$c d_p(f) \le \int_0^{2\pi} |N_\alpha u|^p(\theta)\, d\pi(\theta) \le C d_p(f).$$

6.5 Boundary Limits and Representation of Functions in H^p

Since $H^p \subset N$, it follows from (2) in Section 5.3 that

(1) If $f \in H^p$, then the nontangential limit of f exists at a.e. point of ∂D.

In this section, we will investigate the relationship between f (defined in D) and its nontangential limit (defined on ∂D), which we will also denote by f.

6.5 Boundary Limits and Representation of Functions in H^p

There are two reasons why we want to do this: (a) The mean convergence of H^p functions to their boundary values (3) and the consequent Poisson integral representation (5) are simple consequences of the equivalence established in Section 6.4, and (b) for developments below, it is nice to know that we can consider a function $f \in H^p$ as a function in $L^p(\partial D)$.

Our first topic in this section is L^p convergence to boundary values. As usual, we start with the probabilistic result and then deduce its analytical counterpart.

(2) Let $f \in H^p$ and let $F_t = f(B_t)$, $t < \tau$. If T_n is any sequence of times $\uparrow \tau$, then as $n \to \infty$, $F(T_n) \to f(B_\tau)$ a.s. and in L^p.

Proof The a.s. convergence is a consequence of (2) in Section 5.2. (2) of Section 6.4 shows that $F_\tau^* \in L^p$. Since $|F(T_n) - f(B_\tau)| \leq (2F_\tau^*)^p$ and $F(T_n) \to f(B_\tau)$ a.s., the L^p convergence follows from the dominated convergence theorem.

(3) If $f \in H^p$, then as $r \uparrow 1$,

$$\int |f(re^{i\theta}) - f(e^{i\theta})|^p \, d\pi(\theta) \to 0.$$

Proof It suffices to show that the result holds for any sequence $r_n \uparrow 1$. If we let $T_n = \inf\{t : |B_t| > r_n\}$ and apply (2), it follows that $F(T_n) \to f(B_\tau)$ in L^p, or, if we let v_n denote the distribution of $(B(T_n), B_\tau)$,

$$\int |f(x) - f(y)|^p \, dv_n(x, y) \to 0.$$

This conclusion is similar to the one desired, but it is different enough to make it painful to obtain one from the other. In the face of this hard difficulty, we will take a soft solution. We recall the following result from real analysis:

(4) If $X_n \in L^p$ and $X_n \to X$ a.e., then $X_n \to X$ in L^p if and only if $E|X_n|^p \to E|X|^p$.

We leave the proof as an exercise for the reader (see Chung (1974), page 97), and then we are done, because

$$F(T_n) \to f(B_\tau) \quad \text{in} \quad L^p$$

implies

$$\int |f(r_n e^{i\theta})|^p \, d\pi(\theta) \to \int |f(e^{i\theta})|^p \, d\pi(\theta)$$

implies

$$\int |f(r_n e^{i\theta}) - f(e^{i\theta})|^p \, d\pi(\theta) \to 0.$$

The last result is valid for $p > 0$. When $p \geq 1$, we can use the L^p convergence to express f in terms of its boundary values.

(5) Let $k_\theta(z)$ be the probability density (w.r.t. π) of exiting D at $e^{i\theta}$ starting from z. If $f \in H^p$ and $p \geq 1$, then

$$f(z) = \int k_\theta(z) f(e^{i\theta}) \, d\pi(\theta).$$

Proof From (2) of Section 6.4, it follows that $E|F_\tau^*|^p < \infty$, and since $p \geq 1$, F_t, $t < \tau$, is uniformly integrable. If σ is a stopping time, the optional stopping theorem implies that

$$f(B_\sigma) = E(f(B_\tau)|\mathscr{F}_\sigma),$$

and it follows from the strong Markov property that

$$E(f(B_\tau)|\mathscr{F}_\sigma) = \int k_\theta(B_\sigma) f(e^{i\theta}) \, d\pi(\theta).$$

The last two results imply that the equality in (5) holds for a.e. $z \in D$. The left-hand side, $f(z)$, is clearly a continuous function of z. In Section 3.3, we showed that the right-hand side is harmonic in D, and hence continuous, so the equality holds for all $z \in D$.

Example 1 The equality in (5) does not hold when $p < 1$. The half-plane mapping $\rho(z) = (1 + z)/(1 - z)$ has boundary limit

$$\rho(e^{i\theta}) = i \sin\theta/(1 - \cos\theta),$$

so there are two problems:

(i) $\operatorname{Re} \rho(z) > 0$ in D, but

$$\int k_\theta(z) \operatorname{Re} \rho(e^{i\theta}) \, d\pi(\theta) = 0$$

(ii) $|\sin\theta|/(1 - \cos\theta) \sim 2/|\theta|$ as $\theta \to 0$, so

$$\int k_\theta(z) |\operatorname{Im} \rho(e^{i\theta})| \, d\pi(\theta) = \infty.$$

The resolution of the first problem is obvious. $\operatorname{Re} \rho = k_0$, so

$$\operatorname{Re} \rho(z) = \int k_\theta(z) \, d\delta_0(\theta),$$

where δ_0 is a point mass at 0. This example suggests that we might generalize the representation in (5) to allow measures on ∂D that are not absolutely continuous w.r.t. π. The next result shows that this generalization does not enlarge by very much the class of functions that can be represented.

(6) The following three classes of functions are the same:

(i) the set of u that can be written as

6.5 Boundary Limits and Representation of Functions in H^p

$$u(z) = \int k_\theta(z)\, d\mu(\theta),$$

where μ is a signed measure with finite variation

(ii) the set of u that can be written as a difference of two positive harmonic functions

(iii) $h^1 =$ the set of harmonic functions u with

$$\sup_{r<1} \int |u(re^{i\theta})|\, d\theta < \infty.$$

Proof (i) $\Rightarrow$ (ii) $\Rightarrow$ (iii) is trivial. Since the proof of (iii) $\Rightarrow$ (i) is very similar to the proofs given in Sections 3.3 and 3.5, it is left as an exercise for the reader.

Remark: For developments in Section 6.8, it is useful to know that (6) implies that any $u \in h^1$ can be written as $u_1 - u_2$, where $u_1, u_2 \geq 0$ and $\|u\|_1 = \|u_1\|_1 + \|u_2\|_1$. To prove this, we observe that if

$$u = \int k_\theta(z)\, d\mu(\theta),$$

then $\|u\|_1 =$ the variation $\mu = \sup_A \mu(A) - \mu(A^c)$.

The last result completes our consideration of boundary limits of functions in H^p. Since we will deal with $p \geq 1$ for most of the rest of this chapter, the important results to remember are (1) and (5). They show that a function $f \in H^p$, $p > 0$ has a nontangential limit at a.e. point of ∂D and that if $p \geq 1$, the values of f in D can be recovered from the boundary limits, so we can think of f in H^p as being a function in $L^p(\partial D, \pi)$ (and we will do this when we prove the duality theorem).

A similar viewpoint is possible for local martingales $X \in \mathcal{M}^p_\tau, p \geq 1$. Since this will simplify things below, we will take a few minutes now and spell out the details. $X^*_\tau \in L^p$, so standard arguments imply that if we let

$$Y_t = \begin{cases} X_t & t < \tau \\ \lim_{t \uparrow \tau} X_t & t \geq \tau, \end{cases}$$

then Y is a martingale, $Y \in \mathcal{M}^p$ (the space defined in Section 6.2), and Y can be reconstructed from its limiting value by

$$Y_t = E(Y_\infty | \mathcal{F}_t).$$

The last representation is the probabilistic analogue of the Poisson integral representation. When $X_t = f(B_t)$, $t < \tau$, and $f \in H^p$, $p \geq 1$, the relationship is closer than an analogy, since results above imply that $Y_\infty = f(B_\tau)$ (the right-hand side being the nontangential limit of f evaluated at B_τ).

Notes: Result (3) about the mean convergence of $f(re^{i\theta})$ to $f(e^{i\theta})$ was first proved by F. Riesz (1923) (see, for example, Duren (1970), pages 20–22). The key to Riesz's proof was the following factorization theorem:

(7)   Every function f in H^p can be written as $f(z) = b(z)g(z)$, where $|b(z)| \le 1$ and $g \in H^p$ is a function that does not vanish in D.

Once (7) was established, (1) followed from known results. The function b has boundary limits, since it is bounded, and since g is never 0, we can pick $n > 1/p$ and consider $g^{1/n}$ to reduce the result to the easy case where $p > 1$.

The reader should note that taking nth roots to reduce to the trivial case $p > 1$ was also the key to the proof of (3) in Section 6.2, but in that proof, we used the fact that Brownian motion did not hit zero to construct a pathwise nth root, so we did not have to factor out the zeros.

6.6 Martingale Transforms

Martingale transforms are a natural generalization of the following:

Example 1   Let $f = u + iv$ be an analytic function with $f(0) = 0$, and let B_t be a complex Brownian motion starting at 0. Itô's formula implies that

$$u(B_t) = \int_0^t \nabla u(B_s) \cdot dB_s$$

$$v(B_t) = \int_0^t \nabla v(B_s) \cdot dB_s,$$

and the Cauchy-Riemann equations say that

$$\nabla v = \begin{pmatrix} 0 & 1 \\ -1 & 0 \end{pmatrix} \nabla u,$$

so if we let

$$H_s = \nabla u(B_s) \quad A = \begin{pmatrix} 0 & 1 \\ -1 & 0 \end{pmatrix},$$

then we can write

$$u(B_t) = \int_0^t H_s \cdot dB_s$$

$$v(B_t) = \int_0^t A H_s \cdot dB_s.$$

The last equation obviously makes sense if B is a d-dimensional Brownian motion, H is a locally bounded (R^d-valued) predictable process, and A is any $d \times d$ matrix. To define the transform of a general local martingale X in this setting, we now recall that since we have assumed that our filtration is generated by a Brownian motion, it follows from results in Section 2.14 that

$$X_t = X_0 + \int_0^t H_s \cdot dB_s.$$

6.6 Martingale Transforms

We can therefore define the transform of X by A as

$$(A * X)_t = \int_0^t AH_s \cdot dB_s$$

(whenever this makes sense, e.g., if H is locally bounded).

In this section, we will study properties of martingale transforms as mappings between the $\mathcal{M}^p$ spaces. The results we prove here are analogues of classical results about conjugate functions that we will prove by probabilistic methods in Section 6.7. The first and most basic result is:

(1) If $p > 0$, $X \to A * X$ is a bounded linear transformation from $\mathcal{M}^p$ to $\mathcal{M}^p$.

Proof If

$$X_t = \int_0^t H_s \cdot dB_s,$$

then

$$(A * X)_t = \int_0^t AH_s \cdot dB_s,$$

so

$$\langle A * X \rangle_t = \int_0^t |AH_s|^2 \, ds$$
$$\leq C \int_0^t |H_s|^2 \, ds$$
$$= C \langle X \rangle_t,$$

where

$$C = \sup\{|Ay|^2 : |y| = 1\},$$

so the desired conclusion follows from the equivalence of norms demonstrated in Section 6.3.

When $p > 1$, the norms on $\mathcal{M}^p$ and $\mathcal{H}^p$ are equivalent, so we have:

(2) If $p > 1$, $X \to A * X$ is a bounded linear transformation from $\mathcal{H}^p$ to $\mathcal{H}^p$.

The next example shows that this is false when $p = 1$.

Example 2 Let B_t be a two-dimensional Brownian motion starting at 0, let $\tau = \inf\{t : B_t^1 = -1\}$, and let

$$X_t = B_{t \wedge \tau}^1 = \int_0^{t \wedge \tau} \begin{pmatrix} 1 \\ 0 \end{pmatrix} \cdot dB_s.$$

If we let $A = \begin{pmatrix} 0 & 0 \\ 1 & 0 \end{pmatrix}$ (or, if you want, $\begin{pmatrix} 0 & -1 \\ 1 & 0 \end{pmatrix}$), then

$$(A * X)_t = \int_0^{t \wedge \tau} \binom{0}{1} \cdot dB_s = B_{t \wedge \tau}^2.$$

Now $X_t = B_{t \wedge \tau}^2 \geq -1$ and it is trivial that if $a \geq -1$, $|a| \leq 2 + a$, so

$$E|X_t| \leq 2 + EX_t = 2$$

and $X \in \mathcal{H}^1$. On the other hand, it follows from results in Section 1.9 that $(A * X)_\infty = B_\tau^2$ has a Cauchy distribution

$$P((A * X)_\infty = x) = \frac{1}{\pi} \frac{1}{1 + x^2},$$

so $E(A * X)_\infty^+ = \infty$, and Fatou's lemma implies that

$$\liminf_{t \to \infty} E(A * X)_t^+ \geq \infty.$$

In the last example,

$$\|A * X\|_{\mathcal{H}} = \sup_t E|A * X|_t = \infty,$$

but it doesn't miss by much

$$P(|A * X)_\infty| > y) = \frac{2}{\pi} \int_y^\infty \frac{dx}{1 + x^2} \sim \frac{2}{\pi} y^{-1}$$

as $y \to \infty$. The next result, first proved by Burkholder (1966), shows that this is the worst possible behavior for $X \in \mathcal{H}^1$. The statement and proof given below are from Burkholder (1979a).

(3) If $\langle Y \rangle_t \leq \langle X \rangle_t$ for all $t \geq 0$, then

$$\lambda P(Y^* \geq \lambda) \leq 2 \|X\|_{\mathcal{H}}.$$

Proof Doob's inequality implies that

$$\lambda P(X^* > \lambda) \leq \sup_t E|X_t| = \|X\|_{\mathcal{H}},$$

so it suffices to estimate the probability of $\{Y^* > \lambda, X^* \leq \lambda\}$. To do this, we observe that by stopping at $T_n = \inf\{t : \langle X \rangle_t > n\}$, we can suppose that $E\langle X \rangle_\infty < \infty$, that is, $X \in \mathcal{M}^2$.

Let $\tau = \inf\{t : |X_t| > \lambda\}$. It is trivial that if we let $Y_t^* = \sup\{|Y_s| : s \leq t\}$, then

$$\lambda P(Y^* \geq \lambda, X^* \leq \lambda) \leq \lambda P(Y_\tau^* \geq \lambda).$$

Applying Doob's inequality to the submartingale $Y_{t \wedge \tau}^2$ ($Y \in \mathcal{M}^2$, since $\langle Y \rangle_t \leq \langle X \rangle_t$) gives

$$\lambda^2 P((Y_\tau^*)^2 \geq \lambda^2) \leq EY_\tau^2.$$

Now $X, Y \in \mathcal{M}^2$, so

$$EY_\tau^2 = E\langle Y \rangle_\tau \leq E\langle X \rangle_\tau = EX_\tau^2.$$

6.6 Martingale Transforms

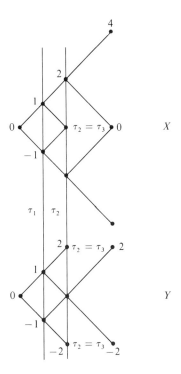

Figure 6.1

To finish up, we observe that $|X_\tau| \leq \lambda$ and $|X_t|$ is uniformly integrable, so
$$EX_\tau^2 \leq \lambda E|X_\tau| \leq \lambda E|X_\infty| = \lambda \|X\|_{\mathscr{H}}.$$
Combining the inequalities above shows that
$$\lambda^2 P(Y^* > \lambda, X^* \leq \lambda) \leq \lambda \|X\|_{\mathscr{H}}$$
and proves the desired result.

A simple example shows that 2 is the best possible constant in (3).

Example 3 Let B_t be a one-dimensional Brownian motion, and make the following definitions (for a picture, see Figure 6.1):

$\tau_1 = \inf\{t : |B_t| = 1\}$
$\tau_2 = \inf\{t \geq \tau_1 : |B_t - B(\tau_1)| = 1\}$
$\tau_3 = \inf\{t \geq \tau_2 : B_t = 0 \text{ or } |B_t - B(\tau_2)| = 2\}$
$B(\tau_1) = 1 \text{ or } -1$
$B(\tau_2) = 2, 0, \text{ or } -2$
$B(\tau_3) = 4, 0, \text{ or } -4$

$$\begin{array}{ccl} \varphi(s,\omega) & \psi(s,\omega) & \\ 1 & 1 & s < \tau_1 \\ 1 & -1 & \tau_1 \le s < \tau_2 \\ 1 & 1 & \tau_2 \le s < \tau_3,\ B(\tau_2) \ne 0 \\ 0 & 0 & \text{otherwise} \end{array}$$

$$X_t = \int_0^t \varphi_s\, dB_s \quad Y_t = \int_0^t \psi_s\, dB_s.$$

Since $\varphi^2 = \psi^2$, we have $\langle X\rangle_t = \langle Y\rangle_t$. A simple calculation shows that if we look at B, X, and Y at times τ_1, τ_2, and τ_3, we have:

B	X	Y	Y^*
1, 2, 4	1, 2, 4	1, 0, 2	2
1, 2, 0	1, 2, 0	1, 0, -2	2
1, 0, 2	1, 0, 0	1, 2, 2	2
1, 0, -2	1, 0, 0	1, 2, 2	2

with a similar table for $a = -1$. From the last computation it follows that

$$\sup_t E|X_t| = E|X_\infty| = 1$$

and

$$P(Y^* \ge 2) = 1,$$

so for $\lambda = 2$,

$$\lambda P(Y^* \ge \lambda) = 2 \sup_t E|X_t|,$$

showing that (3) is sharp.

(1), (2), and (3) are the main results on martingale transforms, but, of course, it is also possible to consider how $X \to A*X$ behaves on other spaces. We will mention only three results and leave the proofs as exercises.

Exercise 1 If $X \in \mathcal{M}^\infty = $ the bounded martingales, then $(A*X)$ may be unbounded. In Chapter 7, we will see that it does not miss by much. $(A*X) \in \mathcal{BMO}$ and, furthermore, $X \to (A*X)$ is a bounded map from $\mathcal{BMO}$ to $\mathcal{BMO}$.

Exercise 2 If $X \in \mathcal{L}^1 = $ the uniformly integrable martingales, then $(A*X)$ need not be in $\mathcal{H}^1$. (Hint: Recall our discussion of $\mathcal{H} \log \mathcal{H}$.)

Exercise 3 If $X \in \mathcal{H} \log \mathcal{H}$, then $(A*X) \in \mathcal{M}^1$. Is $(A*X) \in \mathcal{H} \log \mathcal{H}$?

6.7 Janson's Characterization of $\mathcal{M}^1$

Having seen an example of an $X \in \mathcal{H}^1$ and a matrix A such that $A*X \notin \mathcal{H}^1$, it is natural (if somewhat precocious) to ask which $X \in \mathcal{H}^1$ have $A*X \in \mathcal{H}^1$

6.7 Janson's Characterization of $\mathcal{M}^1$

for all matrices A. If we let J^1 denote this collection, then (1) in Section 6.6 implies that $J^1 \supset \mathcal{M}^1$ and, as you might guess from the title of this section, $J^1 = \mathcal{M}^1$. In fact, more is true: There is a finite set of matrices $A_1, \ldots, A_m$ such that if $A_i * X \in \mathcal{K}^1$ for $i = 1, \ldots, m$, then $X \in \mathcal{M}^1$ (and hence $A * X \in \mathcal{M}^1$). To discover which sets of matrices have this property, we start by making the trivial observation that $A_1, \ldots, A_m$ cannot have a common eigenvector in R^d, for if we have $A_i y = \lambda_i y$ for $i = 1, \ldots, m$, then we can let $\tau = \inf\{t : y \cdot B_t = -1\}$ and $X_t = y \cdot B_{t \wedge \tau}$. X is Example 2 of Section 6.6, so $X \in \mathcal{K}^1 - \mathcal{M}^1$, but for $i = 1, \ldots, m$,

$$(A_i * X)_t = \int_0^{t \wedge \tau} \lambda_i y \cdot dB_s = \lambda_i X_t \in \mathcal{K}^1.$$

Janson's (1977) theorem says that this trivial necessary condition is sufficient.

(1) Let $A_1, \ldots, A_m$ be matrices that do not have a common eigenvector in R^d. Let A_0 be the identity matrix. If the transforms $A_i * X$, $i = 0, \ldots, m$, are all in $\mathcal{K}^1$, then $X \in \mathcal{M}^1$.

The key to the proof is Janson's generalization of the subharmonicity lemma of Chao and Taibleson (1973), which is in turn a generalization of ideas of Stein and Weiss (1971), who attribute the idea to Calderon. ... Using the notation in (1), we can state this result as:

(2) There is a $p_0 < 1$ (that depends only on the matrices $A_1, \ldots, A_m$) such that if $F_t = (1 + \sum_{i=0}^m (X_t^i)^2)^{1/2}$, then F_t^p is a local submartingale for all $p > p_0$.

Once we prove (2), (1) follows immediately. To see this, observe that if we can pick $p < 1$ so that $G_t = F_t^p$ is a local submartingale, then

$$E(G_t^{1/p}) = E\left[\left(1 + \sum_{i=0}^m (X_t^i)^2\right)^{1/2}\right] \leq E\left[1 + \sum_{i=0}^m |X_t^i|\right]$$

(since the L^2 norm of $(1, X_t^0, \ldots, X_t^m) \in R^{m+2}$ is less than its L^1 norm). It follows from our assumption that $\sup_t EG_t^{1/p} < \infty$, and, since $1/p > 1$, it follows from Doob's inequality that

$$\infty > E(\sup_t G_t)^{1/p} = E(\sup_t F_t) \geq E(\sup_t |X_t|),$$

so $X \in \mathcal{M}^1$. If we keep track of the constants, we get

$$E(\sup_t |X_t|) \leq E(\sup_t G_t)^{1/p}$$

$$\leq \left(\frac{1/p}{1/p - 1}\right) \sup_t EG_t^{1/p}$$

$$\leq (1 - p)^{-1} \left(1 + \sum_{i=0}^m \|A_i * X\|_{\mathcal{K}}\right).$$

To get rid of the 1 on the right-hand side and replace p by p_0, we apply the last result to X_t/ε, multiply both sides of the inequality by ε, and let $\varepsilon \to 0$, $p \to p_0$ to get

(3) $$E(\sup_t |X_t|) \leq (1 - p_0)^{-1} \sum_{i=0}^{m} \|A_i * X\|_{\mathscr{H}}.$$

Proof of (2) We start by observing that the result is trivial for $p = 2$ ($F_t = 1 + \sum_{i=0}^{m} (X_t^i)^2$) and for $p > 2$ (since a convex function of a local submartingale is a local submartingale), so we assume $p < 2$. Let $g(x) = (1 + |x|^2)^{p/2}$. A little differentiation gives

$$D_i g = \frac{p}{2}(1 + |x|^2)^{(p-2)/2} 2x_i$$

$$D_{ij} g = \frac{p}{2} \cdot \frac{p-2}{2}(1 + |x|^2)^{(p-4)/2} 4x_i x_j \quad i \neq j$$

$$D_{ii} g = \frac{p}{2} \cdot \frac{p-2}{2}(1 + |x|^2)^{(p-4)/2} 4x_i^2 + \frac{p}{2}(1 + |x|^2)^{(p-2)/2} 2.$$

Applying Itô's formula, we conclude that

$$F_t^p - F_0^p = \sum_{i=0}^{m} \int_0^t p X_s^i F_s^{p-2} dX_s^i$$

$$+ \frac{1}{2} \sum_{i=0}^{m} \sum_{j=0}^{m} \int_0^t \frac{p}{2}\left(\frac{p}{2} - 1\right) 4 X_s^i X_s^j F_s^{p-4} d\langle X^i, X^j \rangle_s$$

$$+ \frac{1}{2} \sum_{i=0}^{m} \int_0^t p F_s^{p-2} d\langle X^i \rangle_s.$$

The first term on the right-hand side is a local martingale, so to prove the result we need to show that we can pick $p < 1$ so that the second term plus the third term is ≥ 0. If $X_t = \int_0^t H_s dB_s$, then

$$\langle X^i \rangle_t = \int_0^t |A_i H_s|^2 ds$$

and

$$\langle X^i, X^j \rangle_t = \int_0^t (A_i H_s, A_j H_s) ds.$$

To complete our proof, we need to show:

(4) $$\frac{1}{2}\left(\frac{p}{2} - 1\right) \sum_{ij} 4 x_i x_j (A_i \varphi, A_j \varphi) + (1 + |x|^2) \sum_i |A_i \varphi|^2 > 0$$

for all $x \in R^{m+1}$ and $\varphi \in R^d$. To do this, we observe that if θ_{ij} is the angle between $x_i A_i \varphi$ and $x_j A_j \varphi$, then

$$\sum_{ij} (x_i A_i \varphi, x_j A_j \varphi) = \sum_{ij} |x_i A_i \varphi| |x_j A_j \varphi| \cos \theta_{ij}$$

$$\leq \sum_i (|x_i A_i \varphi|)^2 \leq |x|^2 \sum_i |A_i \varphi|^2.$$

6.7 Janson's Characterization of $\mathcal{M}^1$

Looking at the first inequality above, we see that there is equality only if $\theta_{ij} = 0$ for all i, j and $|x_i| = c|A_i\varphi|$ for all i. Now if $x, \varphi \neq 0$, it follows from the last equality that $|x_0| \neq 0$, and since $A_0\varphi = \varphi$, the first equality in the last sentence implies that φ is an eigenvector of all the A_i, contradicting our assumption, so we must have

$$\frac{\sum_{ij} x_i x_j (A_i\varphi, A_j\varphi)}{|x|^2 \sum_i |A_i\varphi|^2} < 1$$

for all nonzero $x \in R^{m+1}$ and $\varphi \in R^d$. The value of the left-hand side is not changed by multiplying x or φ by a positive constant, and it is continuous on the compact set $\{(x, \varphi) \in R^{m+1+d} : |x| = 1, |\varphi| = 1\}$, so the supremum of the expression over K is a number $\delta < 1$. Hence the expression in (4) is $\geq$

(5) $\qquad (1 + |x|^2) \sum_i |A_i\varphi|^2 - \frac{1}{2}\left(\frac{p}{2} - 1\right) 4\alpha(|x|^2 \sum_i |A_i\varphi|^2).$

When $p = 1$, the coefficient of the second term is $\alpha < 1$, so we can pick $p_0 < 1$ such that (5) > 0 for all $p > p_0$, and this completes the proof.

Remark: Readers familiar with Janson's (1977) proof (which is for d-adic martingales) should notice that the outline is the same, but two details are different: (a) Itô's formula replaces the computation Janson does for "small jumps" (our jumps have size zero), and then, since we do not have a small/large dichotomy, we need only his first compactness argument. (b) The restriction on the matrices sounds the same, but it is different. In Janson's theorem, the matrices do not have a common eigenvector in $R_0^d = \{x : \sum_i x_i = 0\}$.

Since $\begin{pmatrix} 0 & -1 \\ 1 & 0 \end{pmatrix}$ has no eigenvector in R^2, it follows from Janson's theorem that we have:

(6) $\qquad$ If $d = 2$ and X and its conjugate martingale $\tilde{X}$ are in $\mathcal{H}^1$, then $X \in \mathcal{M}^1$.

If $d = 3$, then any matrix A has a real eigenvalue and, hence, also an eigenvector in R^d (take the real or imaginary part), so it takes at least two matrices to characterize $\mathcal{M}^1$. We leave it to the reader to discover what happens if we take

$$A_1 = \begin{pmatrix} 0 & 1 & 0 \\ -1 & 0 & 0 \\ 0 & 0 & 0 \end{pmatrix} A_2 = \begin{pmatrix} 0 & 0 & 1 \\ 0 & 0 & 0 \\ -1 & 0 & 0 \end{pmatrix}.$$

If you are very clever, you will discover Riesz transforms. If you get stuck, you can find the connection spelled out in Gundy and Varopoulos (1979).

6.8 Inequalities for Conjugate Harmonic Functions

With each harmonic function u, there is associated a unique conjugate harmonic function $\bar{u}$, which has (a) $\bar{u}(0) = 0$ and (b) $u + i\bar{u}$ is analytic in D. In this section, we will investigate conjugation as an operation on h^p, the set of harmonic functions u in D with

$$d_p(u) = \sup_{r<1} \int_0^{2\pi} |u(re^{i\theta})|^p \, d\pi(\theta) < \infty.$$

Let $\|u\|_p = (d_p(u))^{1/p}$. If $p \geq 1$, this equation defines a norm on h^p. Our first result, due to M. Riesz (1927), shows that if $p > 1$, $u \to \bar{u}$ is a bounded linear map from h^p to h^p.

(1) If $p > 1$, then there is a constant C (that depends only on p) such that
$$\|\bar{u}\|_p \leq C \|u\|_p.$$

Proof Let $U_t = u(B_t)$, $t < \tau$, and let $\bar{U}_t = \bar{u}(B_t)$, $t < \tau$. If $\|u\|_p = 1$, then Doob's inequality implies that $E(U^*) \leq (p/(p-1))^p$, and it follows from results in Section 6.4 that

$$\|\bar{u}\|_p \leq E|\bar{U}^*|^p \leq KE|U^*|^p,$$

so (1) holds with $C = K^{1/p}(p/(p-1))$.

To be fair, we should observe that (1) is easy to prove analytically—there is a simple argument using Green's theorem that is due to P. Stein (1933) and that gives a much better value for C_p, namely,

$$C_p = 2(p/(p-1))^{1/p} \quad 1 < p \leq 2$$
$$C_p = C_{p/(p-1)} \quad 2 \leq p < \infty.$$

Since this proof has some interesting probabilistic aspects, we present it here.

If the theorem is true for some $1 < p < \infty$, then it is also true for the conjugate index $q = p/(p-1)$ with $C_q = C_p$ (exercise for the reader: See Zygmund (1959), page 255, for the answer), so it suffices to prove the result when $1 < p \leq 2$. In view of the decomposition of functions in h^1 given in Section 6.5, we can also assume that $u(z) > 0$ in D. To prove the result, we compute

$$\frac{\partial^2}{\partial z_i^2} u^p(z) = \frac{\partial}{\partial z_i}\left(pu^{p-1}(z)\frac{\partial u}{\partial z_i}\right)$$
$$= p(p-1)u^{p-2}(z)\left(\frac{\partial u}{\partial z_i}\right)^2 + pu^{p-1}(z)\frac{\partial^2 u}{\partial z_i^2}.$$

Summing and noticing that $\Delta u = 0$, we get

(a) $\Delta u^p(z) = p(p-1)u^{p-2}(z)|\nabla u(z)|^2.$

Similarly, if we let $v = \bar{u}$, then

6.8 Inequalities for Conjugate Harmonic Functions

$$\frac{\partial^2}{\partial z_i^2}|f|^p = \frac{\partial^2}{\partial z_i^2}(u^2+v^2)^{p/2}$$

$$= \frac{\partial}{\partial z_i}\left(\frac{p}{2}(u^2+v^2)^{(p-2)/2}\left(2u\frac{\partial u}{\partial z_i}+2v\frac{\partial v}{\partial z_i}\right)\right)$$

$$= \frac{p}{2}\cdot\frac{p-2}{2}\cdot(u^2+v^2)^{(p-4)/2}\left(2u\frac{\partial u}{\partial z_i}+2v\frac{\partial v}{\partial z_i}\right)^2$$

$$+\frac{p}{2}(u^2+v^2)^{(p-2)/2}\left(2\left(\frac{\partial u}{\partial z_i}\right)^2+2u\frac{\partial^2 u}{\partial z_i^2}+2\left(\frac{\partial v}{\partial z_i}\right)^2+2v\frac{\partial^2 v}{\partial z_i^2}\right).$$

Summing and using the relationships $\Delta u = 0$, $\Delta v = 0$, $\nabla u \cdot \nabla v = 0$, and $|\nabla u| = |\nabla v|$, we get

(b) $\Delta |f|^p = p(p-2)(u^2+v^2)^{(p-4)/2}(u^2|\nabla u|^2 + v^2|\nabla u|^2)$
$\qquad + p(u^2+v^2)^{(p-2)/2}(|\nabla u|^2 + 0 + |\nabla u|^2 + 0)$
$\qquad = p^2(u^2+v^2)^{(p-2)/2}|\nabla u|^2.$

Since $p-2 \leq 0$, it follows from (a) and (b) that we have

(c) $\Delta |f|^p \leq \dfrac{p}{p-1}\Delta |u|^p.$

This is the key inequality for the proof. From here, completing the proof is easy, and how you do it is a matter of background.

A *probabilist* would reason as follows: Itô's formula says that if h is C^2 in D, then for $t < \tau = \inf\{t : B_t \notin D\}$,

$$h(B_t) - h(B_0) = \int_0^t \Delta h(B_s)\cdot dB_s + \frac{1}{2}\int_0^t \nabla h(B_s)\, ds$$

(here we have written B_t as a real two-dimensional Brownian motion). The first term on the right is a martingale for $t < \tau_r$ (since $\langle h\circ B\rangle_{t\wedge\tau} \leq C(t\wedge\tau_r)$), so

$$E_0 h(B_{\tau_r}) = h(0) + \frac{1}{2}E\int_0^{\tau_r}\Delta h(B_s)\,ds.$$

Letting $h = u^p$, and then $h = |f|^p$, and using (c), gives

$$\int_0^{2\pi}|f(re^{i\theta})|^p\,dm(\theta) \leq \frac{p}{p-1}\int_0^{2\pi}|u(re^{i\theta})|^p\,dm(\theta),$$

since $|u(0)| = |f(0)|$.

An *analyst* would use the divergence theorem

$$\int_0^{2\pi}\frac{\partial\varphi}{\partial r}(re^{i\theta})r\,d\theta = \int_{|z|\leq r}\Delta\varphi(z)\,dA(z),$$

where A denotes Lebesgue measure. This implies that

$$\frac{\partial}{\partial r}\int_0^{2\pi}|f(re^{i\theta})|^p\,d\theta \leq \frac{p}{p-1}\frac{\partial}{\partial r}\int_0^{2\pi}|u(re^{i\theta})|^p\,d\theta,$$

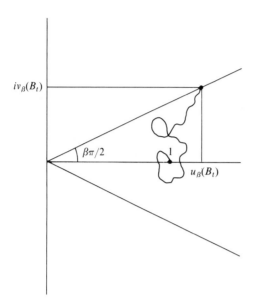

Figure 6.2

so integrating from 0 to r and recalling that $v(0) = 0$, we have a second proof of the inequality

$$\int_0^{2\pi} |f(re^{i\theta})|^p \, d\theta \leq \frac{p}{p-1} \int_0^{2\pi} |u(re^{i\theta})|^p \, d\theta.$$

Since Riesz's theorem is an old and important one, there are many different proofs. Calderon (1950c) has given another proof that you can find in Zygmund (1959), pages 253–255. Pichorides (1972) has given a refinement of Calderon's proof and has found the best possible constant: $C_p = \tan(\pi/2p)$ if $1 < p \leq 2$ (and hence $\cot(\pi/2p)$ if $p \geq 2$). The function that shows that this inequality is sharp is simple. Let $f_\beta(z) = \rho(z)^\beta = ((1+z)/(1-z))^\beta$. Since $\rho \in H^p$ for $p \in (0, 1)$, $f_\beta \in H^p$ for $p \in (0, 1/\beta)$. Since ρ maps D one-to-one onto $\{z : \operatorname{Re} z > 0\} = \{re^{i\theta} : r > 0, \theta < \pi/2\}$, f_β maps D one-to-one onto the cone $\Gamma_\beta = \{re^{i\theta} : r > 0, \theta < \beta\pi/2\}$. From the last observation (and Lévy's theorem), it follows that $f_\beta(B_t)$, $t < \tau$, is a time change of a Brownian motion C_t that starts at 1 and runs until it leaves Γ_β. Drawing a picture (see Figure 6.2) reveals that if $u_\beta = \operatorname{Re} f_\beta$ and $v_\beta = \operatorname{Im} f_\beta$, then

$$\lim_{t \uparrow \tau} |v_\beta(B_t)| = \left|\tan \frac{\beta\pi}{2}\right| \lim_{t \uparrow \tau} |u_\beta(B_t)|.$$

If $1 < p < 1/\beta$, then $f_\beta \in H^p$. Combining the equality above with the L^p convergence results of Section 6.5 shows that

$$\|v_\beta\|_p = \left|\tan \frac{\beta\pi}{2}\right| \|u_\beta\|_p,$$

6.8 Inequalities for Conjugate Harmonic Functions

so the optimal constant $C_p \geq \tan(\pi/2p)$ for all $1 < p < \infty$. Repeating the argument above for $i\rho(z)^\beta$ shows that $C_p \geq \cot(\pi/2p)$ for all $1 < p < \infty$, so Pichorides's constants are the best possible.

As the constants in the last remark might suggest, Riesz's theorem is false for $p = 1$. We have seen the counterexample many times: the half-plane map $\rho(z) = (1+z)/(1-z)$. By computations in Section 6.1, $u(z) = \operatorname{Re} \rho(z) \in h^1$, but $\bar{u}(z) = \operatorname{Im} \rho(z)$ is not, since

$$\pi(\{\theta : |\bar{u}(e^{i\theta})| \geq \lambda\}) \sim 2/\pi\lambda \quad \text{as} \quad \lambda \to \infty.$$

The next result, due to Kolmogorov (1925), shows that (up to a constant multiple) this is the worst behavior we can have for $u \in h^1$.

(2) There is a constant C such that if $u \in h^1$, then

$$\pi(\{\theta : |\bar{u}(e^{i\theta})| \geq \lambda\}) \leq C\lambda^{-1}\|u\|_1.$$

Remark: In the jargon, this is called a weak type $(1, 1)$ inequality. We will give Davis's (1974) proof, because it has the advantage of identifying the best constant $C = 1.347\ldots$.

Proof Let $Z_t = X_t + iY_t$ be a complex Brownian motion, and let $\alpha = \inf\{t : |Y_t| = 1\}$. Since $t \to |X_t|$ is a submartingale and the stopping time α depends only upon the Y component, it is easy to see that if x and $-1 < y < 1$ are real numbers,

(i) $E_0|X_\alpha| \geq E_{iy}|X_\alpha|$
(ii) $E_{x+iy}|X_\alpha| \geq E_{iy}|X_\alpha|$.

The next step is to show:

(iii) Let x be a real number and β a stopping time for Z_t. Then

$$E_x|X_{\alpha \wedge \beta}| \geq E_0|X_\alpha|P_x(\beta \geq \alpha).$$

Proof By (ii) and the triangle inequality,

$$E_0|X_\alpha| \leq E_x|X_\alpha| \leq E_x|X_{\alpha \wedge \beta}| + E_x|X_\alpha - X_{\alpha \wedge \beta}|.$$

By the strong Markov property,

$$E_x|X_\alpha - X_{\alpha \wedge \beta}| = E_x E_{X(\alpha \wedge \beta)}|X_\alpha - X_{\alpha \wedge \beta}|,$$

and we have, using (i),

$$E_{X_{\alpha \wedge \beta}}|X_\alpha - X_{\alpha \wedge \beta}| \begin{cases} = 0 & \alpha \leq \beta \\ \leq E_0|X_\alpha| & \alpha > \beta \end{cases}$$

(the centering makes it look like starting at 0, but $|Y_{\alpha \wedge \beta}| \geq 0$, so there is less time until α). Combining the results above, we have

$$E_0|X_\alpha| \leq E_x|X_{\alpha \wedge \beta}| + E_0|X_\alpha|P_x(\alpha > \beta),$$

proving (iii).

We are now ready to prove Kolmogorov's inequality. Let $f = u + i\bar{u}$. By Lévy's theorem, if we let $\sigma(t) = \int_0^t |f'(B_s)|^2 \, ds$ for $t < \tau$, then $Z_{\sigma(t)} = f(B_t)$, $t < \tau$, defines a Brownian motion run for an amount of time $\gamma = \sigma(\tau)$. If we write $Z_t = X_t + iY_t$ and let $\tau_r = \inf\{t : |B_t| > r\}$, then $|X_{t \wedge \sigma(\tau_r)}| = |u(B_{t \wedge \tau_r})|$ is a bounded submartingale, so

$$E|X_{\alpha \wedge \sigma(\tau_r)}| \leq E|X_{\sigma(\tau_r)}| = E|u(B_{\tau_r})| \leq \|u\|_1.$$

As $r \uparrow 1$, $\tau_r \uparrow \tau$ and $\sigma(\tau_r) \uparrow \sigma(\tau) = \gamma$, so using Fatou's lemma gives

(iv) $E|X_{\alpha \wedge \gamma}| \leq \|u\|_1$.

We have finally assembled all the ingredients to complete the proof:

$$\pi(\{\theta : |\bar{u}(e^{i\theta})| \geq 1\}) = P(|Y_\gamma| \geq 1)$$
$$\leq P(Y_\gamma^* \geq 1) = P(\alpha \leq \gamma),$$

so it follows from (iii) and (iv) that the above

$$\leq (E_0|X_\alpha|)^{-1} E_0|X_{\alpha \wedge \gamma}|$$
$$\leq (E_0|X_\alpha|)^{-1} \|u\|_1.$$

Substituting u/λ for u, we have

$$\pi(\{\theta : |\bar{u}(e^{i\theta})| \geq \lambda\}) \leq (E_0|X_\alpha|)^{-1} \|u\|_1 / \lambda.$$

It is clear from the argument above that the inequality is sharp. If we let $g(z) = (2/\pi) \log \rho(z)$, then g maps D one-to-one onto the strip $\{z : -1 < \operatorname{Im} z < 1\}$, so

$$1 = \pi(\{\theta : |\bar{u}(e^{i\theta})| \geq 1\}) = P(\alpha \leq \gamma)$$
$$= (E_0|X_\alpha|)^{-1} E_0|X_\alpha| = (E_0|X_\alpha|)^{-1} \|u\|_1.$$

To compute the constant, observe that by (4) in Section 5.1, the probability density of X_α is

$$e^{x\pi/2}/(1 + e^{x\pi}),$$

so

$$E|X_\alpha| = 2\int_0^\infty x \frac{e^{x\pi/2}}{1 + e^{x\pi}} \, dx$$
$$= \frac{8}{\pi^2} \sum_{n=0}^\infty \int_0^\infty (-1)^n y e^{(2n+1)y} \, dy$$
$$= \frac{8}{\pi^2} \sum_{n=0}^\infty (-1)^n (2n+1)^{-2}.$$

If you recognize that $\pi^2/8 = \sum_{n=0}^\infty (2n+1)^{-2}$, then you can write the constant as

$$(1 - 3^{-2} + 5^{-2} - 7^{-2} + \cdots)/(1 + 3^{-2} + 5^{-2} + 7^{-2} + \cdots).$$

Although the inequalities (i)–(iv) are important, the two main ideas in the proof of (2) are the observation that: (a) If we time change $f(B_t)$, $t < \tau$, to obtain a Brownian motion $Z_t = X_t + iY_t$, $t < \gamma$, then

6.8 Inequalities for Conjugate Harmonic Functions

$$\pi(\{\theta : \bar{u}(e^{i\theta})| \geq 1\}) = P(|Y_\gamma| \geq 1) \leq P(Y_\gamma^* \geq 1),$$

and (b) for $u \in h^1$, $P(Y_\gamma^* \geq 1)/\|u\|_1$ is largest when $\gamma = \alpha = \inf\{t : |Y_t| = 1\}$. With this philosophy as a guide, the reader should have no problem proving a one-sided version of Kolmogorov's inequality.

Exercise 1 If $u \in h^1$ and $u > 0$, then

$$\pi(\{\theta : \bar{u}(e^{i\theta}) \geq \lambda\}) \leq 2\|u\|_1 \int_\lambda^\infty \frac{dy}{\pi(1+y^2)},$$

and this constant is the best possible.

Hint: The function that shows that this inequality is sharp is

$$f(z) = i\lambda + (1-i\lambda)\left(\frac{i-z}{i+z}\right),$$

which maps D one-to-one onto $\{x + iy : x > 0, y < \lambda\}$ and has $f(0) = 1$. In this case, the reflection principle shows that

$$\pi(\{\theta : \bar{u}(e^{i\theta}) \geq \lambda\}) = 2 \int_\lambda^\infty \frac{dy}{\pi(1+y^2)}.$$

Applying (2) to $u(rz)$ shows that if $\|u\|_1 = 1$, then

$$\pi(\{\theta : |\bar{u}(re^{i\theta})| \geq \lambda\}) \leq C\lambda^{-1},$$

so if $p < 1$

$$\int_0^{2\pi} |\bar{u}(re^{i\theta})|^p \, d\pi(\theta) = \int_0^\infty p\lambda^{p-1} \pi(\{\theta : |\bar{u}(re^{i\theta})| > \lambda\}) \, d\lambda$$

$$\leq \int_0^1 p\lambda^{p-1} \, d\lambda + \int_1^\infty p\lambda^{p-1} C\lambda^{-1} \, d\lambda = 1 + \frac{Cp}{1-p},$$

and we have proved another inequality due to Kolmogorov:

(3) If $u \in h^1$ and $p < 1$, there is a constant C (that depends only on p) such that

$$d_p(\bar{u}) \leq C\|u\|_1^p.$$

The argument above proves (3), but since it is not very interesting and does not give a very good value of the constant, we will give two more proofs of (3). The first is a purely probabilistic one, and because of this, it gives a crude value for the constant. The second is an analytic translation of the probabilistic argument and gives the best constant for positive u.

Proof 2 Again without loss of generality, we can suppose that $u > 0$ in D and $\|u\|_1 = 1$. In this case, $u(0) = 1$ and $\bar{u}(0) = 0$, so $(u(B_t), \bar{u}(B_t))$, $t < \tau$, is a time change of a two-dimensional Brownian motion starting from $(1,0)$ and running for an amount of time γ. Since $u \geq 0$, γ must be smaller than T_1, the time it takes a Brownian motion starting at 1 to hit 0, and it follows that

$$d_p(\bar{u}) \leq E(\bar{U}^*)^p \leq CE\langle \bar{U}\rangle_\infty^{p/2} \leq CET_1^{p/2}.$$

In Chapter 1, we found that the probability density of T_1 is

$$(2\pi)^{-3/2} t^{-3/2} e^{-1/2t},$$

so if $p < 1$,

$$ET_1^{p/2} = (2\pi)^{-3/2} \int_0^\infty t^{(p-3)/2} e^{-1/2t} \, dt.$$

To evaluate the integral, let $u = 1/2t$ to obtain

$$(2\pi)^{-3/2} 2^{(1-p)/2} \int_0^\infty u^{-(1+p)/2} e^{-u} \, du,$$

and observe that the value of the integral is $\Gamma((1-p)/2)$ if $p < 1$.

The reader should observe that if $u(x) = \operatorname{Re} \rho(x)$, then $\gamma = T_1$. This suggests that for nonnegative functions $\|\bar{u}\|_p / \|u\|_p$ should be largest for the Poisson mapping. The proof above cannot be used to prove this fact, since we have used the clumsy estimates $d_p(\bar{u}) \leq E|\bar{U}^*|^p \leq CE\langle \bar{U}\rangle_\infty^{p/2}$. This inaccuracy can be avoided if we abandon the correspondence and translate the proof into analytical terms.

Proof 3 The idea for this proof is due to Littlewood (1926). Again without loss of generality, we can suppose that $u > 0$ in D and $\|u\|_1 = 1$.

A function f analytic in D is said to be subordinate to g if there is an analytic function ω with $|\omega(z)| \leq |z|$ such that $f(z) = g(\omega(z))$.

One reason we are interested in this concept is that

(i) If $u > 0$ and $u(0) = 1$, then $f = u + i\bar{u}$ is subordinate to $\rho(z) = (1+z)/(1-z)$.

Proof Let $\omega(z) = \rho^{-1}(f(z))$; ω is analytic and maps D into D and 0 into $\rho^{-1}(1) = 0$. Let $\sigma(z) = \omega(z)/z$, $z \neq 0$, and $\sigma(z) = \omega'(0)$ at $z = 0$; σ is analytic in D and has $|\sigma(z)| \leq 1$ on ∂D, so it follows from the maximum principle that $|\sigma(z)| \leq 1$ in D, that is, $|\omega(z)| \leq |z|$.

Another reason for our interest in subordination is

(ii) If f is subordinate to g, then

$$\int_0^{2\pi} |f(re^{i\theta})|^p \, d\pi(\theta) \leq \int_0^{2\pi} |g(re^{i\theta})|^p \, d\pi(\theta) \quad \text{for all } r < 1.$$

Proof Let B_t be a complex Brownian motion and $\tau_r = \inf\{t : |B_t| > r\}$. By hypothesis,

$$E|f(B_{\tau_r})|^p = E|g(\omega(B_{\tau_r}))|^p.$$

6.8 Inequalities for Conjugate Harmonic Functions

If we subject $\omega(B_t)$, $t \le \tau_r$, to Lévy's time change, we get a Brownian motion B'_t run for an amount of time γ_r. Since $|\omega(z)| \le |z|$, it follows that $\gamma_r < \tau'_r = \inf\{t : |B'_t| > r\}$. Combining this with the fact that $|g(B'_{t \wedge \tau'_r})|^p$ is a submartingale gives

$$E|g(\omega(B_{\tau_r}))|^p = E|g(B'_{\gamma_r})|^p \le E|g(B'_{\tau'_r})|^p$$

and proves (ii).

Combining (i) and (ii) proves (3) with $C_p = d_p(q)$, where $q(z) = \operatorname{Im} \rho(z)$. To compute the value of the constant, write $\rho(z) = R(z)e^{i\Phi(z)}$, where $R(z) = |\rho(z)|$ and $\Phi(z) \in (-\pi/2, \pi/2)$. Since there is no confusion about which root to take, $F(z) = R(z)^p e^{ip\Phi(z)}$ is analytic in D. By the mean-value theorem,

$$1 = \operatorname{Re} F(0) = \int_{-\pi}^{\pi} \operatorname{Re} F(re^{i\theta}) \, d\pi(\theta)$$

$$= \int_{-\pi}^{\pi} R(re^{i\theta})^p \cos(p\Phi(re^{i\theta})) \, d\pi(\theta).$$

As $r \to 1$, $\Phi(re^{i\theta}) \to \operatorname{sgn}(\theta)\pi/2$ and $R(re^{i\theta}) \to |q(e^{i\theta})|$. Since $q \in h^p$ and $|\cos(p\Phi(re^{i\theta}))| \le 1$, it follows from the dominated convergence theorem that

$$1 = \int_{-\pi}^{\pi} |q(e^{i\theta})|^p \left(\cos \frac{p\pi}{2}\right) dm(\theta)$$

$$= \left(\cos \frac{p\pi}{2}\right) \lim_{r \uparrow 1} \int_{-\pi}^{\pi} |q(re^{i\theta})|^p \, dm(\theta),$$

so $d_p(q) = \sec(p\pi/2)$.

Remark: If at the beginning of the computation of the constant we take an arbitrary f with $u = \operatorname{Re} f > 0$ and use the inequality $|\Phi(re^{i\theta})| \le \pi/2$, we get a purely analytical proof with the constant given above. This argument is due to Hardy (1928).

Proof 3 shows that

(3a) If $u \ge 0$, $d_p(\bar{u}) \le \sec\left(\dfrac{\pi p}{2}\right) \|u\|_1$

(3b) If $u \in h^1$, $d_p(\bar{u}) \le 2 \sec\left(\dfrac{\pi p}{2}\right) \|u\|_1$.

The Poisson kernel shows that (3a) is a sharp result. Since (3b) is obtained from (3a) by using the triangle inequality, we should expect that (3b) is not sharp, and indeed it is not. Burgess Davis (1976) solved the problem of finding the optimal constant in inequality (3b). He showed that the smallest value for C_p is $\|\bar{u}\|_p$, where $u = (k_1 + k_{-1})/2$. In this case, the corresponding analytic function is

$$g(z) = \frac{1}{2}\left(\frac{1+z}{1-z} + \frac{1-z}{1+z}\right) = \frac{2z}{(1-z^2)},$$

which maps D one-to-one onto $S = C - \{x + iy : x = 0, |y| \geq 1\}$. To prove his theorem, Davis uses Lévy's theorem to reduce the result to an optimal stopping problem for Brownian motion, which is solved by considering related discrete time problems. Since the argument is rather lengthy and the improvement on (3b) is rather slight, the reader is referred to Davis's paper for the details or Davis (1979b) for a sketch of the proof. A. Baernstein (1978) has given a purely analytical proof of this result.

Up to this point, we have only discussed $u \in h^p$ for $p \geq 1$. For $1 < p < \infty$, the class h^p was preserved under conjugation. For $p = 1$, this was false, but it was almost true: $\bar{u} \in h^p$ for all $p < 1$. When $p < 1$, things fall apart—there is an analytic function $f = u + iv$ such that $u \in h^p$ for all $p < 1$, and yet $f \notin N$ (hence $v \notin h^p$ for any $p > 0$). The example is a randomly chosen function

(4) $$f(z, \omega) = \sum_{n=1}^{\infty} \xi_n(\omega) g(z^{2^n})$$

where $g(z) = z/(1 - z^2)$

and ξ_n, $n \geq 1$, are independent random variables with

$$P(\xi_n = 1) = P(\xi_n = -1) = 1/2$$

(for analysts, let $\Omega = [0, 1]$ and $\xi_n(\omega) = \text{sgn}(\cos(2^n \omega))$, the nth Rademacher function). I claim that

(4a) For every ω, $\text{Re} f(\cdot, \omega) \in h^p$ for all $p < 1$

and

(4b) With probability 1, $\mathscr{L}_0 = \{\theta : \lim_{r \uparrow 1} f(re^{i\theta}, \omega) \text{ exists}\}$ has Lebesgue measure 0.

The proof of (4a) is a straightforward, but somewhat lengthy, calculation and is therefore left to the reader (see Duren (1970), page 66). We will proceed, then, with the more interesting claim (4b), which is implied by the following result.

(5) Let $g_n(z)$, $n \geq 1$, be complex-valued and continuous in $|z| \leq 1$ except at a finite number of points z with $|z| = 1$.

If (i) $\sum_{n=1}^{\infty} |g_n(z)| < \infty$ and for all $r < 1$ the convergence is uniform on $D_r = \{z : |z| < r\}$, and (ii) for each N, as $r \uparrow 1$ we have

$$\sum_{n=N}^{\infty} |g_n(re^{i\theta})|^2 \to \infty \quad \text{uniformly in } \theta,$$

then with probability 1, $f(z, \omega) = \sum_{n=1}^{\infty} \xi_n(\omega) g_n(z)$ has a radial limit almost nowhere.

Proof Let $E = \{(\theta, \omega) : \lim_{r \to 1} f(re^{i\theta}, \omega) \text{ exists}\}$. It suffices to show that for almost every θ the section $E_\theta = \{\omega : (\theta, \omega) \in E\}$ has probability 0, for then the desired conclusion follows from Fubini's theorem.

6.8 Inequalities for Conjugate Harmonic Functions

Suppose θ is such that $e^{i\theta}$ is not a discontinuity point of any of the g_n, and $P(E_\theta) = \alpha > 0$. We are going to construct a decreasing sequence of events A_n, $n \geq 0$, with $A_0 = E_\theta$ and do this in such a way that $\lim_{n\to\infty} P(A_n) \geq \alpha/2$. For $k, n \geq 1$, we let

$$B_n(k) = \{\omega \in A_{n-1} : |f(re^{i\theta}, \omega)| \leq k \text{ for all } r < 1\}.$$

$B_n(k) \uparrow A_{n-1}$ as $k \uparrow \infty$, so if we have already constructed A_{n-1} with $P(A_{n-1}) \geq (1/2 + 1/(n+1))\alpha$, then we can pick k_n so that $P(B_n(k_n)) \geq (1/2 + 1/(n+2))\alpha$ and let $A_n = B_n(k_n)$. Let $B = \bigcap_{n=1}^{\infty} A_n$. By construction, $P(B) \geq \alpha/2$ and, for all n,

$$E\left(\left|\sum_{m=n}^{\infty} g_m(re^{i\theta})\xi_m(\omega)\right|^2 1_B\right) \leq k_n^2 P(B).$$

The functions $\xi_j \xi_k$, $0 \leq j < k < \infty$, are an (incomplete) orthonormal set in $L^2(\Omega)$, so

$$\sum_{0 \leq m < n < \infty} (E 1_B \xi_n \xi_m)^2 \leq \|1_B\|_2^2.$$

It follows that we can pick M large enough so that

$$\sum_{M \leq m < n < \infty} (E 1_B \xi_n \xi_m)^2 \leq \left(\frac{P(B)}{4}\right)^2.$$

Since $B \subset A_M$, we have (since $\xi_m^2 \equiv 1$)

$$k_M^2 P(B) \geq E\left(\left|\sum_{m=M}^{\infty} g_m(re^{i\theta})\xi_m(\omega)\right|^2 1_B\right)$$

$$= E\left(\sum_{m=M}^{\infty} g_m(re^{i\theta})^2 1_B\right) + 2 \sum_{M \leq m < n} g_m(re^{i\theta})\overline{g_n(re^{i\theta})} E(1_B \xi_n(\omega)\xi_m(\omega)).$$

Cauchy-Schwarz implies that the last expression above

$$\leq 2\left(\sum_{M \leq m \leq n} |g_m(re^{i\theta})|^2 |g_n(re^{i\theta})|^2\right)^{1/2} \left(\sum_{M \leq m \leq n} (E 1_B \xi_m \xi_n)^2\right)^{1/2}$$

$$\leq 2\left(\sum_{m=M}^{\infty} |g_m(re^{i\theta})|^2\right)^{1/2} \left(\sum_{n=M}^{\infty} |g_n(re^{i\theta})|^2\right)^{1/2} \frac{P(B)}{4}.$$

Combining the results above, we see that

$$k_M^2 P(B) \geq \left(\sum_{m=M}^{\infty} |g_m(re^{i\theta})|^2\right) P(B) \left(1 - \frac{1}{2}\right),$$

contradicting (ii) and proving (5).

The next exercise recaptures the main aspects of the last proof in a simpler setting.

Exercise 2 If $\sum_{n=1}^{\infty} |a_n|^2 = \infty$, and $S_N = \sum_{n=1}^{N} a_n \xi_n(\omega)$ then $\lim_{N \to \infty} S_N$ exists with probability 0.

180   6 Hardy Spaces and Related Spaces of Martingales

Remark: Probabilists will recognize this exercise as a special case of the Kolmogorov three-series theorem (see Chung (1974), pages 118–119).

Proof Let $C > 0$ and $B = \{\omega : \sup_N |S_N(\omega)| \leq C\}$. If $\omega \in B$, $|S_n(\omega) - S_m(\omega)| \leq 2C$ for all m, n, and

$$4C^2 P(B) \geq E\left(\left|\sum_{k=m+1}^{n} a_k \xi_k\right|^2 1_B\right)$$

$$= P(B) \sum_{k=m+1}^{n} |a_k|^2 + 2 \sum_{m+1 \leq j < k \leq n} a_j \bar{a}_k E(1_B \xi_j \xi_k).$$

Again, $\xi_j \xi_k$, $j < k$, is an orthonormal system in $L^2(\Omega)$, so we can pick M large enough such that

$$\sum_{M+1 \leq j \leq k} (E(1_B \xi_j \xi_k))^2 \leq \left(\frac{P(B)}{4}\right)^2,$$

and can apply Cauchy-Schwarz to show that

$$4C^2 P(B) \geq \left(\sum_{k=M+1}^{n} |a_k|^2\right) P(B) \left(1 - \frac{1}{2}\right),$$

which is a contradiction, since we have supposed that $\sum |a_k|^2 = \infty$.

There are many other results for conjugate functions. At this point, we have not covered all the inequalities on the first three pages of Chapter VII of Zygmund (1959)! Another inequality that can be proved, using results from Section 6.2, is the following.

Exercise 3 *Zygmund's Inequality.* If $\varphi \in L^1(\partial D)$ and we let $\bar{\varphi}$ denote the conjugate function on ∂D (i.e., the boundary limits of the conjugate of $\mathscr{P}\varphi$), then there is a constant C such that

$$\int_0^{2\pi} |\bar{\varphi}(e^{i\theta})| \, d\theta \leq C \int_0^{2\pi} |\varphi(e^{i\theta})| \log^+ |\varphi(e^{i\theta})| \, d\theta.$$

We will see some more conjugate function inequalities in Section 7.7.

6.9 Conjugate Functions of Indicators and Singular Measures

In this section, we will investigate the conjugate functions of $u = \mathscr{P}1_A$, $A \subset \partial D$, and $u = \mathscr{P}\mu$ where μ is a measure with $\mu(\partial D) < \infty$, which is singular w.r.t. surface measure. In each case, it turns out that

$(*)$ $\bar{\varphi}(\theta) = \lim_{r \uparrow 1} \bar{u}(re^{i\theta})$ exists for a.e. θ

and the value of $\pi(\theta : \bar{\varphi}(\theta) \leq y)$ denotes only on $\pi(A)$ or $\mu(\partial D)$, respectively.

6.9 Conjugate Functions of Indicators and Singular Measures

The result for 1_A was first observed by Stein and Weiss (1959), who proved the result by computing the distribution when $A = \bigcup_i (a_i, b_i)$ is a disjoint union of intervals (see pages 273–276 of their paper) and observing that the distribution depended only on $\sum_i |a_i - b_i|$. The result becomes transparent if we look at $f = u + i\bar{u}$ through the eyes of a Brownian motion B_t starting at 0. Doing this, we observe:

(a) If $\tau = \inf\{t : B_t \notin D\}$, then $f(B_t)$, $t < \tau$, is a time change of a Brownian motion C_t starting at $f(0) = \pi(A)$ and run for an amount of time

$$\gamma = \int_0^\tau |f'(B_s)|^2 \, ds.$$

(b) If $0 < \pi(A) < 1$, then $0 < u < 1$ in D and $u(B_t) \in (0, 1)$ for all $t < \tau$. This implies that $\gamma \leq T = \inf\{t : \operatorname{Re} C_t \notin (0, 1)\}$.

(c) Since $1_A \in L^1(\partial D, \pi)$, it follows from results in Section 6.5 that

$$\lim_{t \uparrow \gamma} \operatorname{Re} C_t = \lim_{t \uparrow \tau} u(B_t) = 1_A(B_\tau) \quad \text{a.s.,}$$

so we have $\gamma = T$.

(d) Since $T < \infty$, it follows that

$$\lim_{t \uparrow \tau} \bar{u}(B_t) = \lim_{t \uparrow \gamma} \operatorname{Im} C_t = \operatorname{Im} C_T,$$

and using the equivalent of Brownian and nontangential convergence in $d = 2$ proved in Section 4.3, we conclude that (∗) holds and that

(1) $\qquad \pi(\theta : \bar{\varphi}(\theta) > y) = P_{\pi(A)}(\operatorname{Im} C_T > y).$

To compute the value of the right-hand side, we observe that $z \to \exp(i\pi(x - 1/2))$ maps the strip $\{0 < \operatorname{Re} z < 1\}$ one-to-one onto the half space $\{\operatorname{Re} z > 0\}$ (see Figure 6.3) and sends

$$1 \to i$$
$$1 + i\lambda \to ie^{-\lambda \pi/2}$$
$$a \to \exp(i\pi(a - 1/2)),$$

so if we let $a = \pi(A)$, $b = \pi(a - 1/2)$, and $c = -\lambda\pi/2$, then it follows from results in Section 1.9 that

$$P_{\pi(A)}(\operatorname{Im} C_T > y) = \int_{-e^c}^{e^c} \frac{1}{\pi} \frac{\cos b}{(\cos b)^2 + (y - \sin b)^2} \, dy,$$

and changing variables $x \cos b = y - \sin b$ shows the last integral

$$= \frac{1}{\pi}\left(\left(\tan^{-1} \frac{e^c - \sin b}{\cos b}\right) - \tan^{-1}\left(\frac{-e^{-c} - \sin b}{\cos b}\right)\right).$$

Remark: To be fair to the analysts, we should say that the proof above is essentially due to Calderon (1966) and was later rediscovered by Davis (1973b), who wrote the proof in probabilistic language. We leave it to the reader to

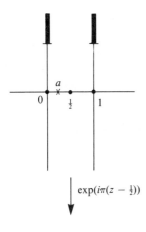

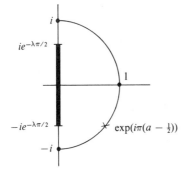

Figure 6.3

show that (1) is consistent with the result given by Stein and Weiss (1959) (see (4.3) on page 273) that

$$\exp\left(\frac{i}{2}|\{\theta:|\bar{\varphi}(\theta)|>y\}|\right) = \frac{\sinh\frac{y}{2} + i\sin\frac{|A|}{2}}{\sinh\frac{y}{2} - i\sin\frac{|A|}{2}}.$$

At this point, the reader can probably guess how we are going to prove the result for $u = \mathscr{P}\mu$ when μ is singular. Let $f = u + i\bar{u}$ and observe that

(a) $f(B_t)$, $t < \tau$, is a time change of a Brownian motion C_t starting at $f(0) = \mu(\partial D)$ and running for an amount of time

$$\gamma = \int_0^\tau |f'(B_s)|^2 \, ds.$$

(b) If $0 < \mu(\partial D) < \infty$, then $0 < u < \infty$ in D and $u(B_t) \in (0, \infty)$ for all $t < \tau$. This implies that $\gamma \leq T' = \inf\{t : \operatorname{Re} C_t \notin (0, \infty)\}$.

(c) Since $\mathscr{P}\mu \in h^1$, it follows from results in Section 6.5 that

6.9 Conjugate Functions of Indicators and Singular Measures

$$\lim_{t \uparrow \gamma} \operatorname{Re} C_t = \lim_{t \uparrow \tau} u(B_t) = 0 \quad \text{a.s.,}$$

and so we have $\gamma = T'$.

(d) Since $T' < \infty$, it follows that

$$\lim_{t \uparrow \tau} \bar{u}(B_t) = \lim_{t \uparrow \gamma} \operatorname{Im} C_t = \operatorname{Im} C(T'),$$

and using the equivalence of Brownian and nontangential convergence in $d = 2$ proved in Section 4.3, we conclude that (∗) holds and that

(2) $\quad \pi(\theta : \bar{\varphi}(\theta) > y) = P_{\mu(\partial D)}(\operatorname{Im} C(T') > y).$

This time it is trivial to compute the right-hand side. If we let $a = \mu(\partial D)$, then

$$P_a(\operatorname{Im} C(T') > y) = \int_{-\infty}^{y} \frac{1}{\pi} \frac{a}{a^2 + x^2} dx.$$

The proof above is due to B. Davis (1973).

7 H^1 and BMO, $\mathcal{M}^1$ and $\mathcal{BMO}$

7.1 The Duality Theorem for $\mathcal{M}^1$

In Chapter 6, we saw that if $p > 1$, $X \to X_\infty$ maps $\mathcal{M}^p$ one-to-one onto $L^p(\mathcal{F}_\infty)$ in such a way that the $\mathcal{M}^p$ norm of X is equivalent to the L^p norm of X_∞. From this observation, it follows immediately that every continuous linear functional φ on $\mathcal{M}^p$ can be written as $\varphi(X) = E(X_\infty Y)$, where $Y \in L^q$, $q = p/(p-1)$. When $p \le 1$, the equivalence of $\mathcal{M}^p$ and L^p breaks down and the reasoning above fails to identify the dual space $(\mathcal{M}^p)^*$. In this section, we will consider the problem of describing $(\mathcal{M}^1)^*$. The first step in the solution is to introduce a decomposition due to Bernard and Maisoneuve (1977), which expresses a general $X \in \mathcal{M}^1$ as a sum of very simple elements of $\mathcal{M}^1$. As will be the case many times below, (a) the probabilistic definition was developed after and imitates the definition invented by analysts (see Coifman (1974), Latter (1977), Coifman and Weiss (1977)) and, more embarrassingly, (b) we will often assume (e.g., in the proof of (4)) that our martingales start at 0 at time 0 and forget to mention this.

(1)   A martingale $A \in \mathcal{M}^1$ is said to be an atom if there is a stopping time T such that

   (i) $A_t = 0$ if $t \le T$
   (ii) $A^* = \sup_t |A_t| \le P(T < \infty)^{-1}$.

   Since $A^* = 0$ on $\{T = \infty\}$, it is easy to see:

(2)   If A is an atom, $\|A\|_1 = EA^* \le 1$.

   It follows from (2) and the triangle inequality that

(3)   If A^n is a sequence of atoms and c_n is a sequence of numbers with $\sum |c_n| < \infty$, then $X = \sum c_n A^n \in \mathcal{M}^1$ and $\|X\|_1 \le \sum |c_n|$.

   Clearly, (3) allows us to construct many examples of martingales in $\mathcal{M}^1$. The next result shows that every $X \in \mathcal{M}^1$ can be built up in this way and furthermore that $\sum |c_n| \le C \|X\|_1$.

7.1 The Duality Theorem for $\mathcal{M}^1$

(4) For all $X \in \mathcal{M}^1$, there is a sequence of atoms A^n, $n \in Z$, and a sequence of constants c_n, $n \in Z$, with $\sum_n |c_n| \leq 6\|X\|_1$ such that as $N \to \infty$,

$$\sum_{n=-N}^{N} c_n A^n \to X \quad \text{in} \quad \mathcal{M}^1.$$

Remark: I think the proof of this result is beautiful. It is a trivial computation, but it is also an ingenious idea. To convince yourself of the latter, you should put the book down for a few minutes and try to construct your own decomposition.

Proof One answer to the problem is: for each $n \in Z$, let

$$T_n = \inf\{t : |X_t| > 2^n\}$$

and let

$$A_t^n = (X(t \wedge T_{n+1}) - X(t \wedge T_n))/c_n.$$

The definition is arranged so that $\sum c_n A_t^n$ is a telescoping series, so we have

$$X_t - \sum_{n=-N}^{N} c_n A_t^n = X(t) - X(t \wedge T_{N+1}) + X(t \wedge T_{-N}).$$

The last term on the right is $\leq 2^{-N}$ and, hence, $\to 0$ in $\mathcal{M}^1$ as $N \to \infty$. To estimate the $\mathcal{M}^1$ norm of $Y_t = X(t) - X(t \wedge T_{N+1})$, we observe that $Y \equiv 0$ on $\{X^* < 2^{N+1}\}$ and $Y^* \leq 2X^*$, so by the dominated convergence theorem, $EY^* \leq E(2X^*; X^* \geq 2^{N+1}) \to 0$ as $n \to \infty$.

Up to this point, the values of the c_n's and the precise form of the stopping times have not entered into the proof. We must now choose the c_n's to make the A^n's atoms. To do this, we observe that

$$|X(t \wedge T_{n+1}) - X(t \wedge T_n)| \leq |2^{n+1} - (-2^n)| = 3 \cdot 2^n,$$

so if we want $|A_t^n| \leq P(T_n < \infty)^{-1}$, we must pick

$$c_n = 3 \cdot 2^n P(T_n < \infty).$$

Having done this, we find that

$$\sum_n |c_n| = \sum_n 3 \cdot 2^n P(X^* > 2^n)$$

$$= 3 \cdot \frac{2^n}{2^{n-1}} \sum_n \int_{2^{n-1}}^{2^n} P(X^* > 2^n)\, dy$$

$$\leq 6 \int_0^\infty P(X^* > y)\, dy = 6\|X\|_1.$$

Remark 1: It is important to observe that we use 2^n, $n \in Z$, and not just $n \geq 0$. We do this (that is, use $n < 0$) and use a sequence that grows geometrically, so that the picture remains the same, if we multiply by 2^m. The last feature is crucial if we are going to prove an estimate like (4), which is unaffected if the quantities under consideration are multiplied by a constant.

Remark 2: While it is important to let $T_n = \inf\{t : |X_t| > a^n\}$ and to pick $c_n = (a+1)a^n P(T_n < \infty)$, the actual choice of a is not crucial. If we repeat the last computation above in this generality, we find that

$$\sum |c_n| \leq \frac{(a+1)}{(1-a^{-1})} EX^*.$$

The constant is optimized by taking $a = 1 + \sqrt{2}$, and for this value of a we get a constant $= 3 + 2\sqrt{2} = 5.828\ldots$, which hardly seems worth the effort.

With the decomposition in (4) established, it is "easy" to find the dual of $\mathcal{M}^1$. A linear functional φ will be continuous if and only if

(5) $$\sup\{|\varphi(X)| : X \text{ is an atom}\} < \infty.$$

To be precise, a linear functional defined on the linear span of the atoms will have a continuous extension to $\mathcal{M}^1$ if and only if (5) holds. As in the case of the decomposition, it is hard to guess the answer (the reader is again invited to try), but if somebody tells you the answer and shows you which atoms to use as test functions, it is not hard to fill in the details.

We say that Y has bounded mean oscillation (and write $Y \in \mathcal{BMO}$) if $Y \in \mathcal{M}^2$ and there is a constant c such that for all stopping times T,

(6) $$E|Y_\infty - Y_T| \leq cP(T < \infty).$$

The infimum of the set of constants for which (6) holds is called the $\mathcal{BMO}$ norm of Y and is denoted as $\|Y\|_*$. This definition may not look very natural now, but it will by the end of the next proof.

(7) Let $\mathcal{A}$ be the set of atoms. For all $Y \in \mathcal{M}^1$,

$$\tfrac{1}{2}\|Y\|_* \leq \sup(|E(X_\infty Y_\infty)| : X \in \mathcal{A}) \leq \|Y\|_*.$$

Proof We will first prove the inequality on the right. If $X \in \mathcal{A}$ and T is a stopping time for which (1) holds, then

$$\begin{aligned} E(X_\infty Y_T) &= EE(X_\infty Y_T | \mathcal{F}_T) \\ &= E(Y_T E(X_\infty | \mathcal{F}_T)) \\ &= E(Y_T X_T) \\ &= 0, \end{aligned}$$

since $X_T = 0$. From this it follows that

$$\begin{aligned} |EX_\infty Y_\infty| &= |EX_\infty (Y_\infty - Y_T)| \\ &\leq E(X^* |Y_\infty - Y_T|) \\ &\leq P(T < \infty)^{-1} E|Y_\infty - Y_T| \leq \|Y\|_*. \end{aligned}$$

7.1 The Duality Theorem for $\mathcal{M}^1$

To prove the other inequality, let T be an arbitrary stopping time, let $Z_\infty = \text{sgn}(Y_\infty - Y_T)$, and let $Z_t = E(Z_\infty | \mathcal{F}_t)$ be the martingale generated by this random variable. Since $|Z_t| \leq 1$, $X_t = (Z_t - Z_{T \wedge t})/2P(T < \infty)$ is an atom. From the definition of Z_∞, it follows that

$$E|Y_\infty - Y_T| = E(Z_\infty(Y_\infty - Y_T)) = EZ_\infty Y_\infty - EZ_\infty Y_T.$$

The first computation in the proof shows that

$$E(Z_\infty Y_T) = E(Z_T Y_T) = E(Z_T Y_\infty),$$

so we have

$$E|Y_\infty - Y_T| = E((Z_\infty - Z_T)Y_T),$$

and it follows that

$$\frac{E|Y_\infty - Y_T|}{2P(T < \infty)} = E\left(\frac{(Z_\infty - Z_T)}{2P(T < \infty)} Y_\infty\right) = E(X_\infty Y_\infty).$$

Taking the supremum over all stopping times now gives the desired result.

With (4) and (7) established, it is now routine to conclude that $(\mathcal{M}^1)^* = \mathcal{BMO}$. To prove $(\mathcal{M}^1)^* \subset \mathcal{BMO}$, we observe:

(a) $\mathcal{M}^2 \subset \mathcal{M}^1$, with $\|X\|_2 = (E|X^*|^2)^{1/2} \geq E|X^*| = \|X\|_1$, so if φ is a continuous linear functional on $\mathcal{M}^1$, φ induces a continuous linear functional on $\mathcal{M}^2$ with $\|\varphi\|_2 = \sup\{|\varphi(X)| : \|X\|_2 \leq 1\} \leq \|\varphi\|_1$.
(b) from the duality theorem for $\mathcal{M}^2$, it follows that there is a $Y \in \mathcal{M}^2$ such that $\varphi(X) = EX_\infty Y_\infty$ for all $X \in \mathcal{M}^2$. Since $\mathcal{A} \subset \mathcal{M}^2$, it follows from (7) that $Y \in \mathcal{BMO}$.
(c) from the atomic decomposition, it follows that $\mathcal{M}^2$ is dense in $\mathcal{M}^1$, so the correspondence $\varphi \to Y$ defined in (b) is one-to-one.

To prove that $(\mathcal{M}^1)^* = \mathcal{BMO}$, we now have to prove that all the linear functionals given above are continuous. This follows from the next result (Fefferman's inequality).

(8) If $X, Y \in \mathcal{M}^2$, then

$$|E(X_\infty Y_\infty)| \leq 6\|X\|_1 \|Y\|_*.$$

Proof From (4), it follows that X can be written as $\sum_n c_n A^n$, where $A^n, n \in \mathbb{Z}$, is a sequence of atoms and

$$\left|\sum_{n=-N}^{N} c_n A_t^n\right| \leq X^* + 1 \quad \text{for all } N, t.$$

Since $X \in \mathcal{M}^2$, we have $X^* \in L^2$, and it follows from the Cauchy-Schwarz inequality and the dominated convergence theorem that

$$EX_\infty Y_\infty = \sum_n c_n E(A_\infty^n Y_\infty).$$

Using the triangle inequality now with the results of (7) and (4) gives the desired conclusion:

$$|EX_\infty Y_\infty| \leq \sum_n |c_n| |E(A_\infty^n Y_\infty)|$$
$$\leq \sum_n |c_n| \|Y\|_* \leq 6\|X\|_1 \|Y\|_*.$$

Remark: What we have shown above is that if $\varphi \in (\mathcal{M}^1)^*$, then there is a $Y \in \mathcal{BMO}$ such that $\varphi(X) = E(X_\infty Y_\infty)$ for all $X \in \mathcal{M}^2$. In Section 7.2, when we give the "classical" proof of the duality result, we improve this conclusion slightly by showing that $\varphi(X) = E\langle X, Y\rangle_\infty$ for all $X \in \mathcal{M}^1$.

(9) If there is a constant c such that for all stopping times T

(∗) $E|Y_\infty - Y_T| \leq cP(T < \infty),$

then it follows that we have

(∗∗) $E(|Y_\infty - Y_T| \big| \mathcal{F}_T) \leq c$ a.s.

for all stopping times.

Proof Applying (∗) to the stopping time

$$T' = \begin{cases} T & \text{if } E(|Y_\infty - Y_T| \big| \mathcal{F}_T) > c \\ \infty & \text{otherwise,} \end{cases}$$

we see that if $P(T' < \infty) > 0$, then

$$cP(T' < \infty) \geq E(|Y_\infty - Y_T|1_{(T' < \infty)})$$
$$= E(E(|Y_\infty - Y_T| \big| \mathcal{F}_T)1_{(T' < \infty)})$$
$$> cP(T' < \infty),$$

a contradiction, so $P(T' < \infty) = 0$.

7.2 A Second Proof of $(\mathcal{M}^1)^* = \mathcal{BMO}$

In this section, we will give a second proof of the duality theorem for $\mathcal{M}^1$ following Meyer (1976). This approach starts with a somewhat different definition of $\mathcal{BMO}$.

Let $X \in \mathcal{M}^2$ with $X_0 = 0$. We say that $X \in \mathcal{BMO}_2$ if there is a constant c such that, for all stopping times T,

(1) $E(X_\infty - X_T)^2 \leq c^2 P(T < \infty).$

The infimum of the constants with this property is called the $\mathcal{BMO}_2$ norm of X and is denoted by $\langle\!\langle X \rangle\!\rangle_*$.

7.2 A Second Proof of $(\mathcal{M}^1)^* = \mathcal{BMO}$

Remark 1: From Jensen's inequality for conditional expectations, it follows that if $X \in \mathcal{BMO}_2$, then $X \in \mathcal{BMO}$ and $\langle\!\langle X \rangle\!\rangle_* \geq \|X\|_*$. An inequality in the other direction, $\langle\!\langle X \rangle\!\rangle_* \leq C\|X\|_*$, is also true and is a consequence of the two proofs of the duality theorem. We will also give a direct proof of the second inequality in Section 7.6.

Remark 2: Since $E((X_\infty - X_T)^2 | \mathscr{F}_T) = E(X_\infty^2 - X_T^2 | \mathscr{F}_T) = E(\langle X \rangle_\infty - \langle X \rangle_T | \mathscr{F}_T)$, the definition of $\mathcal{BMO}_2$ can be written as

(1') $\quad E(\langle X \rangle_\infty - \langle X \rangle_T) \leq c^2 P(T < \infty)$

or, in view of (9) in Section 7.1, as

(1'') $\quad E(\langle X \rangle_\infty - \langle X \rangle_T | \mathscr{F}_T) \leq c^2$ a.s.

The first step in our second proof of the duality theorem is the same as in the first proof. We show that every continuous linear functional on $\mathcal{M}^1$ comes from a $Y \in \mathcal{BMO}_2$.

(2) If φ is a continuous linear functional on $\mathcal{M}^1$, then there is a $Y \in \mathcal{BMO}_2$ such that for all $X \in \mathcal{M}^2$

$$\varphi(X) = E X_\infty Y_\infty.$$

Proof It suffices to prove the result when $\langle\!\langle \varphi \rangle\!\rangle_1 = \sup\{|\varphi(X)| : \langle\!\langle X \rangle\!\rangle_1 \leq 1\} = 1$. Jensen's inequality implies that $\langle\!\langle X \rangle\!\rangle_2 = (E\langle X \rangle_\infty)^{1/2} \geq E(\langle X \rangle_\infty^{1/2}) = \langle\!\langle X \rangle\!\rangle_1$, so φ induces a continuous linear functional on $\mathcal{M}^2$, and it follows as in Section 7.1 that there is a $Y \in \mathcal{M}^2$ such that $\varphi(X) = E X_\infty Y_\infty$ for all $X \in \mathcal{M}^2$. To show that $Y \in \mathcal{BMO}_2$, let T be a stopping time and let $X_t = Y_t - Y_{T \wedge t}$. X is the stochastic integral $J \cdot Y$ where $J = 1_{[T, \infty)}$, so using our formula for the covariance of two stochastic integrals ((3) of Section 2.6), we have

$$\langle X, Y \rangle_t = \int_0^t J_s \, d\langle Y, Y \rangle_s$$

$$\langle X, X \rangle_t = \int_0^t J_s^2 \, d\langle Y, Y \rangle_s.$$

Since $J_s^2 = J_s$, it follows that

$$\langle X, X \rangle_\infty = \langle X, Y \rangle_\infty = \langle Y, Y \rangle_\infty - \langle Y, Y_T \rangle.$$

Let Z denote the common value of the three expressions. Since $X, Y \in \mathcal{M}^2$, it follows from the inequalities of Doob, Cauchy-Schwarz, and Kunita-Watanabe that

$$EZ \equiv E\langle X, Y \rangle_\infty = E X_\infty Y_\infty.$$

On the other hand, we have

$$E X_\infty Y_\infty = \varphi(X) \leq \langle\!\langle X \rangle\!\rangle_1 \equiv E(Z^{1/2}),$$

and since $Z = 0$ on $\{T = \infty\}$, it follows from the Cauchy-Schwarz inequality applied to $Z 1_{(T<\infty)}$ that

$$E(Z^{1/2}) \leq (EZ)^{1/2} P(T < \infty)^{1/2}.$$

Combining the last three results proves that

$$EZ \leq (EZ)^{1/2} P(T < \infty)^{1/2},$$

that is,

$$EZ \equiv E(\langle Y \rangle_\infty - \langle Y \rangle_T) \leq P(T < \infty).$$

Since this result holds for all stopping times, it shows that $Y \in \mathscr{BMO}$ with $\langle\!\langle Y \rangle\!\rangle_* \leq 1$.

At this point, we have shown that $(\mathscr{M}^1)^* \subset \mathscr{BMO}_2$. To prove the other inclusion, we need another version of Fefferman's inequality.

(3) If $X \in \mathscr{M}^1$ and $Y \in \mathscr{BMO}_2$, then

$$E \int_0^\infty |d\langle X, Y \rangle_s| \leq \sqrt{2} \langle\!\langle X \rangle\!\rangle_1 \langle\!\langle Y \rangle\!\rangle_*.$$

Proof By stopping, it suffices to prove the result when X, Y, $\langle X \rangle$, and $\langle Y \rangle$ are bounded. Since $\langle X \rangle_t^{-1/4} \langle X \rangle_t^{1/4} = 1$, it follows from the Kunita-Watanabe inequality that

$$E \int_0^\infty |d\langle X, Y \rangle_s|^2 \leq E \int_0^\infty \langle X \rangle_t^{-1/2} d\langle X \rangle_t \cdot E \int_0^\infty \langle X \rangle_t^{1/2} d\langle Y \rangle_t.$$

Since $\langle X \rangle_t$ has bounded variation, ordinary (Riemann-Stieljes) integration gives

$$\int_0^\infty \langle X \rangle_t^{-1/2} d\langle X \rangle_t = 2 \langle X \rangle_\infty^{1/2},$$

so

$$E \int_0^\infty \langle X \rangle_t^{-1/2} d\langle X \rangle_t = 2 \langle\!\langle X \rangle\!\rangle_1.$$

To estimate the second integral, we fix ω and integrate by parts to obtain

$$E \int_0^\infty \langle X \rangle_t^{1/2} d\langle Y \rangle_t = E \langle X \rangle_\infty^{1/2} \langle Y \rangle_\infty - \int_0^\infty \langle Y \rangle_t d\langle X \rangle_t^{1/2}$$

$$= E \int_0^\infty \langle Y \rangle_\infty - \langle Y \rangle_t d\langle X \rangle_t^{1/2}.$$

At this point, it is very easy to complete the proof if we leave out one detail. I claim that since $\langle X \rangle_t$ is adapted to $\mathbb{F}$, the last expression

7.2 A Second Proof of $(\mathcal{M}^1)^* = \mathcal{BMO}$

$$= E\int_0^\infty E(\langle Y\rangle_\infty - \langle Y\rangle_t | \mathscr{F}_t)\,d\langle X\rangle_t^{1/2}.$$

If you accept this, then there is nothing left to show, for definition (1″) of $\mathcal{BMO}_2$ implies that the above is

$$\leq \langle\!\langle Y\rangle\!\rangle_*^2 E(\langle X\rangle_\infty^{1/2}) = \langle\!\langle X\rangle\!\rangle_1 \langle\!\langle Y\rangle\!\rangle_*^2.$$

Combining the inequalities above shows that

$$E\int_0^\infty |d\langle X, Y\rangle_s| \leq (2\langle\!\langle X\rangle\!\rangle_1)^{1/2}(\langle\!\langle X\rangle\!\rangle_1 \langle\!\langle Y\rangle\!\rangle_*^2)^{1/2},$$

which is the desired inequality.

To complete the proof of (3), it remains only to justify the equality claimed above. To do this, we will prove a general result:

(4) If Z is a bounded random variable and A is a bounded increasing process adapted to $\mathbb{F}$ with $A_0 = 0$, then

$$E(ZA_\infty) = E\int_0^\infty E(Z|\mathscr{F}_t)\,dA_t.$$

Technical Remark: Some care is needed in defining the integrand, since A may be singular and, for each t, $E(Z|\mathscr{F}_t)$ is defined only up to a null set. In our situation, there is no problem. The Brownian filtration admits only continuous martingales, so we take versions of $E(Z|\mathscr{F}_t)$ that are continuous in t for each ω. In the language of the general theory, we are taking the optional projection of the process $Y_t = Z$ (which is constant in time), but in our situation we do not need this notion, since there is only one reasonable way to define $E(Z|\mathscr{F}_t)$ for all t simultaneously.

Proof Suppose without loss of generality that $Z \geq 0$.

$$EZA_\infty = E\sum_{k=1}^\infty Z\left(A\left(\frac{k}{n}\right) - A\left(\frac{k-1}{n}\right)\right)$$

$$= E\sum_{k=1}^\infty E\left(Z\left(A\left(\frac{k}{n}\right) - A\left(\frac{k-1}{n}\right)\right)\bigg|\mathscr{F}_{k/n}\right)$$

$$= E\sum_{k=1}^\infty \left(A\left(\frac{k}{n}\right) - A\left(\frac{k-1}{n}\right)\right)E(Z|\mathscr{F}_{k/n})$$

$$\to E\int_0^\infty E(Z|\mathscr{F}_t)\,dA_t$$

as $n \to \infty$, by the dominated convergence theorem.

Taking $Z = \langle Y \rangle_\infty$ in (4), we see that

$$\int_0^\infty \langle Y \rangle_\infty \, d\langle X \rangle_t^{1/2} - \int_0^\infty \langle Y \rangle_t \, d\langle X \rangle_t^{1/2}$$

$$= \int_0^\infty E(\langle Y \rangle_\infty | \mathcal{F}_t) \, d\langle X \rangle_t^{1/2} - \int_0^\infty \langle Y \rangle_t \, d\langle X \rangle_t^{1/2}$$

$$= \int_0^\infty E(\langle Y \rangle_\infty - \langle Y \rangle_t | \mathcal{F}_t) \, d\langle X \rangle_t^{1/2},$$

which completes both the proof of (3) and the proof of the duality theorem. As we mentioned in Section 7.1, one advantage of the new proof is that it gives a formula for the linear functional that is valid on the whole space and not just on a dense subset.

7.3 Equivalence of *BMO* to a Subspace of $\mathcal{BMO}$

Our next goal is to prove the duality theorem $(H^1)^* = BMO$. One-third of the work for the proof of this result was done in Sections 6.4 and 6.5, when we showed:

(a) $f \to \text{Re} f(B_t)$, $t < \tau$, maps H_0^1 one-to-one into $\mathcal{M}_\tau^1$ and, furthermore, the H^1 norm of f is equivalent to the $\mathcal{M}_\tau^1$ norm of its image

(b) if we let $X_t = \text{Re} f(B_{t \wedge \tau})$, then $X \in \mathcal{M}^1$, so if we let $M : f \to \text{Re} f(B_{t \wedge \tau})$, then the results in (a) hold when the τ is erased.

It follows from (b) that $M(H_0^1)$ is a closed subspace of $\mathcal{M}^1$ and that all continuous linear functionals on H_0^1 have the form $\Lambda(Mf)$, where $\Lambda \in (\mathcal{M}^1)^*$. Since $(\mathcal{M}^1)^* = \mathcal{BMO}$ (the second third of the work), it then remains to identify $M(H_0^1)^* \subset \mathcal{BMO}$ and to show that we can map BMO one-to-one onto $M(H_0^1)^*$ in such a way that the BMO norm of a function is equivalent to the $\mathcal{BMO}$ norm of its image.

The answer to the first question can be found by very naive reasoning: $M(H_0^1) = \{X \in \mathcal{M}^1 : X_0 = 0, X_t = h(B_t), t < \tau, \text{ and } X \text{ is constant for } t \geq \tau\}$, a space we call $\mathcal{M}_h^1$, so if $\mathcal{BMO}_h$ is defined in the obvious way, we should have $(\mathcal{M}_h^1)^* = \mathcal{BMO}_h$. When we prove that $(H^1)^* = BMO$ in Section 7.4, we will show that this reasoning is correct. To prepare for that result, this section is devoted to determining which harmonic functions h give rise to martingales in $\mathcal{BMO}_h$ (the last third of the work). The first baby step in doing this is to observe:

(1) If we let

$$\mathcal{P}f = \int k_\theta(z) f(e^{i\theta}) \, d\pi(\theta),$$

then $h = \mathcal{P}\varphi$ where $\varphi \in L^2(\partial D, \pi)$.

7.3 Equivalence of BMO to a Subspace of $\mathscr{BMO}$

Proof $\mathscr{BMO}_h \subset \mathscr{M}_h^2$, so the result follows from (2) in Section 7.3.

The last result is not much but, in view of the remarks at the end of Section 6.5, it allows us to think of h as being a function $\varphi \in L^2(\partial D, \pi)$ and to conclude that

$$X_t = \begin{cases} h(B_t) & t < \tau \\ \varphi(B_\tau) & t \geq \tau \end{cases}$$

is a continuous martingale $\in \mathscr{M}^2$.

A more substantive conclusion results if we use the definition of $\mathscr{BMO}$: for all stopping times T,

$$E((X_\infty - X_T)^2 | \mathscr{F}_T) \leq c^2.$$

So it follows from the strong Markov property that

$$E((X_\infty - X_T)^2 | \mathscr{F}_T) = w(B_{T \wedge \tau}),$$

where $w(z) = 0$ on ∂D and

$$w(z) = \int k_\theta(z)(\varphi(e^{i\theta}) - \mathscr{P}\varphi(z))^2 2\pi(\theta)$$

when $z \in D$. So in terms of w, the condition for x to be in $\mathscr{BMO}$ is

(2)   For all $z \in D$, $w(z) \leq c^2$.

The aim of this section is to show that the last definition is equivalent to the following notion in analysis.

(3) φ is in BMO if there is a constant c such that for all intervals I,

$$\int_I |\varphi - \varphi_I|^2 \frac{d\theta}{|I|} \leq c^2,$$

where

$$\varphi_I = \int_I \varphi \frac{d\theta}{|I|}$$

is the average value of φ on I. The smallest positive constant with this property is denoted by $\|\varphi\|_*$.

Conditions (2) and (3) have a very similar form. To emphasize this we will introduce some notation. Let

$$j_\theta(re^{i\psi}) = \begin{cases} 1/(1-r) & \text{if } |\psi - \theta| < (1-r)\pi \\ 0 & \text{otherwise} \end{cases}$$

$$\bar{\mathscr{P}}\varphi(z) = \int j_\theta(z) \varphi(e^{i\theta}) \, d\pi(\theta)$$

$$\bar{w}(z) = \int j_\theta(z)(\varphi(e^{i\theta}) - \bar{\mathscr{P}}\varphi(z))^2 \, d\pi(\theta).$$

If we recall that $d\pi(\theta) = d\theta/2\pi$ and observe that as z ranges over D, the supports of the maps $\theta \to j_\theta(z)$ run through all possible intervals I, then we see that (3) says simply that $\bar{w}(z) \leq c^2$ for all $z \in D$.

The first step in understanding the relationship between (2) and (3) is to understand the relationship between $k_\theta(z)$ and $j_\theta(z)$. To begin, we will look at the asymptotic behavior of $k_\theta(r)$ as $r \to 1$.

(4) $$k_\theta(r) = \frac{1-r^2}{|r - e^{i\theta}|^2}$$
$$= \frac{1-r^2}{(r-\cos\theta)^2 + (\sin\theta)^2}$$
$$= \frac{1-r^2}{1+r^2 - 2r\cos\theta}$$
$$= \frac{(1+r)(1-r)}{(1-r)^2 + 2r(1-\cos\theta)}$$
$$= \frac{1+r}{1-r}\left(1 + \frac{2r(1-\cos\theta)}{(1-r)^2}\right)^{-1},$$

so if we let $\theta = y(1-r)$, then

$$(1-r)k_{y(1-r)}(r) \to 2(1+y^2)^{-1}$$

2π times the density of a Cauchy distribution with parameter 1.

The last result should be no surprise. If we approach the boundary of the disk and rescale the picture so that we are at a distance 1 from the boundary, then the boundary will approach a straight line, so the rescaled exit distribution will approach the Cauchy distribution and the factor of 2π arises since on ∂D we are looking at the density w.r.t. $d\pi = d\theta/2\pi$.

The computations above show that the natural place for viewing $k_\theta(r)$ is at $\theta = y(1-r)$. Since this is the width of the support of $j_\theta(r)$, the strict positivity of the Cauchy density leads easily to:

(5) There is a constant A such that, for all z and θ,

$$j_\theta(z) \leq Ak_\theta(z).$$

Proof From (4) we see that $j_\theta(z) \leq C_r k_\theta(z)$ for all $|z| \leq r$, so it suffices to consider what happens when $r \to 1$. From (4), it follows that

$$k_\theta(r) = \frac{1+r}{1-r}\left(1 + \frac{2r(1-\cos\theta)}{(1-r)^2}\right)^{-1}.$$

Now $1 - \cos\theta = \varepsilon(\theta)$, where $\varepsilon(\theta) \sim \theta^2/2$ as $\theta \to 0$, so $|\varepsilon(\theta)| \leq C\theta^2$ for $\theta \in [0, 2\pi]$ and it follows that if $r < 1$ and $|\theta| < (1-r)\pi$, then

$$1 + \frac{2r(1-\cos\theta)}{(1-r)^2} \leq 1 + \frac{2}{(1-r)^2} C(1-r)^2 \pi^2,$$

proving (5).

7.3 Equivalence of BMO to a Subspace of $\mathscr{BMO}$

With (5) established, it is easy to prove:

(6)   For all $z \in D$, $\bar{w}(z) \leq Aw(z)$.

Proof By (5),

$$Aw(z) \geq \int (\varphi(e^{i\theta}) - \mathscr{P}\varphi(z))^2 j_\theta(z)\,d\pi(\theta)$$

$$\geq \int (\varphi(e^{i\theta}) - \bar{\mathscr{P}}\varphi(z))^2 j_\theta(z)\,d\pi(\theta),$$

since $a \to \int (\varphi(e^{i\theta}) - a)^2 j_\theta(z)\,d\pi(\theta)$ is minimized at $a = \bar{\mathscr{P}}\varphi$. In more familiar terms, the mean $\mu = EX$ minimizes $E(X-a)^2$, since

$$E(X-a)^2 = E(X-\mu+\mu-a)^2 = E(X-\mu)^2 + (\mu-a^2).$$

Having proved (6), we turn our attention to the other comparison. From the proof of (6), it is clear that it would be enough to show that there is a constant B such that

(*) $$B\bar{w}(z) \geq \int (\varphi(e^{i\theta}) - \bar{\mathscr{P}}\varphi(z))^2 k_\theta(z)\,d\pi(\theta),$$

for this would imply, by the argument above, that $B\bar{w}(z) \geq w(z)$.

The proof of (6) was easy, because all we had to do was show that some multiple of $k_\theta(z)$ was $\geq j_\theta(z)$. This statement is false if we interchange the roles of j_θ and k_θ, so we will have to work harder to prove the result we want—we have to add up a large number of functions of the form $c1_{[a,b]}$ to make something $\geq k_\theta$. Because of this difficulty, we will prove a slightly weaker result than (*) that is still sufficient to prove that the norms are equivalent.

(7)   There is a constant B such that for all $z \in D$,

$$B\left(\sup_z \bar{w}(z)\right) \geq \int (\varphi(e^{i\theta}) - \bar{\mathscr{P}}\varphi(z))^2 k_\theta(z)\,d\pi(\theta).$$

Proof As we mentioned above, we prove this by adding up multiples of $1_{[a,b]}$ to make something $\geq k_\theta(z)$. The first step is to introduce the intervals. Let

$$I_0 = [-(1-r)\pi, (1-r)\pi]$$

and for $n \geq 1$, $I_n = 2^n I_0$, let N be the largest integer such that $2^N(1-r) \leq 1$, and write

$$\int_{-\pi}^{\pi} = \int_{I_0} + \sum_{n=1}^{N} \int_{I_n - I_{n-1}} + \int_{[-\pi,\pi]-I_N}.$$

For a picture of the decomposition, see Figure 7.1.

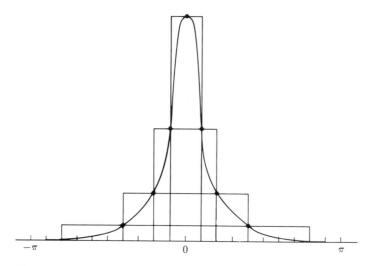

Figure 7.1 $k_\theta(.9) = 19/(1 + 180(1 - \cos\theta))$, $I_0 = [-.1\pi, .1\pi]$, $N = 3$

Estimating the integral over I_0 is easy. If $\theta \in I_0$, then (4) implies that

$$(a_0)\quad k_\theta(r) = \frac{1-r^2}{(1-r)^2 + 2r(1-\cos\theta)} \leq \frac{1-r^2}{(1-r)^2}$$
$$= \frac{1+r}{1-r} \leq 2j_\theta(r),$$

so we have

$$(b_0)\quad 2\bar{w}(r) \geq \int_{I_0} (\varphi(e^{i\theta}) - \bar{\mathscr{P}}\varphi(r))^2 k_\theta(r)\, d\pi(\theta).$$

(The reason for this strange numbering will become apparent as the proof goes on.)

Estimating the rest of the integrals requires more work. The first step in estimating the integral over $I_n - I_{n-1}$ is to observe

(a_n) There is a constant C such that if $r \geq 1/2$ and $n \leq N$, $(1-r)k_\theta(r) \leq C4^{-n}$ for all $\theta \in I_n - I_{n-1}$.

Proof From (4), it follows that

$$(1-r)k_\theta(r) = (1+r)\left(1 + \frac{2r(1-\cos\theta)}{(1-r)^2}\right)^{-1}.$$

Now if $n \leq N$ and $\theta \in I_n - I_{n-1}$, then

$$1 - \cos\theta \geq 1 - \cos(2^{n-1}\pi(1-r)),$$

and we have from calculus that

$$\inf_{x \in (0,1)} \frac{1 - \cos x}{x^2} = \varepsilon > 0,$$

7.3 Equivalence of BMO to a Subspace of $\mathscr{BMO}$

so
$$(1-r)k_\theta(r) \le 2(1+\varepsilon(2^{n-1}\pi)^2)^{-1},$$
proving (a_n).

The estimate in (a_n) takes care of $k_\theta(r)$ on $I_n - I_{n-1}$. To estimate the rest of the integral in (7), we let
$$\|f\|_2^n = \left(\int_{I_n} |f|^2\, d\theta\right)^{1/2}$$
and
$$a_n = \int_{I_n} \varphi\, \frac{d\theta}{|I_n|}.$$
Since $\|\cdot\|_2^n$ is a norm,
$$\|\varphi - a_0\|_2^n \le \|\varphi - a_n\|_2^n + \sum_{k=1}^n \|a_k - a_{k-1}\|_2^n.$$

The first term
$$= \left(\int_{I_n} (\varphi - a_n)^2\, d\theta\right)^{1/2} \le \|\varphi\|_* |I_n|^{1/2}.$$

The second term
$$= \sum_{k=1}^n |a_k - a_{k-1}| |I_n|^{1/2}.$$

To estimate $|a_k - a_{k-1}|$, we observe that
$$|a_k - a_{k-1}| = ((a_k - a_{k-1})^2)^{1/2} = \left(\int_{I_{k-1}} (a_k - a_{k-1})^2\, \frac{d\theta}{|I_{k-1}|}\right)^{1/2}$$
$$\le \left(\int_{I_{k-1}} (\varphi - a_{k-1})^2\, \frac{d\theta}{|I_{k-1}|}\right)^{1/2} + \left(\int_{I_{k-1}} (a_k - \varphi)^2\, \frac{d\theta}{|I_{k-1}|}\right)^{1/2}$$
$$\le \|\varphi\|_* + \left(2\int_{I_k} (a_k - \varphi)^2\, \frac{d\theta}{|I_k|}\right)^{1/2} \le 3\|\varphi\|_*.$$

Summing the estimates above, we find that
$$\|\varphi - a_0\|_2^n \le (3n+1)\|\varphi\|_* |I_n|^{1/2},$$
so
$$\int_{I_n} (\varphi - a_0)^2\, d\theta \le 16 n^2 \|\varphi\|_*^2 2^n \pi(1-r).$$

Combining this with the estimate (a_n) and recalling that $a_0 = \bar{\mathscr{P}}\varphi$, shows that for $n \le N$,

(b_n)
$$\int_{I_n - I_{n-1}} (\varphi - \bar{\mathscr{P}}\varphi)^2 k_\theta(z)\, d\theta \le C'\|\varphi\|_*^2 n^2 / 2^{-n}.$$

The last detail is to estimate the integral over $[-\pi, \pi] - I_N$. To do this, we observe that $I_N \supset [-\pi/2, \pi/2]$, so $\cos\theta \leq 0$ on $[-\pi, \pi] - I_N$, and it follows from the proof of (a_n) that

(a_{N+1}) if $r \geq 1/2$ and $\theta \in [-\pi, \pi] - I_N$, then

$$(1-r)k_\theta(r) \leq 2(1 + (2^N\pi)^2)^{-1},$$

and repeating the arguments used to prove (b_n) proves:

(b_{N+1}) $\displaystyle\int_{[-\pi,\pi]-I_N} (\varphi - \bar{\mathscr{P}}\varphi)^2 k_\theta(z)\, d\theta \leq C' \|\varphi\|_*^2 N^2 / 2^{-N}.$

Adding up the estimates $(b_0) + (b_1) + \cdots + (b_{N+1})$ and recalling that $\|\varphi\|_*^2 = \sup_z \bar{w}(z)$, proves (7).

With (7) proved, it follows immediately (for reasons given after the proof of (6)) that we have

(8) For all $z \in D$,

$$B\|\varphi\|_*^2 \geq w(z).$$

Combining this result with (6) shows that if we let

$$\langle\!\langle \varphi \rangle\!\rangle_*^2 = \sup_z w(z)$$

(= the $\mathscr{BMO}$ norm of the associated martingale, $\mathscr{P}\varphi(B_{t\wedge\tau})$), then

$$\frac{1}{A}\|\varphi\|_*^2 \leq \langle\!\langle \varphi \rangle\!\rangle_*^2 \leq B\|\varphi\|_*^2.$$

For some arguments in Section 7.4 and beyond, it is useful to write the definition of $\langle\!\langle\ \rangle\!\rangle_*$ in a slightly different way. Let $\varphi \in L^2(\partial D, \pi)$ and let $u = \mathscr{P}\varphi$. Since $u(z) = E_z\varphi(B_\tau)$, then

$$w(z) = E_z(u(B_\tau) - u(z))^2$$
$$= E_z u(B_\tau)^2 - u(z)^2.$$

Since $U_t = u(B_{t \wedge \tau}) \in \mathscr{M}^2$, it follows that

$$w(z) = E_z \langle U \rangle_\tau = E_z \int_0^\tau |\nabla u(B_s)|^2\, ds,$$

so in view of the results of Section 1.11, the condition for $\varphi \in BMO$ can be written as

(9) $\displaystyle\sup_z \int_D G_D(z,w) |\nabla u(w)|^2\, dw < \infty,$

where

$$G_D(z,w) = \frac{2}{\pi} \log\left|\frac{1 - \bar{w}z}{z - w}\right|$$

is the Green's function for D.

The last result is very similar to a classical analytical characterization of BMO. A positive measure λ on D is said to be a Carleson measure if there is a constant c such that

$$\lambda(S) \leq K \cdot h$$

for every sector

$$S = \{re^{i\theta} : 1 - h \leq r < 1, |\theta - \theta_0| \leq h\}.$$

In his solution of the "Corona problem," Carleson proved that

(10) $\varphi \in BMO$ if and only if the measure defined by

$$|\nabla u(z)|^2 \log \frac{1}{|z|} dx\, dy,$$

where $u = \mathscr{P}\varphi$, is a Carleson measure and, furthermore, the constant K in the definition can be chosen so that

$$C_1 \|\varphi\|_*^2 \leq K(\varphi) \leq C_2 \|\varphi\|_*^2.$$

For an analytic proof see Garnett (1980), Chapter 6, Section 3. Even though (9) and (10) are very similar, I do not know how to get the second result from the first.

7.4 The Duality Theorem for H^1, Fefferman-Stein Decomposition

In this section, we will identify $(H_0^1)^*$ as BMO. There is a standard way to recover a complex linear functional Λ from the real part of the corresponding functional on the associated real Banach space: $\Lambda(f) = \operatorname{Re} \Lambda(f) - i \operatorname{Re} \Lambda(if)$, so, to make the transition to martingales easier, we consider only real linear functionals. The identification of $(H_0^1)^*$ is a four-step procedure.

(1)   If $\Lambda \in (H_0^1)^*$, then there are functions g_1 and g_2 in $L^\infty(\partial D, \pi)$ such that if $f = u + iv \in H^1$, then

$$\Lambda(f) = \int_{\partial D} ug_1 + vg_2\, d\pi.$$

Proof of (1) By results in Sections 7.2 and 7.3, the mapping $f \to (u, v)|_{\partial D}$ identifies H_0^1 with a subspace K^1 of $L^1(\partial D) \times L^1(\partial D)$ in such a way that $\|f\|_{H^1}$ and $\|u\|_{L^1} + \|v\|_{L^1}$ are equivalent norms. If Λ is a continuous linear functional on H_0^1, then it gives rise in an obvious way to a linear functional on K^1, which by the Hahn-Banach theorem extends to a continuous linear functional on all of $L^1 \times L^1$. Since $(L^1 \times L^1)^* = L^\infty \times L^\infty$, it follows that there exist g_1 and $g_2 \in L^\infty$ such that (1) holds.

The next step is to show:

(2)   If $f_1, f_2 \in H_0^2$ and u_i, v_i are the boundary limits of $\mathrm{Re} f_i$, $\mathrm{Im} f_i$, then
$$\int_{\partial D} u_1 u_2 \, d\pi = \int_{\partial D} v_1 v_2 \, d\pi.$$

Proof Let $U_t^i = u_i(B_t)$ for $t < \tau$. Itô's formula implies
$$u_i(B_t) = \int_0^t \nabla u_i(B_s) \cdot dB_s,$$
so the formula for the covariance of two stochastic integrals gives
$$\langle U^1, U^2 \rangle_t = \int_0^t \nabla u_1(B_s) \cdot \nabla u_2(B_s) \, ds.$$
A similar argument shows
$$\langle V^1, V^2 \rangle_t = \int_0^t \nabla v_1(B_s) \cdot \nabla v_2(B_s) \, ds,$$
and the Cauchy-Riemann equations imply
$$\nabla v_i = \begin{pmatrix} 0 & -1 \\ 1 & 0 \end{pmatrix} \nabla u_i.$$
so we have $\nabla u_1 \cdot \nabla u_2 = \nabla v_1 \cdot \nabla v_2$ and $\langle U^1, U^2 \rangle_t = \langle V^1, V^2 \rangle_t$.

Since we have assumed the $f_i \in H_0^2$, we have $U^i, V^i \in \mathcal{M}^2$ with $U_0^i = V_0^i = 0$, so the usual domination argument shows
$$E_0 U_\infty^1 U_\infty^2 = E_0 \langle U^1, U^2 \rangle_\infty = E_0 \langle V^1, V^2 \rangle_\infty = E_0 V_\infty^1 V_\infty^2,$$
proving (2).

(2) allows us to write
$$\Lambda(f) = \int_{\partial D} u g_1 + u \bar{g}_2 \, d\pi,$$
where $\bar{g}_2$ has the obvious meaning: it is the boundary limit of the conjugate function of $\mathscr{P} g_2$. The rest of the proof is very easy. Recalling that conjugation is a martingale transform, we see that

(3) $g_1 + \bar{g}_2 \in BMO$.

Proof Since $g_1, g_2 \in L^\infty$, they are in BMO. Let $g_3 = \bar{g}_2$, $u_i = \mathscr{P} g_i$. By results in Section 7.3, g_i is in BMO if and only if
$$\sup_z E_z \int_0^\tau |\nabla u_i(B_s)|^2 \, ds < \infty.$$
The Cauchy-Riemann equations imply that $|\nabla u_2| = |\nabla u_3|$, so it follows that $g_3 = \bar{g}_2 \in BMO$.

7.4 The Duality Theorem for H^1, Fefferman-Stein Decomposition

Given the martingale duality theorem and the results of the last two sections, the last step is trivial.

(4) If $h \in BMO$ and we let

$$\Lambda(f) = \int_{\partial D} uh \, d\pi$$

for all $f \in H_0^2$, then Λ has an extension that is a continuous linear functional on H_0^1.

Proof Let $h \in BMO$, $f \in H_0^2$, $u = \text{Re} f$, and let $H_t = h(B_{t \wedge \tau})$, $U_t = u(B_{t \wedge \tau})$. Since $U_0 = 0$ and $BMO \subset H^2$, the usual domination argument shows

$$\int_{\partial D} uh \, d\pi = EU_\infty H_\infty.$$

Using the martingale duality theorem and the equivalence of the two norms shows

$$EU_\infty H_\infty \leq 6\|U\|_1 \|H\|_* \leq C\|f\|_1 \|h\|_*$$

and completes the proof of (4).

Remark: The theorem above makes no mention of Carleson measures, but it is otherwise the same as the one given by Fefferman and Stein (1972), on page 145. The only difference is that we have carried out the two main steps, (3) and (4), using the correspondences developed in the last two sections to reduce these steps to results about martingales.

As a corollary of the duality theorem, we get the following result from BMO.

(5) If $\varphi \in BMO$, then φ can be written as $g_1 + \bar{g}_2 + c$, where c is a constant and $g_1, g_2 \in L^\infty$, and furthermore, this representation can be done in such a way that

$$\|g_1\|_\infty + \|g_2\|_\infty \leq C\|\varphi\|_*.$$

Proof The first conclusion is trivial. If $\varphi \in BMO$, then

$$f \to \int_{\partial D} (\text{Re} f) \varphi \, d\pi$$

defines a continuous linear functional on H_0^1, so it follows from the proof of the duality theorem that these are $g_1, g_2 \in L^\infty$ with

$$\int_{\partial D} (\text{Re} f) \varphi \, d\pi = \int_{\partial D} (\text{Re} f)(g_1 + \bar{g}_2) \, d\pi$$

for all $f \in H_0^2$, and taking test functions of the form $\mathscr{P}\psi$, where $\psi \in L^2$ and $\int \psi \, d\pi = 0$, shows that $\varphi - (g_1 + \bar{g}_2)$ is constant.

To prove the second conclusion, we observe that, given a $\varphi \in BMO$, the first equation in this proof defines a continuous linear functional Λ on H_0^1 with $\|\Lambda\|_1 = \sup\{|\Lambda f| : \|f\|_1 \leq 1\} \leq C\|\varphi\|_*$. Now H_0^1 is equivalent to a subspace of $L^1 \times L^1$, so it follows from the Hahn-Banach theorem that we can extend Λ to be a linear functional Γ on $L^1 \times L^1$ in such a way that

$$\|\Gamma\|_{1 \times 1} \leq C'\|\varphi\|_*,$$

where $\|\Gamma\|_{1 \times 1} = \sup\{|\Gamma(u_1, u_2)| : \|u_1\|_1 + \|u_2\|_1 \leq 1\}$ (the constant changes only because we change from the L^1 norm to the $L^1 \times L^1$ norm). Now any $\Gamma \in (L^1 \times L^1)^*$ comes from a pair $(q_1, q_2) \in L^\infty$ with

$$\|q_1\|_\infty, \|q_2\|_\infty \leq \|\Gamma\|_{L^1 \times L^1}$$

(consider functions of the form $(u_1, 0), (0, u_2)$), so the proof is complete.

It is easy to generalize the last proof to prove a result about martingales. To do this, we observe that if $A_1, \ldots, A_m$ are matrices without a common eigenvector in R^d, then Janson's theorem ((1) in Section 6.7) implies that if we let A_0 be the identity matrix, then the mapping $X \to (A_0 * X, \ldots, A_m * X)$ embeds $\mathcal{M}^1$ in $\mathcal{L}^1 \times \cdots \times \mathcal{L}^1$ in such a way that

(6) $$\|X\|_1 \leq C \sum_i \|A_i * X\|_{\mathcal{L}}$$

(recall that if $Y \in \mathcal{L}^1$, then $\|Y\|_{\mathcal{L}} = \|Y\|_{\mathcal{L}^1}$). If $\varphi(X) = E\langle X, Y \rangle$ defines a continuous linear function on $\mathcal{M}^1$, then (6) implies that

$$\psi(A_0 * X, \ldots, A_m * X) = \varphi(X)$$

defines a continuous linear functional on a subspace of $\mathcal{L}^1 \times \cdots \times \mathcal{L}^1$, which the Hahn-Banach theorem asserts can be extended to the whole space. Since $\mathcal{L}^1$ is isomorphic to $L^1(\mathcal{F}_\infty)$, we have $(\mathcal{L}^1)^* = \mathcal{M}^\infty$ and, hence, $(\mathcal{L}^1 \times \cdots \times \mathcal{L}^1)^* = \mathcal{M}^\infty \times \cdots \times \mathcal{M}^\infty$, that is, there are martingales $Y^0, \ldots, Y^m \in \mathcal{M}^\infty$ such that if $X^0, \ldots, X^m \in \mathcal{L}^1$, then

$$\psi(X^0, \ldots, X^m) = \sum_{i=0}^m E(X_\infty^i Y_\infty^i) = \sum_{i=0}^m E\langle X^i, Y^i \rangle_\infty,$$

and hence we have

$$E\langle X, Y \rangle = \varphi(X) = \psi(A_0 * X, \ldots, A_m * X) = \sum_{i=0}^m E\langle A_i * X, Y^i \rangle_\infty.$$

If

$$X_t = \int_0^t H_s \cdot dB_s \quad \text{and} \quad Y_t^i = \int_0^t K_s^i \cdot dB_s,$$

then

$$\langle A_i * X, Y^i \rangle_\infty = \int_0^\infty (A_i H_s, K_s^i) \, ds$$

$$= \int_0^\infty (H_s, A_i^T K_s^i) \, ds = \langle X, A_i^T * Y^i \rangle_\infty,$$

7.4 The Duality Theorem for H^1, Fefferman-Stein Decomposition

so we have

$$E\langle X, Y\rangle_\infty = \sum_{i=0}^m E\langle X, A_i^T * Y^i\rangle_\infty$$

and, since this equality holds for all $X \in \mathcal{M}^1$,

$$Y = \sum_{i=0}^m A_i^T * Y^i \quad Y^i \in \mathcal{M}^\infty.$$

Now in the arguments above we started with an arbitrary $Y \in \mathcal{BMO}$ so it follows from the last equation that we have:

(7) If $A_1, \ldots, A_m$ do not have a common eigenvector in R^d, then any $Y \in \mathcal{BMO}$ can be written as

$$Y^0 + \sum_{i=1}^m A_i^T * Y^i,$$

with $Y^i \in \mathcal{M}^\infty$ for $i = 0, 1, \ldots, m$, and furthermore this can be done in such a way that

$$\|Y^i\|_\infty \le C\|Y\|_*.$$

Since the proof of (7) uses the Hahn-Banach theorem, it does not tell us how to find the martingales $Y^0, \ldots, Y^m$. In simple cases, this can be done "by hand."

Example 1 Let $d = 2$, $A = \begin{pmatrix} 0 & 1 \\ -1 & 0 \end{pmatrix}$, and let $X_t = B_{t \wedge \tau}^1$ where $\tau = \inf\{t : |B_t \cdot \theta| = 1\}$. If we let $Y_t = B_{t \wedge \tau} \cdot \theta$ and pick a, b so that $(1, 0) = a\theta + bA\theta$, then

$$X_t = a\int_0^{t \wedge \tau} \theta \cdot dB_s + b \int_0^{t \wedge \tau} A\theta \cdot dB_s$$
$$= (aY_t) + A * (bY_t).$$

Finding a decomposition in general, however, seems to be quite a difficult problem. A. Uchiyama succeeded in doing this recently for d-adic martingales (1982a) and has generalized the construction to *BMO* functions (1982b), but I do not know how to prove the result for $\mathcal{BMO}$, and I invite the reader to consider this problem.

The key to deducing the duality theorem for H^p from its probabilistic analogue was the fact that the mapping M (mnemonic for *M*artingale) defined by

$$M : \varphi \to \mathscr{P}\varphi(B_{t \wedge \tau}) \quad t \ge 0$$

has $\|M\varphi\|_* \le C\|\varphi\|_*$. For some applications, we will need to know that the inverse N, defined by

$$N : X \to \varphi(e^{i\theta}) = E(X_\infty | B_\tau = e^{i\theta}),$$

is also continuous.

(8) There is a constant C such that, for all $X \in \mathscr{BMO}$,

$$\|NX\|_* \leq C\|X\|_*.$$

Proof From the duality theorem for H_0^1 and the open mapping theorem (see Dunford and Schwarz (1957), Chapter II, Section 2), it follows that

$$\|f\|_* \leq C \sup_{\substack{\varphi \in H_0^2 \\ \|\varphi\|_1 \leq 1}} \left|\int f\varphi \, d\pi\right|.$$

Since B_τ has a uniform distribution,

$$\int (NX)\varphi \, d\pi = E(E(X_\infty|B_\tau)\varphi(B_\tau))$$

$$= EE(X_\infty \varphi(B_\tau)|B_\tau) = E(X_\infty \varphi(B_\tau))$$

(all the expectations above exist, since $X \in \mathscr{M}^2$ and $\varphi \in H^2$). If we let $U_t = \mathscr{P}\varphi(B_{t \wedge \tau})$, then it follows from the martingale duality theorem that

$$|E(X_\infty \varphi(B_\tau))| \leq C\|X\|_* \|U\|_1 \leq C'\|X\|_*.$$

From (8) and duality, it follows that if N is extended in the obvious way it maps $\mathscr{M}^1$ continuously into H^1. Combining this result with a trivial result for $p > 1$ gives the following:

(9) If $X \in \mathscr{M}^1$ and $NX = \varphi$, then there is a unique analytic function f with $\operatorname{Re} f = \mathscr{P}\varphi$ and $\operatorname{Im} f(0) = 0$. If $p \geq 1$ and we denote this function by FX, then there is a constant $C < \infty$ that depends only on p such that

$$\|FX\|_p \leq C\|X\|_p.$$

Proof For $p > 1$, this result is trivial, since

$$\|FX\|_p \leq C\|\operatorname{Re} FX\|_{L^p(\partial D)}$$

and

$$\|\operatorname{Re} FX\|_{L^p(\partial D)}^p = E|E(X_\infty|B_\tau)|^p$$
$$\leq EE(|X_\infty|^p|B_\tau)$$
$$= E|X_\infty|^p$$
$$\leq E|X^*|^p$$
$$= \|X\|_p^p.$$

To prove the result for $p = 1$, we observe that if $X \in \mathscr{M}_0^2$, then $FX \in H_0^2$, so the duality result and the Hahn-Banach theorem imply that

$$\|f\|_1 \leq C \sup_{\|\varphi\|_*=1} \int (\operatorname{Re} f)\varphi \, d\pi.$$

If we let $u = \mathscr{P}\varphi$, then we have

$$\int (\operatorname{Re} FX)\varphi \, d\pi = E(E(X_\infty | B_\tau) u(B_\tau))$$
$$= EE(X_\infty U_\infty | B_\tau)$$
$$= E(X_\infty U_\infty)$$
$$\leq C \|X\|_1 \|U\|_*$$
$$\leq C' \|X\|_1,$$

so it follows that $\|f\|_1 \leq C\|X\|_1$ for all $X \in \mathscr{M}^2$. Since $\mathscr{M}^2$ is dense in $\mathscr{M}^1$, the desired result follows.

Remark: The last result is false when $p < 1$. It is known that there are nontrivial continuous linear functionals on H^p for $p < 1$ (see Duren, Romberg, and Shields (1969)). In Section 7.8, we will show that there are no nontrivial continuous linear functionals on $\mathscr{M}^p$, $p < 1$, so $X \to FX$ cannot be continuous.

Note: We learned the results on M and N, (8) and (9) above, from Varopoulos (1979), who attributes the results to Maurey.

Exercise 1 Sharpen (3) by showing that if $\varphi \in BMO$ and $\bar{\varphi}$ is the conjugate function, then $\bar{\varphi} \in BMO$ and

$$\langle\!\langle \bar{\varphi} \rangle\!\rangle_* = \langle\!\langle \varphi \rangle\!\rangle_*, \quad \|\bar{\varphi}\|_* \leq C \|\varphi\|_*.$$

As a corollary of the last result, we see that if $\varphi \in L^\infty$, then $\bar{\varphi} \in BMO$, a result first proved independently by Spanne (1966) and Stein (1967). The function $f(z) = i \log z$ shows that we may have $\varphi \in L^\infty$ but $\bar{\varphi} \notin L^\infty$.

7.5 Examples of Martingales in $\mathscr{BMO}$

In Section 7.1, we defined the $\mathscr{BMO}$ norm $\|X\|_*$ of a martingale to be the smallest number c such that for all stopping times T

(1) $\qquad E(|X_\infty - X_T|) \leq c P(T < \infty).$

In this section, we will try to explain what kind of martingales are in $\mathscr{BMO}$ by giving four examples and describing some of their properties.

Example 1 If $X \in \mathscr{M}^\infty = \{X : X^* \in L^\infty\}$, then it is immediate that $X \in \mathscr{BMO}$ and $\|X\|_* \leq 2\|X\|_\infty$. With a little thought, we can replace the 2 by a 1:

$$(E|X_\infty - X_T|)^2 \leq E(X_\infty - X_T)^2$$
$$= EX_\infty^2 - EX_T^2$$
$$\leq \|X\|_\infty^2.$$

There are also unbounded martingales in $\mathcal{BMO}$. Perhaps the simplest example is the following.

Example 2 Let $X_t = B_{t \wedge 1}$. If T is a stopping time, then $X_\infty - X_T = B_1 - B_{T \wedge 1}$ is independent of $\mathcal{F}_T$ and has a normal distribution with mean 0 and variance $(1-T)^+$. Since $B_c \stackrel{d}{=} c^{1/2} B_1$ and

$$E|B_1| = 2 \int_0^\infty (2\pi)^{-1/2} x e^{-x^2/2} \, dx = (2/\pi)^{1/2},$$

it follows that

$$E(|X_\infty - X_T| | \mathcal{F}_T) \le (2/\pi)^{1/2} P(T < \infty)$$

and, taking $T \equiv 0$, that $\|X\|_* = (2/\pi)^{1/2}$.

Once you get started, it is easy to use Brownian motion to construct numerous examples of unbounded martingales in $\mathcal{BMO}$. An obvious modification of the last example is to let S be a random variable that is independent of B_t, $t \ge 0$, and has an exponential distribution $P(S > t) = e^{-\lambda t}$. If we let $X_t = B_{S \wedge t}$, then the lack-of-memory property, $P(S > t + u | S > t) = P(S > u)$, of the exponential implies that X is in $\mathcal{BMO}$. Another possibility is to introduce the stopping times $R_n = \inf\{t > R_{n-1} : |B(t) - B(R_{n-1})| > 1\}$ and let $X_t = B(t \wedge R_N)$, where N is an independent random variable with a geometric distribution $P(N > n) = (1-p)^n$, $n = 0, 1, 2, \ldots$. The next example is a variation of this—N depends on B_t, $t \le T_N$, and is chosen to try to produce a large maximum.

Example 3 Let $R_0 = 0$, and for $n \ge 1$, let $R_n = \inf\{t > R_{n-1} : |B(t) - B(R_{n-1})| > 1\}$. Let $N = \inf\{n : B(R_n) - B(R_{n-1}) = -1\}$ and let $X_t = B(t \wedge R_N)$. See Figure 7.2 for a sample path. It is easy to get a bound on the $\mathcal{BMO}$ norm of X. If T is a stopping time, then on $\{R_k \le T < R_{k+1}, N > k\}$, $X_T = k + a$ for some $a \in (-1, 1)$, and $P(N > k + j | \mathcal{F}_T) \le (1/2)^{j-1}$ (the worst case is $a \to 1$). On the other hand, a look at Figure 7.2 shows that $|X_\infty - X_T| \le N - k$ (the worst case is $a \to -1$), so combining the estimates gives that

$$E(|X_\infty - X_T| | \mathcal{F}_T) \le E(N - k | \mathcal{F}_t) \le 2,$$

and we have $\|X\|_* \le 2$. A look at the parenthetical remarks above shows that this is not the right answer, but we are not far from it. It is easy to compute $f(a) = E(|X_\infty - X_T| | X_T = a, T < R_1)$ and maximize to find $\|X\|_* = 9/8$ (details of this computation are sketched in Exercise 1 at the end of this section).

We have taken the trouble to compute the $\mathcal{BMO}$ norm of the last example so that we can check the accuracy of some of the inequalities below. By abstracting the construction above, you can produce many examples of martingales in $\mathcal{BMO}$. The next construction is so general that we can think of it as giving the typical element of $\mathcal{BMO}$ (see the proofs of (1) and (5) in the next section).

7.5 Examples of Martingales in $\mathcal{BMO}$

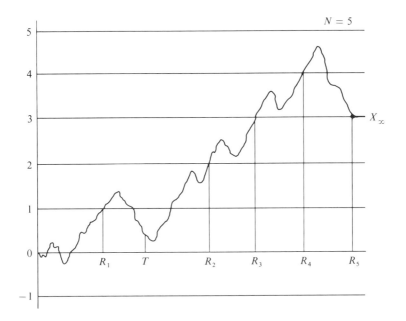

Figure 7.2

Example 4 Let T_n be an increasing sequence of stopping times with $T_0 = 0$ and $|B(t) - B(T_{n-1})| \leq A$ for $t \in [T_{n-1}, T_n]$, $n \geq 1$. Let N be a stopping time for $\mathcal{F}(T_n)$, that is, $\{N \leq n\} \in \mathcal{F}(T_n)$, and suppose that for all $n \geq 1$, $P(N > n | \mathcal{F}(T_{n-1})) \leq \theta < 1$ on $\{N > n - 1\}$. If we let $X_t = B(t \wedge T_N)$, then $X \in \mathcal{BMO}$ and as the reader can easily show

$$\|X\|_* \leq A(1 + (1 - \theta)^{-1}).$$

The four examples above should give you an idea of what type of martingales are in $\mathcal{BMO}$ and, hopefully, make the results in the next two sections more obvious. To lead you in the direction of the John-Nirenberg inequality ((1) in the next section), we will now compute $P(X^* > \lambda)$ for the three unbounded examples.

Example 2. $\quad P(\sup X_t > \lambda) = 2P(B_1 > \lambda) = 2 \int_\lambda^\infty \frac{1}{\sqrt{2\pi}} e^{-y^2/2} \, dy.$

$$P(X^* > \lambda) \leq 2P(\sup_t X_t > \lambda) = \frac{4}{\sqrt{2\pi}} e^{-\lambda^2/2} \int_0^\infty e^{-\lambda x} e^{-x^2/2} \, dx$$

$$\sim \frac{4}{\sqrt{2\pi}} e^{-\lambda^2/2} \lambda^{-1}. \text{ As } \lambda \to \infty.$$

Example 3. If n is a positive integer, then

$$P(X^* > n) = (1/2)^n.$$

Example 4. If n is a positive integer, then
$$P(X^* > nA) \le P(N > n) \le (1-\theta)^n.$$

Exercise 1 Compute the $\mathscr{BMO}$ norm of the martingale in Example 3.

(a) By the strong Markov property and independent increments of Brownian motion, it suffices to consider the case $X_T = a \in (-1, 1)$ a.s. on $\{T < \infty\}$.

(b) When $X_T = a$, the distribution of $|X_\infty - X_T|$ is given by the following table:

X_∞	$\lvert X_\infty - X_T \rvert$	Probability
-1	$a+1$	$\dfrac{(1-a)}{2}$
0	$\lvert a \rvert$	$\dfrac{(a+1)}{2} \cdot \dfrac{1}{2}$
n	$n-a$	$\dfrac{(a+1)}{2} \cdot \left(\dfrac{1}{2}\right)^{n+1}$

Summing over the possibilities, we get
$$\frac{E(|X_\infty - X_T|)}{P(T < \infty)} = \begin{cases} (a+1)(2-a)/2 & \text{if } a \ge 0 \\ (a+1)(1-a) & \text{if } a \le 0. \end{cases}$$

The maximum value occurs for $a = 1/2$, and here the value is $9/8$.

7.6 The John-Nirenberg Inequality

In this section, we will prove a classical result on *BMO* functions due to John and Nirenberg (1961), which was discovered almost a decade before it was known that $(H^1)^* = BMO$. Following our usual inclination, we will prove the probabilistic analogue first and then use this result to prove the analytical result.

(1) There is a constant $C \in (0, \infty)$ such that if $\|X\|_* \le 1$, then
$$P(X^* > \lambda) \le Ce^{-\lambda/e}.$$

Proof of (1) The first step is to prove:

(2) If $Y \in \mathscr{BMO}$ and $S \ge T$ are stopping times, then
$$E|Y_S - Y_T| \le \|Y\|_* P(T < \infty).$$

Proof If T is a stopping time, then $Z_t = Y_t - Y_{t \wedge T}$ is a martingale and $|Z_t|$ is a submartingale which is dominated by an integrable r.v. (recall that $Y \in \mathscr{M}^1$),

7.6 The John-Nirenberg Inequality

so the optional stopping theorem implies that

$$E|Y_S - Y_T| = E|Z_S| \leq E|Z_\infty| = E|Y_\infty - Y_T|,$$

proving the result, since, by definition of $\|Y\|_*$,

$$E|Y_\infty - Y_T| \leq \|Y\|_* P(T < \infty).$$

With (2) established, it is easy to prove (1). Let $a > 1$ and define a sequence of stopping times $R_0 = 0$ and, for $n \geq 1$, $R_n = \inf\{t > R_{n-1} : |X_t - X(R_{n-1})| > a\}$. We have assumed that $\|X\|_* \leq 1$, so applying the lemma with $S = R_{i+1}$ and $T = R_i$ gives

$$P(R_i < \infty) \geq E|X_{R_{i+1}} - X_{R_i}| \geq a P(R_{i+1} < \infty).$$

Therefore, $P(R_{i+1} < \infty | R_i < \infty) \leq 1/a$ and, by induction, $P(R_k < \infty) \leq (1/a)^k P(R_0 < \infty)$. Let n be an integer and let $an \leq \lambda < a(n+1)$. Then

$$P(X^* > \lambda) \leq P(X^* > an) \leq P(R_n < \infty) \leq (1/a)^n P(R_0 < \infty)$$
$$\leq a(1/a)^{\lambda/a} = a e^{-\lambda (\log a)/a}.$$

Setting $a = e$, to maximize $(\log a)/a$, gives (1).

Remark: To summarize the proof in loose language: The amount of wiggling a martingale in $\mathscr{BMO}$ can do has a geometric upper bound because of (2), and, hence, even if all the movement occurs in the same direction, a large maximum will not be produced.

To translate (1) into a result about *BMO* functions, we recall that in Section 7.3 we showed that if $\varphi \in BMO$ and we let $u = \mathscr{P}\varphi$ and $U_t = u(B_{t \wedge \tau})$, then $U \in \mathscr{BMO}$ and there are constants $c, C \in (0, \infty)$ such that

(∗) $$c\|\varphi\|_* \leq \|U\|_* \leq C\|\varphi\|_*.$$

Combining the last observation with (1) gives:

(3) *The John-Nirenberg Inequality*. There are constants $C, \gamma \in (0, \infty)$ such that if $\|\varphi\|_* \leq 1$ and $\int \varphi \, d\pi = 0$, then

$$|\{\theta : |\varphi(e^{i\theta})| > \lambda\}| \leq C e^{-\gamma \lambda}.$$

Proof (∗) implies that $\|U\|_* \leq C$, so

$$|\{\theta : |\varphi(e^{i\theta})| > \lambda\}| \leq 2\pi P(U^* > \lambda) \leq C' e^{-\lambda/Ce}.$$

It is easy to improve (3) to the usual result.

(4) Let C, γ be the constants in (3). If $\|\varphi\|_* \leq 1$, then for all intervals I,

$$|\{\theta \in I : |\varphi(e^{i\theta}) - \varphi_I| > \lambda\}| \leq C e^{-\gamma \lambda} |I|.$$

Proof If $I = (a - h\pi, a + h\pi)$, then $\psi(e^{i\theta}) = \varphi(\exp(i(a + \theta h))) - \varphi_I$ is in *BMO* with $\|\psi\|_* \leq 1$ and $\int \psi = 0$, so (4) follows from applying (3) to ψ.

The conclusion of (1) can be improved in a similar way.

(5) Let C be the constant in (1). If $\|X\|_* \leq 1$, then for all stopping times T,
$$P\left(\sup_{t \geq T}|X_t - X_T| > \lambda\right) \leq Ce^{-\lambda e}P(T < \infty).$$

Proof This could be proved in the same manner as (4), but it seems easier to observe that repeating the proof of (1) with $R_0 = T$ proves (5).

From (5), we see that the definition of $\mathscr{BMO}$ as
$$\sup_T E(|X_\infty - X_T||T < \infty) < \infty$$

implies the much stronger conclusion that

(6) For all $\alpha < (e\|X\|_*)^{-1}$,
$$\sup_T E(e^{\alpha|X_\infty - X_T|}|T < \infty) < \infty$$

(in both cases, the supremum is over all stopping times).

From (6), it follows that $\mathscr{BMO}$ can be defined as the set of all martingales with
$$\sup_T E(|X_\infty - X_T|^p|T < \infty) < \infty$$

for some or all $p \geq 1$. In particular, when $p = 2$, $\|X\|_* < \infty$ if and only if $\langle\!\langle X \rangle\!\rangle_* < \infty$. By integrating (5), we get a sharper result.

(7) There is a constant $C < \infty$ such that
$$\|X\|_* \leq \langle\!\langle X \rangle\!\rangle_* \leq C\|X\|_*.$$

Proof We proved the left-hand inequality in Section 7.2. To prove the right-hand inequality, we observe that it follows from (5) that if $\|X\|_* \leq 1$, then
$$E(X_\infty - X_T)^2 \leq \int_0^\infty 2\lambda P\left(\sup_{t \geq T}|X_t - X_T| > \lambda\right) d\lambda$$
$$\leq \left(c\int_0^\infty 2\lambda e^{-\lambda/e} d\lambda\right) P(T < \infty),$$

proving (7).

As a corollary of the John-Nirenberg inequality, we get the following striking result due to Zygmund (1929):

(8) If φ is continuous on ∂D, then its conjugate function $\bar{\varphi}$ has
$$\int \exp(\lambda|\bar{\varphi}(\theta)|) d\theta < \infty$$

for all $\lambda < \infty$.

Proof Since every continuous φ is the uniform limit of trigonometric polynomials, we can, for any $\varepsilon > 0$, write $\varphi = \varphi_1 + \varphi_2$, where φ_1 is a trigonometric polynomial (and hence has $\bar{\varphi}_1 \in L^\infty$) and φ_2 has $\|\varphi_2\|_* \le \|\varphi_2\|_\infty < \varepsilon$, so the desired result follows from (6).

On page 253 of Zygmund (1959), you can find a continuous φ whose conjugate $\bar{\varphi}$ is unbounded:

$$\varphi(e^{i\theta}) = \sum_{n=2}^\infty \frac{\sin n\theta}{n \log n} \quad \bar{\varphi}(e^{i\theta}) = -\sum_{n=2}^\infty \frac{\cos n\theta}{n \log n}.$$

Exercise 1 *Calderon's Proof of the John-Nirenberg Inequality.* Let

$$a(x) = \sup\{P(X^* > x) : \|X\|_* \le 1\}.$$

By stopping at the first time, $|X_t| > x$, we see that $a(x) \le 1/x$ and $a(x)a(y) \ge a(x+y)$. Taking logs gives

$$\log a(x) + \log a(y) \ge \log a(x+y),$$

and one easily concludes from this that

$$\limsup_{n \to \infty} \frac{1}{n} \log a(n) = \inf_{m \ge 1} \frac{1}{m} \log a(m) < 0,$$

proving the result with a constant that is not explicit, but is trivially the best possible.

7.7 The Garnett-Jones Theorem

In this section, we will determine which martingales in $\mathscr{BMO}$ are almost bounded, or, to be precise, we will find the closure of $\mathscr{M}^\infty$ in $\mathscr{BMO}$. The solution to this problem is again the probabilistic analogue of a previously proven analytical result, and the analytical result can be recovered from the probabilistic one.

To state the result, we first need some notation. Let $\alpha_0(X)$ be the supremum of the set of all α so that

$$\sup_T E(e^{\alpha|X_\infty - X_T|} | T < \infty) < \infty$$

where the sup is taken over all stopping times (by (6) in Section 7.6, $\alpha_0(X) \ge (e\|X\|_*)^{-1} > 0$). Intuitively, $\alpha_0(X)$ is the exponential rate at which $P(|X_\infty - X_T| > \lambda)$ goes to zero for the worst choice of T. Given this interpretation, it should not be surprising that $\alpha_0(X)$ can be used to measure how well an X in $\mathscr{BMO}$ can be approximated by a $Y \in \mathscr{M}^\infty$.

(1) There are constants $c, C \in (0, \infty)$ such that

$$\frac{c}{\alpha_0(X)} \le \inf_{Y \in \mathscr{M}^\infty} \|X - Y\|_* \le \frac{C}{\alpha_0(X)},$$

and, consequently, the closure of $\mathscr{M}^\infty$ in $\mathscr{BMO}$ is $\{X : \alpha_0(X) = \infty\}$.

Remark: This result and the proof we will give below are due to Varopoulos (1979).

Proof The inequality on the left is an easy consequence of the John-Nirenberg inequality. If $Y \in \mathcal{M}^\infty$ and $\|X - Y\|_* = a$, it follows from (5) in Section 7.6 that if $Z = X - Y$ and $\alpha > 1/e$, then

$$E(e^{\alpha|Z_\infty - Z_T|/a}) \leq C_\alpha P(T < \infty),$$

so

$$P(|Z_\infty - Z_T| > \lambda) \leq C_\alpha e^{-\lambda \alpha / a}.$$

Let $\lambda_0 = \|Y\|_\infty$. If $K > 2$ and $\lambda > K\lambda_0$, then

$$P(|X_\infty - X_T| > \lambda) \leq P(|Z_\infty - Z_T| > \lambda - 2\lambda_0)$$
$$\leq P\left(|Z_\infty - Z_T| > \left(1 - \frac{2}{K}\right)\lambda\right)$$
$$\leq C_\alpha e^{-\alpha\lambda(K-2)/Ka} P(T < \infty).$$

At this point, we have shown that $\alpha_0(X) \geq \alpha(K-2)/Ka$. Letting $\alpha \to 1/e$ and $K \to \infty$ gives that $\alpha_0(X) \geq 1/ea$, so the left-hand inequality in (1) holds with $c = 1/e$ (the constant is inherited from the John-Nirenberg inequality).

To prove the other inequality is more difficult. Given a large α and a martingale $X \in \mathcal{BMO}$ with $\alpha_0(X) > \alpha$, we need to construct $Z \in \mathcal{M}^\infty$ with $\|X - Z\|_\infty \leq C/\alpha$. This construction is accomplished in two stages. First, we construct an approximating martingale of the form described in Example 4 of Section 7.5, but we are forced to take λ large to make $\theta < 1$, and we end up with a $\mathcal{BMO}$ norm that is too large. Then we use an ingenious construction, due to Varopoulos, to introduce a sequence of stopping times that smooths the transitions between the times constructed in our first attempt and reduces the $\mathcal{BMO}$ norm to the right size.

The first part of the construction is straightforward. We let λ be fixed and define R_n inductively by letting $R_0 = 0$ and $R_n = \inf\{t > R_{n-1} : |X(t) - X(R_{n-1})| > \lambda\}$ for $n \geq 1$. If we let $\xi_n = X(R_n) - X(R_{n-1})$ on $\{R_n < \infty\}$ and $Z = \sum_{n=1}^\infty \xi_n 1_{(R_n < \infty)}$, then we can construct a martingale by letting $Z_t = E(Z|\mathcal{F}_t)$. Z_t will differ from X_t by a martingale that is bounded. To see this, observe that when $R_n \leq t \leq R_{n-1}$,

$$X_t - Z_t = X_t - E\left(\sum_{m=1}^\infty \xi_m 1_{(R_m < \infty)} \Big| \mathcal{F}_t\right)$$
$$= X_t - X(R_n) + E\left(\sum_{m=n+1}^\infty \xi_m 1_{(R_m < \infty)} \Big| \mathcal{F}_t\right).$$

From the definition of the R_n, it follows that $|X_t - X(R_n)| \leq \lambda$. To get a bound on the other term, observe that if

$$\sup_T E(e^{\alpha|X_\infty - X_T|}|T < \infty) \leq K,$$

7.7 The Garnett-Jones Theorem

then $P(R_{m+1} < \infty | R_m < \infty) \leq K/e^{\alpha\lambda}$. Therefore, if $Ke^{-\alpha\lambda} < 1$, then

(*) $\quad E\left(\sum_{m=n+1}^{\infty} \xi_m 1_{(R_m < \infty)} \Big| \mathscr{F}_t\right) \leq \lambda(1 - Ke^{-\alpha\lambda})^{-1}$

and we have $Y_t = X_t - Z_t \in \mathscr{M}^\infty$.

$Y_t = X_t - Z_t$ is *not* the martingale in $\mathscr{M}^\infty$ that we want. From Example 4 in Section 7.5, we see that the $\mathscr{BMO}$ norm of Z satisfies $\|Z\|_* \leq \lambda(1 + (1 - Ke^{-\alpha\lambda})^{-1})$, but this is too big because we have no control over K and, hence, over the choice of λ. To circumvent this difficulty, we use the construction referred to above. Pick $\theta < 1$ and then λ so large that $Ke^{-\alpha\lambda} < e^{-\alpha\lambda\theta}$ and (for convenience) $\lambda\alpha = M$ is an integer. Let $\gamma = e^{-\theta}$, let $S_0 = 0$, and, for $n \geq 0$, $1 \leq j \leq M$, let

$$S_{nM+j} = \inf\{t : P(R_{n+1} < \infty | \mathscr{F}_t) \geq \gamma^{M-j}\}.$$

Since $P(R_{n+1} < \infty | \mathscr{F}(R_n)) \leq Ke^{-\alpha\lambda} < \gamma^M$, we have

$$R_n \leq S_{nM+1} \leq \cdots \leq S_{(n+1)M} \leq R_{n+1}.$$

Let

$$U_n = \frac{1}{M} \sum_{j=1}^{M} 1_{(S_{nM+j} < \infty)}.$$

U_n is a staircase that allows us to climb from $\{U_n = 0\} = \{S_{nM+1} = \infty\}$ to $\{U_n = 1\} = \{S_{(n+1)M} < \infty\}$. Let

$$Z' = \sum_{m=1}^{\infty} \xi_m U_m$$

$$Z'_t = E(Z'|\mathscr{F}_t).$$

If $R_n \leq t < R_{n+1}$, then

$$|Z_t - Z'_t| = \left| E\left(\sum_{m=n}^{\infty} \xi_m (1_{(R_m < \infty)} - U_m) \Big| \mathscr{F}_t\right)\right|$$

$$\leq \lambda + \left| E\left(\sum_{m=n+1}^{\infty} \xi_m (1_{(R_m < \infty)} - U_m) \Big| \mathscr{F}_t\right)\right|$$

$$\leq \lambda + \lambda(1 - Ke^{-\alpha\lambda})^{-1}$$

by the arguments used to prove (*) (observe that $|\xi_m(1_{(R_m < \infty)} - U_m)| \leq \lambda$), so the triangle inequality implies that $X_t - Z'_t \in \mathscr{M}^\infty$. To estimate the $\mathscr{BMO}$ norm of Z'_t, we observe that since the Brownian filtration admits only continuous martingales, $P(R_{n+1} < \infty | \mathscr{F}(S_{nM+j})) = \gamma^{m-j}$ on $\{S_{nM+j} < \infty\}$, and it follows that $P(S_{k+1} < \infty) \leq \gamma$. If we let $m(k)$ be the integer part of k/M, then we can write

$$Z' = \sum_{k=1}^{\infty} \xi_{m(k)} \cdot \frac{1}{M} \cdot 1_{(S_k < \infty)},$$

and we see from Example 4 in Section 7.5 that for all $\theta < 1$,

$$\|Z'\|_* \le \frac{\lambda}{M}(1 + (1 - e^{-\theta})^{-1}).$$

Letting $\theta \uparrow 1$ proves the desired result with $C = (2e - 1)/(e - 1) = 2.42$.

Example 1 Let X_t be the martingale in Example 3 of Section 7.5. It is easy to see that $\alpha_0(X) = \log 2 = .693$, so (1) gives

$$.5307 = \frac{1}{e(.693)} \le \inf_{Z \in \mathcal{M}^\infty} \|X - Z\|_* \le \frac{2.42}{.693} = 3.49$$

The upper bound is hardly informative, since $\|X\|_* = 9/8 = 1.125$, but the computation suggests the following question I could not answer.

Problem In Example 3, do we have

$$\inf_{Z \in \mathcal{M}^\infty} \|X - Z\|_* = \|X\|_*?$$

An even simpler question is: Can you construct an example with this property?

With (1) established, we turn our attention to proving the analytical result. Since this is a thankless job and requires more work than a direct analytical proof (see Garnett and Jones (1982)), we just sketch the details.

(2) *The Garnett-Jones Theorem.* Let $\alpha_0(\varphi)$ be the supremum of the set of α that has

$$\sup_I \frac{1}{|I|} \int_I \exp(\alpha|\varphi - \varphi_I|)\, d\theta < \infty,$$

where

$$\varphi_I = \int_I \varphi\, d\theta.$$

There are constants $c, C \in (0, \infty)$ such that

$$\frac{c}{\alpha_0(\varphi)} \le \inf_{\psi \in L^\infty} \|\varphi - \psi\|_* \le \frac{C}{\alpha_0(\varphi)}.$$

Proof As in the proof of (1), the left-hand inequality follows easily from the John-Nirenberg inequality, so we leave the details as an exercise for the reader and turn now to the proof of the more difficult right-hand inequality. From the developments above, the plan of attack should be clear. We pass from φ to $U_t = \mathcal{P}\varphi(B_{t\wedge\tau}) \equiv M\varphi$, decompose $U_t = X_t^1 + X_t^2$ according to the construction in the proof of (1), and then let $\psi_j(e^{i\theta}) = E(X_\infty^j|B_\tau = e^{i\theta}) \equiv NX^j$. From results about the maps M and N defined in Section 7.4, we see that to prove (2), it suffices to show that

(3) $\qquad \alpha_0(\varphi) \le C\alpha_0(M\varphi).$

for then it will follow that

$$\|\varphi - \psi_1\|_* \leq C \|U - X^i\|_* \leq \frac{C'}{\alpha_0(U)} \leq \frac{C''}{\alpha_0(\varphi)}.$$

The proof of (3), however, is tedious and very similar to the proof of (7) in Section 7.3, so the reader is referred to Varopoulos (1979) for the details.

It is inevitable when we use the correspondence between BMO and $\mathcal{BMO}$ (or H^p and $\mathcal{M}^p$) that the resulting inequalities will not be very precise. In the case of (2), an analytical proof gives a much stronger result. As Garnett and Jones (1978) remark, if we norm BMO by

$$\|f\|_{**} = \inf\{\|\varphi\|_\infty + \|\psi\|_\infty : f - \varphi - \tilde{\psi} \text{ is constant}\}$$

(which we know from Section 7.4 to be an equivalent norm), then it follows from results of Helson-Szegö (1960) and Hunt, Muckenhoupt, and Wheeden (1973) that

$$\inf_{g \in L^\infty} \|\varphi - g\|_{**} = \frac{\pi}{2} \frac{1}{\alpha_0(\varphi)}$$

(see Garnett (1980), Section 6). Varopoulos (1980) has shown (see his Theorem 4.2) that there is an analogous result for $\mathcal{BMO}$,

$$\inf_{Y \in \mathcal{M}^\infty} \langle\langle X - Y \rangle\rangle_* \leq \frac{\pi}{2} \frac{1}{\alpha_0(X)},$$

and that the constant is the best possible. The reader is referred to his paper for the details of this and many other interesting developments (including a probabilistic proof of the Corona theorem).

7.8 A Disappointing Look at $(\mathcal{M}^p)^*$ When $p < 1$

In this section, we will consider the martingale spaces $\mathcal{M}^p$, $0 < p < 1$, and concentrate in particular on identifying the dual space $(\mathcal{M}^p)^*$. The motivation for doing this is the hope of using probability to identify the dual of the corresponding H^p space, which in the case $0 < p < 1$ was first found by Duren, Romberg, and Shields (1969) (see Duren (1970), pages 115–118). In this section, we will show that this approach is not feasible—there are no nontrivial continuous linear functionals on $\mathcal{M}^p$, $0 < p < 1$. This result, attributed to P. A. Meyer in a footnote in Getoor and Sharpe (1972), is in sharp contrast with the results for dyadic martingales obtained by Herz (1974b). In the dyadic case, a martingale is in $(\mathcal{M}^p)^*$ if and only if there is a constant C such that, for all stopping times T, $E(|Y_\infty - Y_T||\mathcal{F}_T) \leq CP(T < \infty)^{1/p}$. In this section, we use the techniques of Section 7.1 to show that the same result is true for the Brownian filtration and that, unfortunately, no nonconstant martingale has this property.

The first step in our study of the martingales in $\mathcal{M}^p$ is to obtain an atomic decomposition. The arguments given in Section 7.1 generalize with very little work to the case $p < 1$.

(1) A martingale $A \in \mathcal{M}^p$ is said to be an atom if there is a stopping time T such that

(i) $A_t = 0$ if $t \leq T$
(ii) $|A_t|^p \leq P(T < \infty)^{-1}$ for all t.

If we let $d_p(X) = E(X^*)^p$, then it follows from the definition that

(2) If A is an atom, $d_p(A) = E(A^*)^p \leq 1$.

(3) If A^n is a sequence of atoms and $\sum_n |c_n|^p < \infty$, then as $N \to \infty$, $X^N = \sum_{n=-N}^{N} c_n A^n$ converges to a limit $X \in \mathcal{M}^p$, and $d_p(X) \leq \sum_n |c_n|^p$.

The proof of (4) in Section 7.1 leads to the following:

(4) For all $X \in \mathcal{M}^p$, there is a sequence of atoms A^n, $n \in \mathbb{Z}$, and a sequence of constants c_n, $n \in \mathbb{Z}$, such that

(i) as $N \to \infty$, $\sum_{n=-N}^{N} c_n A^n \to X$ in $\mathcal{M}^p$

and

(ii) $\sum_n |c_n|^p \leq 2(2^{1/p} + 1)^p d_p(X)$.

The proof is the same as the proof in Section 7.1. Let

$$T_n = \inf\{t : |X_t| > 2^{n/p}\}$$
$$A^n = (X(t \wedge T_{n+1}) - X(t \wedge T_n))/c_n$$
$$c_n = (2^{1/p} + 1) 2^{n/p} P(T < \infty)^{1/p}.$$

It follows from computations in Section 7.1 that (i) holds. To check (ii), we observe that

$$d_p\left(\sum_n c_n A^n\right) \leq \sum_n |c_n|^p$$
$$= (2^{1/p} + 1)^p \sum_n 2^n P(T < \infty)$$
$$= (2^{1/p} + 1)^p \sum_n 2^n P((X^*)^p > 2^n)$$
$$\leq 2(2^{1/p} + 1) E(X^*)^p.$$

Remark: In the argument above, the fact that $p \leq 1$ is important. If $p > 1$, the triangle inequality gives that

$$\left\|\sum_n c_n A^n\right\|_p \leq \sum_n |c_n| = (2^{1/p} + 1) \sum_n (2^n P((X^*)^p > 2^n))^{1/p},$$

and the right-hand side $> (2^{1/p} + 1)(\sum_n 2^n P((X^*)^p > 2^n))^{1/p}$. Since the spaces $\mathcal{M}^p$, $p > 1$, are easy to handle directly, we leave it to the reader to figure out how to define atoms for $p > 1$.

7.8 A Disappointing Look at $(\mathcal{M}^p)^*$ When $p < 1$

With the atomic decomposition established, it is easy to find the dual of $\mathcal{M}^p$. From results in Section 7.2, if φ is a continuous linear functional on $\mathcal{M}^p$, then there is a $Y \in \mathcal{M}^2$ such that if $X \in \mathcal{M}^2$, $\varphi(X) = EX_\infty Y_\infty$. To see what properties Y must have, we follow (7) of Section 7.1. Let T be a stopping time, let $Z_\infty = \operatorname{sgn}(Y_\infty - Y_T)$, and let $Z_t = E(Z_\infty | \mathscr{F}_t)$. Since $|Z_t| \leq 1$, $X_t = (Z_t - Z_{T \wedge t})/2P(T < \infty)^{1/p}$ is an atom. By a computation in Section 7.1, $E|Y_\infty - Y_T| = E(Z_\infty(Y_\infty - Y_T)) = E((Z_\infty - Z_T)Y_\infty)$, so

$$E(X_\infty Y_\infty) = \frac{E|Y_\infty - Y_T|}{2P(T < \infty)^{1/p}}.$$

Taking the supremum over all stopping times, we see that there must be a constant c such that for all T,

(5) $\quad E|Y_\infty - Y_T| \leq cP(T < \infty)^{1/p}.$

If $p = 1$, this inequality reduces to (6) of Section 7.1, so on the surface everything is fine. We have a family of spaces, which we might call $\mathscr{B}_p$, $0 < p \leq 1$, that generalize $\mathscr{BMO}$ and have $(\mathcal{M}_p)^* \subset \mathscr{B}_p$. Unfortunately condition (5) is too strong when $p < 1$:

(6) If $p < 1$, then the only martingales $Y \in \mathcal{M}^2$ that satisfy (5) are constant.

Proof The idea of the proof is simple. Suppose that $C = 1$. If we can find a T such that

$$E(|Y_\infty - Y_T| | \mathscr{F}_T) \not\equiv 0$$

and if we can pick an $\varepsilon > 0$ and define another stopping time

$$T' = \begin{cases} T & \text{if } E(|Y_\infty - Y_T| | \mathscr{F}_T) > \varepsilon \\ \infty & \text{otherwise} \end{cases}$$

with $P(T' < \infty) = \delta$, a small number chosen to have $\delta^{1/p-1} < \varepsilon$, then we are done, since if (5) holds with $c = 1$, we have

$$P(T' < \infty)^{1/p} \geq E(|Y_\infty - Y_{T'}|) \geq \varepsilon P(T' < \infty),$$

which implies that $\delta^{1/p-1} \geq \varepsilon$, a contradiction.

The technical problems we now face are (a) to find a stopping time with $E(|Y_\infty - Y_T| | \mathscr{F}_T) \not\equiv 0$ (easy) and (b) to shrink $P(T < \infty)$ to an appropriate size. To solve these problems, we observe that since every continuous local martingale is a time change of Brownian motion, it suffices to prove the result when $Y_t = B_{t \wedge \tau}$ and τ is a stopping time. Let $T_a = \inf\{t : |B_t| > a\}$. Since $Y \in \mathcal{M}^2$ and $|Y_\infty - Y_t| \leq 2Y^*$, it follows from Theorem 9.5.4 in Chung (1974) that as $a \to 0$,

$$E(|Y_\infty - Y(T_a)| | \mathscr{F}(T_a)) \to E(|Y_\infty| | \bigcap_{a > 0} \mathscr{F}(T_a)).$$

The right-hand side is $E|Y_\infty|$ (exercise), but we do not need this. The expectation of the limit is $E|Y_\infty|$, which is > 0 if $Y \not\equiv 0$ (recall that $Y_t = E(Y_\infty|\mathscr{F}_t)$), so it must be positive with positive probability, and hence, if a is small enough and $T = T_a$, then $E(|Y_\infty - Y_T| \, | \, \mathscr{F}_T) \not\equiv 0$, and we have accomplished (a). Since T_a has a continuous distribution, the solution of (b) is trivial. We let

$$T' = \begin{cases} T & \text{if } E(|Y_\infty - Y_T| \, | \, \mathscr{F}_T) \geq \varepsilon \text{ and } 0 < T < \lambda \\ \infty & \text{otherwise} \end{cases}$$

and vary λ to make $P(T' < \infty)$ the right size.

Remark: The last part of the proof obviously breaks down for dyadic martingales. In that setting, if you want a fixed value for the stopping time, say $T = 1$, then the probability of taking on that value cannot be arbitrarily small. It is this curiosity that allows nontrivial examples on the dyadic filtration.

8 PDE's That Can Be Solved by Running a Brownian Motion

A Parabolic Equations

In the first half of this chapter, we will show how Brownian motion can be used to construct (classical) solutions of the following equations:

$$u_t = \tfrac{1}{2}\Delta u$$
$$u_t = \tfrac{1}{2}\Delta u + g$$
$$u_t = \tfrac{1}{2}\Delta u + cu$$
$$u_t = \tfrac{1}{2}\Delta u + b \cdot \nabla u$$

in $(0, \infty) \times R^d$ subject to the boundary condition: u is continuous in $[0, \infty) \times R^d$ and $u(0, x) = f(x)$ for $x \in R^d$.

The solutions to these equations are (under suitable assumptions) given by

$$E_x(f(B_t))$$

$$E_x\left(f(B_t) + \int_0^t g(t - s, B_s)\, ds\right)$$

$$E_x\left(f(B_t) \exp\left(\int_0^t c(B_s)\, ds\right)\right)$$

$$E_x\left(f(B_t) \exp\left(\int_0^t b(B_s) \cdot dB_s - \frac{1}{2}\int_0^t |b(B_s)|^2\, ds\right)\right).$$

In words, the solutions may be described as follows:

(i) To solve the heat equation, run a Brownian motion and let $u(t, x) = E_x f(B_t)$.

(ii) To solve the inhomogeneous equation $u_t - \tfrac{1}{2}\Delta u = g$, add the integral of g along the path.

219

(iii) To introduce cu, multiply $f(B_t)$ by $\exp(\int_0^t c(B_s)\,ds)$ before taking expected values. In more picturesque terms, we think of the Brownian particle as having mass 1 at time 0 and changing size according to $m_t' = c(B_t)\,m_t$, and when we take expected values, we take the particle's weight into account.

(iv) To introduce $b \cdot \nabla u$, we multiply $f(B_t)$ by what may now seem to be a very strange-looking factor. By the end of Section 8.4, this factor will look very natural, that is, it will be clear that this is the only factor that can do the job.

In the first four sections of this chapter, we will say more about why the expressions we have written above solve the indicated equations. In order to bring out the similarities and differences in these equations, we have adopted a rather robotic style. Formulas (2) through (6) and their proofs have been developed in parallel in the four sections, and at the end of each section we discuss what happens when something becomes unbounded.

8.1 The Heat Equation

In this section, we will consider the following equation:

(1) (a) $u_t = \tfrac{1}{2}\Delta u$ in $(0, \infty) \times R^d$
     (b) u is continuous in $[0, \infty) \times R^d$ and $u(0, x) = f(x)$.

The equation derives its name from the fact that if the units of measurement are chosen suitably and if we let $u(t, x)$ be the temperature at the point $x \in R^d$ at time t when the temperatures at time 0 were given by $f(x)$, then u satisfies (1).

The first step in solving (1), as it will be many times below, is to prove:

(2) If u satisfies (a), then $M_s = u(t - s, B_s)$ is a local martingale on $[0, t)$.

Proof Applying Itô's formula gives

$$u(t - s, B_s) - u(t, B_0) = \int_0^s -u_t(t - r, B_r)\,dr$$
$$+ \int_0^s \nabla u(t - r, B_r) \cdot dB_r$$
$$+ \frac{1}{2}\int_0^s \Delta u(t - r, B_r)\,dr,$$

which proves (2), since $-u_t + \tfrac{1}{2}\Delta u = 0$ and the second term is a local martingale.

If we now assume that u is bounded, then M_s, $0 \le s < t$, is a bounded martingale. The martingale convergence theorem implies that as $s \uparrow t$, M_s converges to a limit. If u satisfies (b), this limit must be $f(B_t)$, and since M_s is uniformly integrable, it follows that

$$M_s = E_x(f(B_t)|\mathcal{F}_s).$$

Taking $s = 0$ in the last equation gives us a uniqueness theorem.

8.1 The Heat Equation

(3) If there is a solution of (1) that is bounded, it must be

$$v(t, x) \equiv E_x f(B_t).$$

Now that (3) has told us what the solution must be, the next logical step is to find conditions under which v is a solution. It is (and always will be) easy to show that v is a "generalized solution," that is, we have

(4) Suppose f is bounded. If v is smooth (i.e., it has enough continuous derivatives so that we can apply Itô's formula in the form given at the end of Section 2.9), then it satisfies (a).

Proof The Markov property implies that

$$E_x(f(B_t)|\mathscr{F}_s) = E_{B(s)}(f(B_{t-s})) = v(t-s, B_s).$$

Since the left-hand side is a martingale, $v(t-s, B_s)$ is also. If v is smooth, then repeating the calculation in the proof of (2) shows that

$$v(t-s, B_s) - v(t, B_0) = \int_0^s (-v_t + \tfrac{1}{2}\Delta v)(t-r, B_r)\,dr$$

$$+ \text{ a local martingale},$$

so it follows that the integral on the right-hand side is a local martingale. Since this process is continuous and locally of bounded variation, it must be $\equiv 0$, and hence, $-v_t + \tfrac{1}{2}\Delta v = 0$ in $(0, \infty) \times R^d$ (v_t and Δv are continuous, so if $-v_t + \tfrac{1}{2}\Delta v \neq 0$ at some point (t, x), then it is $\neq 0$ on an open neighborhood of that point, and, hence, with positive probability the integral is $\neq 0$, a contradiction).

It is easy to give conditions that imply that v satisfies (b). In order to keep the exposition simple, we first consider the situation when f is bounded. In this situation, the following condition is necessary and sufficient:

(5) If f is bounded and continuous, then v satisfies (b).

Proof $(B_t - B_0) \stackrel{d}{=} t^{1/2} N$, where N has a normal distribution with mean 0 and variance 1, so if $t_n \to 0$ and $x_n \to x$, the bounded convergence theorem implies that

$$v(t_n, x_n) = Ef(x_n + t_n^{1/2} N) \to f(x).$$

The last step in showing that v is a solution is to find conditions that guarantee that it is smooth. In this case, the computations are not very difficult.

(6) If f is bounded, then $v \in C^\infty$ and hence satisfies (a).

Proof We will show only that $v \in C^2$, since that is all we need to apply Itô's formula. By definition,

$$v(t, x) = E_x f(B_t) = \int (2\pi t)^{-d/2} e^{-|x-y|^2/2t} f(y)\,dy.$$

A little calculus gives

$$D_i e^{-|x-y|^2/2t} = -(x_i - y_i)e^{-|x-y|^2/2t}t$$
$$D_{ii} e^{-|x-y|^2/2t} = ((x_i - y_i)^2 - t)e^{-|x-y|^2/2t}t^2$$
$$D_{ij} e^{-|x-y|^2/2t} = (x_i - y_i)(x_j - y_j)e^{-|x-y|^2/2t}t^2 \quad i \neq j.$$

If f is bounded, then it is easy to see that for $\alpha = i$ or ij,

$$\int |D_\alpha e^{-|x-y|^2/2t}| \, |f(y)| \, dy < \infty$$

and is continuous in R^d, so the result follows from our result on differentiating under the integral sign (an exercise at the end of Section 1.10).

For some applications, the assumption that f is bounded is too restrictive. To see what type of unbounded f we can allow, we observe that, at the bare minimum, we need $E_x|f(B_t)| < \infty$ for all t. Since

$$E_x|f(B_t)| = \int \frac{1}{(2\pi t)^{d/2}} e^{-|x-y|^2/2t} |f(y)| \, dy,$$

a condition that guarantees this is

$$(*) \quad |x|^{-2} \log^+ |f(x)| \to 0 \quad \text{as } x \to \infty.$$

By repeating the proofs of (5) and (6) above and doing the estimates more carefully, it is not hard to show:

(7) If f is continuous and satisfies (*), then v satisfies (1).

Note: All we have done in this section is rewrite well-known results in a different language. An analyst (see, for example, Folland (1976), Section 4A) would write the first equation as

$$\partial_t u - \Delta u = 0$$
$$u(0, x) = f(x)$$

and take Fourier transforms to get

$$\partial_t \hat{u}(\xi, t) - 4\pi^2 |\xi| \hat{u}(\xi, t) = 0$$
$$\hat{u}(\xi, t) = \hat{f}(\xi) e^{-4\pi^2 |\xi|^2 t}.$$

Now

$$K_t(x) = (4\pi t)^{-d/2} e^{-|x|^2/4t}$$

has

$$\hat{K}_t(\xi) = e^{-4\pi^2 |\xi|^2 t},$$

so it follows that

$$u(t, x) = \int K_t(x - y) f(y) \, dy,$$

and we have derived the result without reference to Brownian motion.

Given the simplicity of the derivation above, we would be foolish to claim that Brownian motion is the best way to study the heat equation in $(0, \infty) \times R^d$. The situation changes, however, if we turn our attention to $(0, \infty) \times G$, where G is an open set (e.g., $G = \{z : |z| < 1\}$), and try to solve

(1')   (a) $u_t = \frac{1}{2}\Delta u$ in $(0, \infty) \times G$
       (b) u is continuous in $[0, \infty) \times \bar{G}$ and

$$u(0, x) = f(x) \quad x \in G$$
$$u(t, x) = 0 \quad t > 0, x \in \partial G.$$

In this context, the analyst (for example, Folland (1976), Sections 4B and 7E) must look for solutions of the form $f_j(x) \exp(\lambda_j t)$, "separation of variables," and show that the initial condition can be written as

$$f(x) = \sum_j a_j f_j(x).$$

Proving this even in the special case $G = \{z : |z| < 1\}$ requires a lot more work than when $G = R^d$, but for Brownian motion the amount of work is almost the same in both cases. We let $\tau = \inf\{t : B_t \notin G\}$ and let

$$v(t, x) = E_x(f(B_t); t < \tau).$$

Repeating the proofs above shows

(2')   If u satisfies (a), then $M_s = u(t - s, B_s)$ is a local martingale on $[0, \tau \wedge t)$.

(3')   If there is a solution of (1) that is bounded, it must be $v(t, x)$.

(4')   If v is smooth, then it satisfies (a).

(5')   If f is bounded and continuous, then u is continuous in $[0, \infty) \times G$ and $u(0, x) = f(x)$.

If the reader is patient, he or she can also show that

(6')   If f is bounded, then v_t, $D_i v$, and $D_{ij} v$, $1 \leq i, j \leq d$, all exist and are continuous, so v satisfies (a).

Note: v will not necessarily satisfy the other boundary condition $u(t, y) = 0$ for $y \in \partial G$. We will discuss this point when we consider the Dirichlet problem in Section 8.5.

8.2 The Inhomogeneous Equation

In this section, we will consider what happens when we add a function $g(t, x)$ to the equation we considered in the last section, that is, we will study

(1)   (a) $u_t = \frac{1}{2}\Delta u + g$ in $(0, \infty) \times R^d$
       (b) u is continuous in $[0, \infty) \times R^d$ and $u(0, x) = f(x)$.

8 PDE's That Can Be Solved by Running a Brownian Motion

The first step is to observe that we know how to solve the equation when $g \equiv 0$, so we can restrict our attention to the case $f \equiv 0$. Having made this simplification, we will now solve the equation above by blindly following the procedure used in the last section. The first step is to prove

(2) If u satisfies (a), then

$$M_s = u(t-s, B_s) + \int_0^s g(t-r, B_r)\, dr$$

is a local martingale on $[0, t)$.

Proof Applying Itô's formula gives

$$u(t-s, B_s) - u(t, B_0) = \int_0^s (-u_t + \tfrac{1}{2}\Delta u)(t-r, B_r)\, dr + \int_0^s \nabla u(t-r, B_r) \cdot dB_r,$$

which proves (2), since $-u_t + \tfrac{1}{2}\Delta u = -g$ and the second term is a local martingale.

If g is bounded and u is bounded on $[0, t] \times R^d$ and satisfies (a), then M_s, $0 \le s < t$, is a bounded martingale. By the argument in the last section, if u satisfies (b), then

$$\lim_{s \uparrow t} M_s = \int_0^t g(t-s, B_s)\, ds$$

and

$$M_s = E_x\!\left(\int_0^t g(t-s, B_s)\, ds \,\Big|\, \mathcal{F}_s\right).$$

Taking $s = 0$ gives

(3) Suppose g is bounded. If there is a solution of (1) that is bounded on $[0, t] \times R^d$, it must be

$$v(t, x) = E_x\!\left(\int_0^t g(t-s, B_s)\, ds\right).$$

Again, it is easy to show

(4) Suppose g is bounded. If v is smooth, then it satisfies (a) a.e. in $(0, \infty) \times R^d$.

Proof The Markov property implies that

$$E_x\!\left(\int_0^t g(t-r, B_r)\, dr \,\Big|\, \mathcal{F}_s\right)$$

$$= \int_0^s g(t-r, B_r)\, dr + E_{B(s)}\!\left(\int_0^{t-s} g(t-s-u, B_u)\, du\right).$$

8.2 The Inhomogeneous Equation

Since the left-hand side is a martingale, it follows that

$$v(t-s, B_s) + \int_0^s g(t-r, B_r)\, dr$$

is also. If v is smooth, then repeating the calculation in the proof of (2) shows that

$$v(t-s, B_s) - v(t, B_0) + \int_0^s g(t-r, B_r)\, dr$$
$$= \int_0^s (-v_t + \tfrac{1}{2}\Delta v + g)(t-r, B_r)\, dr$$
$$+ \text{a local martingale},$$

Again, we conclude that the integral on the right-hand side is a local martingale and, hence, must be $\equiv 0$, so we have $(-v_t + \tfrac{1}{2}\Delta v + g) = 0$ a.e.

The next step is to give a condition that guarantees that v satisfies (b). As in the last section, we will begin by considering what happens when everything is bounded.

(5)   If g is bounded, then v satisfies (b).

Proof   If $|g| \leq M$, then

$$E_x \left| \int_0^t g(t-s, B_s)\, ds \right| \leq Mt \to 0.$$

The last step in showing that v is a solution is to check that it is smooth enough. In this case,

$$v(t, x) = \int_0^t ds \int (2\pi s)^{-d/2} e^{-|y-x|^2/2s} g(t-s, y)\, dy,$$

and the normal density

$$p_s(x, y) = (2\pi s)^{-d/2} e^{-|y-x|^2/2s}$$

$\to \infty$ if $x = y$ and $s \to 0$, so things are not as simple as they were in the last section.

The expression we have written for v above is what Friedman (1964) would call a volume potential and would write as

$$V(x, t) = \int_{T_0}^t \int_D Z(x, t; \xi, \tau) f(\xi, \tau)\, d\xi\, d\tau.$$

To translate between notations, set $T_0 = 0$, $D = R^d$,

$$Z(x, t; \xi, \tau) = p_{t-\tau}(x, \xi),$$

and change variables $s = t - \tau$, $y = \xi$. Because of their importance for the parametrix method, the differentiability properties of volume potentials are well known. Since the calculations necessary to establish these properties are quite tedious, we will content ourselves to state what the results are and indicate why they are true. The results we will state are just Theorems 2 to 5 of Friedman (1964), so the reader who is interested in knowing the whole story can find the missing details there.

(6a) If g is a bounded measurable function, then $v(t, x)$ is continuous on $(0, \infty) \times R^d$.

Proof This result follows easily from the bounded convergence theorem.

(6b) If g is bounded and measurable, then the partial derivatives $D_i v = \partial v/\partial x_i$ are continuous, and

$$D_i v = \int_0^t \int D_i p_s(x, y) g(t - s, y) \, dy \, ds.$$

Proof The right-hand side is

$$-\int_0^t \int (2\pi s)^{-d/2} \frac{(x_i - y_i)}{s} e^{-|x-y|^2/2s} g(t - s, y) \, dy \, ds$$

$$= -\int_0^t \frac{ds}{s} E_x[(x_i - B_s^i) g(t - s, B_s)].$$

Although the last formula looks suspicious because we are integrating s^{-1} near 0, everything is really all right. If $|g| \leq M$, then

$$E_x|(x_i - B_s^i) g(t - s, B_s)| \leq M E_x |x_i - B_s^i| = CMs^{1/2},$$

so

$$\int_0^t \frac{ds}{s} E_x|(x_i - B_s^i) g(t - s, B_s)| < \infty$$

and it follows from the exercise in Section 1.10 that the partial derivatives $D_i v$ exist and are continuous.

The computations above are not hard to make rigorous, but you should save your strength. Things get very nasty when we take second derivatives.

(6c) Suppose that g is a bounded continuous function and that for any $N < \infty$ there are constants C, $\alpha \in (0, \infty)$ such that $|g(t, x) - g(t, y)| \leq C|x - y|^\alpha$ whenever $|x|$, $|y|$, and $t \leq N$. Then the partial derivatives $D_{ij} v = \partial^2 v/\partial x_i \partial x_j$ are continuous, and

$$D_{ij} v = \int_0^t \int D_{ij} p_s(x, y) g(t - s, y) \, dy \, ds.$$

Proof Suppose for simplicity that $i = j$. In this case, the right-hand side is

8.2 The Inhomogeneous Equation

$$\int_0^t \int (2\pi s)^{-d/2} \left(\frac{(x_i - y_i)^2}{s^2} - \frac{1}{s} \right) e^{-|x-y|^2/2s} g(t-s, y) \, dy \, ds$$

$$= \int_0^t E_x \left[\frac{(x_i - B_s^i)^2 - s}{s^2} g(t-s, B_s) \right] ds,$$

but this time, however,

$$E_x |(x_i - B_s^i)^2 - s| = s E_0 |(B_1^i)^2 - 1|$$

so

$$\int_0^t ds \, E_x \left| \frac{(x_i - B_s^i)^2 - s}{s^2} \right| = \infty.$$

We can overcome this problem if f is Hölder continuous at x, because we can write

$$E_x \left[\left(\frac{(x_i - B_s^i)^2 - s}{s^2} g(t-s, B_s) \right) \right]$$

$$= E_x \left(\frac{(x_i - B_s^i)^2 - s}{s^2} (g(t-s, B_s) - g(t-s, x)) \right).$$

The second expression $\leq C s^{-1+\alpha}$, so its integral from $s = 0$ to t converges absolutely, and with a little work (6c) follows. (See Friedman (1964), pages 10–12, for more details.)

The last detail now is:

(6d) Let g be as in (6c). Then $\partial v / \partial t$ exists, and

$$\frac{\partial v}{\partial t}(t, x) = g(t, x) + \int_0^t dr \int \frac{\partial}{\partial t} p_{t-r}(x, y) g(r, y) \, dy.$$

Proof To take the derivative w.r.t. t, we rewrite v as

$$v(t, x) = \int_0^t \int p_{t-r}(x, y) g(r, y) \, dy.$$

Differentiating the left-hand side w.r.t. t gives two terms. Differentiating the upper limit of the integral gives $g(t, x)$. Differentiating the integrand gives

$$\int_0^t \int \frac{\partial}{\partial t} p_{t-r}(x, y) g(r, y) \, dy \, dr$$

$$= \int_c^t \int \frac{-d}{2} (2\pi(t-r))^{-(d+2)/2} \exp\left(\frac{-|x-y|^2}{2(t-r)} \right) g(r, y) \, dy \, dr$$

$$+ \int_0^t \int (2\pi(t-r))^{-d/2} \left(\frac{+|x-y|^2}{2(t-r)^2} \right) \exp\left(\frac{-|x-y|^2}{2(t-r)} \right) g(r, y) \, dy \, dr$$

$$= \frac{-d}{2} \int_0^t \frac{dr}{t-r} E_x g(r, B_{t-r}) \, dr + \int_0^t \frac{dr}{2(t-r)^2} E_x (|x - B_{t-r}|^2 g(r, B_{t-r})).$$

In the second integral, we can use the fact that $E_x(|x - B_{t-r}|^2) = C(t - r)$ to cancel one of the $t - r$'s and make the second expression like the first, but even if we do this,

$$\int_0^t \frac{dr}{t - r} = \infty.$$

This is the difficulty that we experienced in the proof of (6c), and the remedy is the same: We can save the day if g is locally uniformly Hölder continuous in x. For further details, see pages 12–13 of Friedman (1964).

As in the last section we can generalize our results to unbounded g's. To see what type of unbounded g's can be allowed, we will restrict our attention to the homogeneous case $g(t, x) = f(x)$. At the bare minimum, we need

$$E_x \int_0^t f(B_s) \, ds < \infty,$$

and if we want (b) to hold, we need to know that if $t_n \downarrow 0$ and $x_n \to x$, then

$$E_{x_n} \left| \int_0^{t_n} f(B_s) \, ds \right| \to 0.$$

If we strengthen the last result to uniform convergence for $x \in R^d$, then we get a definition that is essentially due to Kato (1973).

A function f is said to be in K_d if

(*) $\quad \lim\limits_{t \downarrow 0} \sup\limits_x E_x \left(\int_0^t |f(B_s)| \, ds \right) = 0.$

By Fubini's theorem, we can write the above as

$$\lim\limits_{t \downarrow 0} \sup\limits_x \int k_t(x, y) |f(y)| \, dy = 0,$$

where

$$k_t(x, y) = \int_0^t (2\pi s)^{-d/2} e^{-|x-y|^2/2s} \, ds.$$

By considering the asymptotic behavior of $k_t(x, y)$ as $t \to 0$ and $|x - y|^2/t \to c$, we can cast this condition in a more analytical form as

(**) $\quad \lim\limits_{\alpha \downarrow 0} \sup\limits_x \int_{|x-y| \le \alpha} \varphi(|x - y|) f(y) \, dy = 0,$

where

$$\varphi(z) = \begin{cases} |z|^{-(d-2)} & d \ge 3 \\ -\log|z| & d = 2 \\ 1 & d = 1. \end{cases}$$

This is Theorem 4.5 of Aizenman and Simon (1982). The details of the proof,

8.3 The Feynman-Kac Formula

In this section, we will consider what happens when we add cu to the right-hand side of the equation we considered in Section 8.1, that is, we will study

(1) (a) $u_t = \frac{1}{2}\Delta u + cu$ in $(0, \infty) \times R^d$
(b) u is continuous in $[0, \infty) \times R^d$ and $u(0, x) = f(x)$.

If $c(x) \leq 0$, then this equation describes heat flow with cooling. As in Section 8.1, the solution $u(t, x)$ gives the temperature at the point $x \in R^d$ at time t, but here we do not assume that there is perfect conduction of heat. Instead, we assume that heat at x dissipates at the rate $k(x) = -c(x)$. We will see below that this corresponds to Brownian motion with killing at rate k, that is, the probability that a particle survives until time t is $\exp(-\int_0^t k(B_s) \, ds)$.

The first step in solving (1) is to prove:

(2) If u satisfies (a), then

$$M_s = u(t - s, B_s) \exp\left(\int_0^s c(B_r) \, dr\right)$$

is a local martingale on $[0, t)$.

Proof Let $c_t = \int_0^t c(B_s) \, ds$. Applying Itô's formula gives that
$u(t - s, B_s) \exp(c_s) - u(t, B_0)$

$$= \int_0^s -u_t(t - r, B_r) \exp(c_r) \, dr + \int_0^s \exp(c_r) \nabla u(t - r, B_r) \cdot dB_r$$
$$+ \int_0^s u(t - r, B_r) \exp(c_r) \, dc_r + \frac{1}{2} \int_0^s \Delta u(t - r, B_r) \exp(c_r) \, dr,$$

which proves (2), since $-u_t + cu + \frac{1}{2}\Delta u = 0$ and the second term is a local martingale.

If c is bounded and u is bounded on $[0, t] \times R^d$ and satisfies (a), then M_s, $0 \leq s < t$, is a bounded martingale, so by an argument we have used in the last two sections

$$\lim_{s \uparrow t} M_s = f(B_t) \exp(c_t)$$

and

$$M_s = E_x(f(B_t)\exp(c_t)|\mathcal{F}_s),$$

so taking $s = 0$ gives

(3) Suppose that c is bounded. If there is a solution of (1) that is bounded on $[0, t] \times R^d$, it must be

$$v(t, x) = E_x(f(B_t)\exp(c_t)).$$

As before, it is easy to show

(4) Suppose that c is bounded. If v is smooth, then it satisfies (a) a.e. in $(0, \infty) \times R^d$.

Proof The Markov property implies that

$$E_x(f(B_t)\exp(c_t)|\mathcal{F}_s) = \exp(c_s)E_{B(s)}(f(B_{t-s})\exp(c_{t-s})),$$

so if we let $v(t, x) = E_x(f(B_t)\exp(c_t))$, then the last equality shows that $v(t - s, B_s)$ is a martingale. If v is smooth, then repeating the calculation in the proof of (2) shows that

$$v(t - s, B_s)\exp(c_s) - v(t, B_0)$$
$$= \int_0^s (-v_t + cv + \tfrac{1}{2}\Delta v)(t - r, B_r)\exp(c_r)\,dr + \text{a local martingale}.$$

so again, we conclude that the integral on the right-hand side is a local martingale and, hence, must be $\equiv 0$, so we have that $-v_t + cv + \tfrac{1}{2}\Delta v = 0$ a.e.

The next step is to give a condition that guarantees that v satisfies (b). As before, we begin by considering what happens when everything is bounded.

(5) If c is bounded and f is bounded and continuous, then v satisfies (b).

Proof If $|c| \le M$, then $e^{-Mt} \le \exp(c_t) \le e^{Mt}$, so $\exp(c_t) \to 1$ as $t \to 0$. Since f is bounded, this result implies that

$$E_x \exp(c_t)f(B_t) - E_x f(B_t) \to 0,$$

and so the desired result follows from (5) in Section 8.1.

This brings us to the problem of determining when v is smooth enough to be a solution. To solve the problem in this case, we use a trick to reduce our result to the previous case. We observe that

$$\exp\left(\int_0^t c(B_s)\,ds\right) = 1 + \int_0^t c(B_s)\exp\left(\int_s^t c(B_r)\,dr\right)ds,$$

so taking expected values gives that

$$v(t, x) = 1 + \int_0^t E_x\left[c(B_s)\exp\left(\int_s^t c(B_r)\,dr\right)f(B_t)\right]ds.$$

Conditioning on $\mathcal{F}_s$ and using the Markov property, we can write the equation above as

8.3 The Feynman-Kac Formula

$$v(t,x) = 1 + \int_0^t E_x c(B_s) v(t-s, B_s)\, ds.$$

The second term on the right-hand side is of the form considered in the last section. If we start with the trivial observation that if c and f are bounded, then v is bounded on $[0, t] \times R^d$, and if we apply (6a) and (6b) from the last section, we see that v is continuous and the derivatives $\partial v/\partial x_i$ are continuous. This implies that, for each N, $|v(t, x) - v(t, y)| \leq C_N |x - y|$ whenever $|x|, |y|$, and $t \leq N$. If we assume that c is bounded and locally Hölder continuous, then it follows from (6c) and (6d) that we have

(6) Suppose that f is bounded. If c is bounded and locally Hölder continuous, then v is smooth and, hence, satisfies (a).

As in the last two sections, we can generalize the results above to unbounded c's. Given the formula above, which expresses v as a volume potential, it is perhaps not too surprising that the appropriate assumption is $c \in K_d$. The key to working in this generality is what Simon (1982) calls Khasmin'skii's lemma:

(7) Let $f \geq 0$ be a function on R^d with

$$\alpha \equiv \sup_x E_x \left(\int_0^t f(B_s)\, ds \right) < 1.$$

Then

$$\sup_x E_x \exp\left(\int_0^t f(B_s)\, ds \right) \leq (1 - \alpha)^{-1}.$$

Proof The Markov property and nonnegativity of f imply that

$$\sup_x E_x \int\cdots\int_{0 < s_1 < \cdots < s_n < t} ds_1 \ldots ds_n f(B_{s_1}) \ldots f(B_{s_n}) \leq \alpha^n,$$

so the desired result follows by noticing that

$$\int\cdots\int_{0 < s_1 < \cdots < s_n < t} = \frac{1}{n!} \int_0^t \cdots \int_0^t$$

and summing on n.

From the last result, it should be clear why assuming $c \in K_d$ is natural in this context. This condition guarantees that

$$\sup_x E_x \left(\int_0^t |c(B_s)|\, ds \right) \to 0$$

and, hence, that

$$\sup_x E_x \exp\left(\int_0^t |c(B_s)|\, ds \right) \to 1.$$

With these two results in hand, we can proceed with developing the theory much as we did in the case of bounded coefficients. Since Simon (1982) has written a lengthy and very readable account of how to develop the theory in this generality, we will content ourselves just to briefly describe a few of the results given in his paper. Part of our motivation for doing this is to establish the connection between the notation we use and the way in which mathematical physicists write things. To make it easy for the reader to find the results in Simon's paper, we have used his theorem numbers below.

Let $H_0 = -\frac{1}{2}\Delta$ and $V = -c$

$$H = H_0 + V = -\frac{\Delta}{2} + V,$$

and define a linear operator e^{-tH} by setting

$$(e^{-tH}f)(x) = E_x\left(\exp\left(-\int_0^t V(B_s)\,ds\right)f(B_t)\right).$$

THEOREM B.1.1 Let $V_- \in K_d$ and $V_+ \in K_d^{\text{loc}}$. Then for every $t > 0$ and $p \leq q \leq \infty$, e^{-tH} is bounded from L^p to L^q.

Proof Since the proof relies on some things that we have not explained above, we simply sketch the proof and refer the reader to Simon (1982) for details. Let $\|e^{-tH}\|_{p,q}$ denote the norm of e^{-tH} as a map from L^p to L^q.

Step 1: $p = \infty$, $q = \infty$. The Feynman-Kac formula shows that if t is small, then

$$\|e^{-tH}f\|_\infty \leq C\|f\|_\infty,$$

so the semigroup property $e^{-(s+t)H} = e^{-sH}e^{-tH}$ implies that

$$\|e^{-tH}\|_{\infty,\infty} \leq Ce^{At},$$

where $A = T^{-1}\ln C$.

Step 2: $p = 2$, $q = \infty$. Using the Cauchy-Schwarz inequality in the Feynman-Kac formula gives

$$|e^{-tH}f| \leq (e^{-t(H_0+2V)}1)^{1/2}(e^{-tH_0}|f|^2)^{1/2}.$$

Applying Step 1 to $H_0 + 2V$ gives

$$\|e^{-t(H_0+2V)}\|_{\infty,\infty} \leq C'e^{A't},$$

and an easy estimate shows that

$$\|e^{-tH_0}g\|_\infty = \sup_x \int (2\pi t)^{-d/2} p_t(x,y) g(y)\,dy$$

$$\leq (2\pi t)^{-d/2}\|g\|_1,$$

8.3 The Feynman-Kac Formula

so

$$\|e^{-tH}f\|_\infty \leq C''\|f\|_2.$$

Step 3: $p = 1$, $q = 2$. Since $\langle f, e^{-tH}g\rangle = \langle e^{-tH}f, g\rangle$, we have

$$\|e^{-tH}\|_{1,2} = \|e^{-tH}\|_{2,\infty}.$$

Step 4: $p = 1$, $q = \infty$. By the semigroup property,

$$\|e^{-tH}\|_{1,\infty} \leq \|e^{-tH/2}\|_{1,2}\|e^{-tH/2}\|_{2,\infty}.$$

Step 5: Steps 1 and 4 show that e^{-tH} is bounded from L^∞ to L^∞ and from L^1 to L^∞. The result now follows by "duality and interpolation."

The next result gives another reason why the spaces K_d are well suited for studying Shrödinger semigroups.

PROPOSITION B.1.4 If $V \leq 0$ and $\lim_{t\downarrow 0}\|e^{-tH}\|_{\infty,\infty} = 1$, then $V \in K_d$.

Proof This is an easy consequence of Jensen's inequality and is left as an exercise for the reader.

Theorem B.1.1. shows e^{-tH} maps L^∞ into L^∞. With a little work this can be improved considerably.

THEOREM B.3.1 Let $V_- \in K_d$, $V_+ \in K_d^{\text{loc}}$. If $f \in L^\infty$, then $e^{-tH}f$ is a continuous function.

THEOREM B.7.1 Let $V_- \in K_d$, $V_+ \in K_d^{\text{loc}}$. Then

$$e^{-tH}f(x) = \int e^{-tH}(x, y)f(y)\,dy,$$

where $e^{-tH}(x, y)$ is jointly continuous in x, y, and t in the region $t > 0$.

THEOREM B.3.4 Let $V_- \in K_d$, $V_+ \in K_d^{\text{loc}}$. If $f \in L^\infty$, then, for any $t > 0$, $e^{-tH}f$ has a distributional gradient in L^2_{loc}.

To get more smoothness, one has to assume more boundedness. Suppose for simplicity that $d \geq 2$ and, for $0 < \alpha < 2$, let K_d^α be the set of all functions that satisfy:

(i) $\displaystyle\sup_x \int_{D(x,1)} |x - y|^{-(d-2+\alpha)}|f(y)|\,dy < \infty$ if $\alpha \neq 1$

(ii) $\displaystyle\lim_{r\downarrow 0}\sup_x \int_{D(x,r)} |x - y|^{-(d-2+\alpha)}|f(y)|\,dy = 0$ if $\alpha = 1$.

THEOREM B.3.5 Let $\alpha < 2$. Let $V_- \in K_d$, $V_+ \in K_d^{\text{loc}}$. Suppose that the restriction of V to some bounded open set G lies in K_d^α. If $f \in L^\infty$, then for each $t > 0$, $e^{-tH} f \in C^\alpha(G) =$ the set of functions whose derivatives of order $[\alpha]$ are Hölder continuous of order $\alpha - [\alpha]$.

Remarks: The fact that V is not supposed to be smooth restricts the last result to $\alpha < 2$. The reader should also observe that by writing $e^{-tH} = e^{-tH/2} e^{-tH/2}$ and applying Theorem B.1.1, we can conclude that the results above hold if $f \in L^\infty$ is replaced by $f \in \bigcup_{p=1}^\infty L^p$.

Note: Inspired by the work of Feynman (1948), Kac (1949) proved the first version of what is now known as the Feynman-Kac formula. He proved his result in $d = 1$ for potentials $V = -c$, which are bounded below, by discretizing time, passing to the limit, and ignoring a few details along the way. Rosenblatt (1951) extended Kac's work to $d \geq 2$ and filled in the missing details (e.g., Hölder continuity is needed if one wants the solution to be C^2). Since that time, there have been a number of papers extending the result to more general processes and potentials. The results we have mentioned above are only a small sample of what is known. If the reader would like to see more examples of how probability can be used to study these problems, he should look at McKean (1977), Berthier and Gaveau (1978), and at recent work of Carmona and Simon. Perhaps the best place to begin is Simon's (1982) survey paper.

8.4 The Cameron-Martin Transformation

In this section, we will consider what happens when we add $b \cdot \nabla u$ to the right-hand side of the equation considered in Section 8.1, that is, we will study:

(1)   (a) $u_t = \frac{1}{2} \Delta u + b \cdot \nabla u$ in $(0, \infty) \times R^d$
      (b) u is continuous in $[0, \infty) \times R^d$ and $u(0, x) = f(x)$.

Physically, the extra term corresponds to a force field. In this section, we will see that the probabilistic effect is to add an infinitesimal drift b to our Brownian motion.

The first step in solving (1) is to prove:

(2)   If u satisfies (a), then

$$M_s = u(t-s, B_s) \exp\left(\int_0^s b(B_r) \cdot dB_r - \frac{1}{2} \int_0^s |b(B_r)|^2 dr\right)$$

is a local martingale on $[0, t)$.

Proof Let $Z_s = \int_0^s b(B_r) \cdot dB_r - \frac{1}{2} \int_0^s |b(B_r)|^2 dr$. Applying Itô's formula to

$$f(x_0, x_1, \ldots, x_{d+1}) = u(t - x_0, x_1, \ldots, x_d) \exp(x_{d+1})$$
$$X_s^0 = s,\ X_s^i = B_s^i \text{ for } 1 \leq i \leq d,\ X_s^{d+1} = Z_s$$

8.4 The Cameron-Martin Transformation

gives:

$$u(t-s, B_s)\exp(Z_s) - u(t, B_0) = \int_0^s -u_t(t-r, B_r)\exp(Z_r)\,dr$$

$$+ \int_0^s \exp(Z_r)\nabla u(t-r, B_r)\cdot dB_r$$

$$+ \int_0^s u(t-r, B_r)\exp(Z_r)\,dZ_r$$

$$+ \frac{1}{2}\int_0^s \Delta u(t-r, B_r)\exp(Z_r)\,dr$$

$$+ \sum_{i=1}^d \int_0^s D_i u(t-r, B_r)\exp(Z_r)\,d\langle B^i, Z\rangle_r$$

$$+ \frac{1}{2}\int_0^s u(t-r, B_r)\exp(Z_r)\,d\langle Z\rangle_r.$$

To check this result, observe that the first three lines are the terms involving first derivatives of f. The last three lines are the terms with $D_{ij}f$ where (a) $1 \le i, j \le d$, (b) $1 \le i \le d$, $j = d+1$, or $i = d+1$, $1 \le j \le d$, and (c) $i = j = d+1$, respectively. The terms with i or $j = 0$ vanish, because X^0 is locally b.v.

Applying the associative law and the formula for the covariance of two stochastic integrals to the mess above gives

$$\int_0^s -u_t(t-r, B_r)\exp(Z_r)\,dr + 2 \text{ local martingales}$$

$$+ \int_0^s u(t-r, B_r)\exp(Z_r)(-\tfrac{1}{2}|b(B_r)|^2)\,dr$$

$$+ \frac{1}{2}\int_0^s \Delta u(t-r, B_r)\exp(Z_r)\,dr$$

$$+ \sum_{i=1}^d \int_0^s D_i u(t-r, B_r)\exp(Z_r)b^i(B_r)\,dr$$

$$+ \frac{1}{2}\int_0^s u(t-r, B_r)\exp(Z_r)|b(B_r)|^2\,dr.$$

The third and sixth terms cancel, and if u satisfies (a), the sum of the first, fourth, and fifth is 0, proving (2).

At this point, the reader can probably anticipate the next step.

(3) Suppose that b is bounded. If there is a solution of (1) that is bounded on $[0, t] \times R^d$, it must be

$$v(t, x) = E_x(f(B_t)\exp(Z_t)).$$

This time, however, we cannot simply define our problems away, because $X_s \equiv \int_0^s b(B_r) \cdot dB_r$ may be unbounded. Let $T_n = \inf\{t : |X_t| > n\}$. The exponential formula implies that $\exp(Z_t) \equiv \exp(X_t - \frac{1}{2}\langle X \rangle_t)$ is a local martingale. So observing that $\langle X \rangle_t \geq 0$ and stopping at $s \wedge T_n$ gives

$$E \exp(Z_{s \wedge T_n}) = E \exp(Z_0) = 1.$$

Letting $n \to \infty$ now and using Fatou's lemma shows $E \exp(Z_s) \leq 1$, that is, if we let $Y_s = \exp(Z_s)$, then Y is an L^1 bounded martingale.

Applying the last result with b replaced by $2b$ gives

$$1 \geq E \exp\left(2 \int_0^s b(B_r) \cdot dB_r - 4 \cdot \frac{1}{2} \int_0^s |b(B_r)|^2 \, dr\right)$$

$$\geq \exp(-sb^*) E Y_s^2,$$

where $b^* = \sup|b(x)|$, and we can use the martingale convergence theorem to conclude that if u satisfies (1), then

$$\lim_{s \uparrow t} M_s = f(B_t) \exp(Z_t)$$

and

$$M_s = E_x(f(B_t) \exp(Z_t) | \mathscr{F}_s),$$

so taking $s = 0$ proves (3).

Exercise 1 There are many other ways to prove (3). Let

$$X_s = \int_0^s b(B_r) \cdot dB_r$$

and observe that

$$\langle X \rangle_s = \int_0^s |b(B_r)|^2 \, dr \leq Cs,$$

so it follows from Lévy's theorem (see Section 2.11) that

$$E\left(\sup_{0 \leq s \leq t} \exp(X_s)\right) \leq E\left(\exp\left(\sup_{0 \leq s \leq b^*t} B_s\right)\right),$$

and the right-hand side is finite since we have (see Section 1.5)

$$P_0(B_t^* > a) = 2P_0(B_t > a) \quad \text{for } a > 0$$

Note: This proof was originally in the text. I would like to thank Tom **Liggett** for pointing out the simpler proof used above.

As before, it is easy to show the following:

(4) If v is smooth, it satisfies (a) a.e. in $(0, \infty) \times R^d$.

Proof The Markov property implies that

$$E_x(f(B_t) \exp(Z_t) | \mathscr{F}_s) = \exp(Z_s) E_{B(s)}(f(B_{t-s}) \exp(Z_{t-s})),$$

8.4 The Cameron-Martin Transformation

so if we let $v(t, x) = E_x(f(B_t)\exp(Z_t))$, then the last equality shows that $v(t-s, B_s)$ is a martingale. If v is smooth, then repeating the calculation in the proof of (2) shows that

$$v(t-s, B_s)\exp(Z_s) - v(t, B_0)$$
$$= \int_0^s (-v_t + b \cdot \nabla v + \tfrac{1}{2}\Delta v)(t-r, B_r)\exp(Z_r)\,dr + \text{a local martingale.}$$

Again, we conclude that the integral on the right-hand side is a local martingale and, hence, must be $\equiv 0$, so we have $-v_t + b \cdot \nabla v + \tfrac{1}{2}\Delta v = 0$ a.e.

The next step is to give a condition that guarantees that v satisfies (ii). As before, we begin by considering what happens when everything is bounded.

(5) If b is bounded and f is bounded and continuous, then v satisfies (b).

Proof Let $Y_t = \exp(Z_t)$. As $t \to 0$, $Y_t \to 1$ almost surely, and we have $E_x Y_1^* < \infty$. Since f is bounded, it follows from the dominated convergence theorem that

$$E_x f(B_t)\exp(Z_t) - E_x f(B_t) \to 0,$$

and so the desired result follows from (5) in Section 8.1.

This brings us last, but not least, to the problem of determining when v is smooth enough to be a solution. As you might guess by extrapolating from the last three sections, this is a very difficult problem. The reader is invited to think about how he might try to solve this problem. We do not know a very simple way of doing this, so we will put off consideration of this point until Chapter 9, when we will confront the problem in a more general situation.

Having skipped smoothness, the last item on our outline is: What happens if b is unbounded? To answer this question and to prepare for developments in Chapter 9, we will look at our solution through the eyes of Cameron and Martin (1949). Let P_x be the measure on $(C, \mathcal{C})$ that makes the coordinate maps $B_s(\omega) = \omega_s$ a Brownian motion starting at x, and define a new measure on $(C, \mathcal{F}_t)$ by setting

$$Q_x(A) = \int_A \exp(Z_t)\,dP_x \quad \text{for } A \in \mathcal{F}_t,$$

where

$$Z_t = \int_0^t b(B_s) \cdot dB_s - \frac{1}{2}\int_0^t |b(B_s)|^2\,ds.$$

Since $X_t \equiv \int_0^t b(B_s) \cdot dB_s$ is a local martingale with $\langle X \rangle_t = \int_0^t |b(B_s)|^2\,ds$, it follows from the exponential formula that $\exp(Z_t)$ is a local martingale/P_x. Let $T_n = \inf\{t : |Z_t| > n\}$. By stopping at $T_n \wedge t$ and letting $n \to \infty$, we see that

$$Q_x(C) = E_x \exp Z_t \le 1.$$

In some situations (e.g., b bounded), we will have $Q_x(C) = 1$. The next result shows that when this occurs, Q_x makes the coordinate maps behave like a Brownian motion plus a drift.

(6) If $Q_x(C) = 1$, then under Q_x

$$W_t = B_t - \int_0^t b(B_s)\,ds$$

is a Brownian motion starting at x.

Proof Let W_t^j be the jth component of W_t. Our first goal is to prove that W^j is a one-dimensional Brownian motion starting at x_j. The first step in doing this is to observe that it suffices to show that if for all θ, $U_t = \exp(i\theta W_t^j + \theta^2 t/2)$ is a local martingale under Q_x, because then

$$E_{Q_x}(\exp(i\theta(W_t^j + \theta^2 t/2))|\mathscr{F}_s) = \exp(i\theta W_t^j + \theta^2 s/2),$$

and

$$E_{Q_x}(\exp(i\theta(W_t^j - W_s^j))|\mathscr{F}_s) = \exp(-\theta^2(t-s)/2).$$

In other words, $W_t^j - W_s^j$ is independent of $\mathscr{F}_s$ and has a normal distribution with mean 0 and variance $t - s$. By (2) from Section 2.13, U_t is a local martingale/Q_x if and only if $U_t \exp(Z_t)$ is a local martingale/P_x. Unscrambling the definitions gives

$$U_t \exp(Z_t) = \exp\left(i\theta\left(B_t^j - \int_0^t b^j(B_s)\,ds\right) + \frac{\theta^2 t}{2}\right)$$
$$\cdot \exp\left(\int_0^t b(B_s)\cdot dB_s - \frac{1}{2}\int_0^t |b(B_s)|^2\,ds\right)$$
$$= \exp(C_t - D_t),$$

where

$$C_t = i\theta B_t^j + \int_0^t b(B_s)\cdot dB_s$$

and

$$D_t = \frac{-\theta^2 t}{2} + i\theta \int_0^t b^j(B_s)\,ds + \frac{1}{2}\int_0^t |b(B_s)|^2\,ds.$$

Now, if we let $X_t = \int_0^t b(B_s)\cdot dB_s$, then

$$\langle B^j \rangle_t = t,$$
$$\langle B^j, X \rangle_t = \int_0^t b^j(B_s)\,ds,$$

and $\langle X \rangle_t = \int_0^t |b(B_s)|^2\,ds,$

so $D_t = \frac{1}{2}\langle C \rangle_t$ and it follows from the exponential formula that $U_t \exp(Z_t)$ is a local martingale.

8.4 The Cameron-Martin Transformation

The computation above shows that each component W_t^j is a local martingale. To complete the proof, we observe that (5) in Section 2.8 implies that $\langle W^i, W^j \rangle_t$ is the same under Q_x as it was under P_x, so $\langle W^i, W^j \rangle_t = \delta_{ij} t$. The desired result follows from the characterization of multidimensional Brownian motion given in Section 2.12.

(6) shows that under Q_x, the coordinate maps, which we will now call $X_t(\omega) = \omega_t$, satisfy

$$X_t = W_t + \int_0^t b(X_s)\, ds$$

where W is a Brownian motion, an equation that we can write formally as

(∗) $dX_t = dB_t + b(X_t)\, dt$

(here, to facilitate comparison with the results above, we have replaced the W (for Wiener process) with the letter B, which we usually use to denote Brownian motion).

There is an obvious connection between (∗) and (1):

(2′) If u satisfies (a) and X satisfies (∗), then $M_s = u(t - s, X_s)$ is a local martingale on $[0, t)$.

Proof Applying Itô's formula gives

$$u(t - s, X_s) - u(0, X_0) = \int_0^s -u_t(t - r, X_r)\, dr + \int_0^s \nabla u(t - r, X_r) \cdot dX_r$$
$$+ \frac{1}{2} \sum_{ij} \int_0^s D_{ij} u(t - r, X_r)\, d\langle X^i, X^j \rangle_r.$$

Since $\langle X^i, X^j \rangle_r = \delta_{ij} r$, it follows that the right-hand side

$$= \int_0^s -u_t(t - r, X_r)\, dr + \text{a local martingale}$$
$$+ \sum_{i=1}^d \int_0^s D_i u(t - r, X_r) b^i(X_r)\, dr$$
$$+ \frac{1}{2} \int_0^s \Delta u(t - r, X_r)\, dr,$$

which proves (2′) since $-u_t + b \cdot \nabla u + \frac{1}{2} \Delta u = 0$.

From the discussion above, it should be clear how to approach the problem of solving (∗) and (1) when b is only locally bounded. Let $b_n = b 1_{(|x| \le n)}$. Since b_n is bounded, we can solve (∗) and get a process X^n that satisfies the original equation for $t < T_n = \inf\{t : |X_t| > n\}$. The measures μ_n on $(C, \mathscr{C})$ that give rise to X^n have the property that if $m < n$, then μ_n and μ_m agree on $\mathscr{F}(T_m)$, so we can let $n \to \infty$ and construct a process that solves the original equation for $t < T_\infty =$

8 PDE's That Can Be Solved by Running a Brownian Motion

$\lim T_n$. When $\{T_\infty < \infty\}$ has positive probability, we say that the process explodes. The next result gives a simple condition that rules out explosion.

(7) If $x \cdot b(x) \le C(1 + |x|^2)$, then the process does not explode starting from any $x \in R^d$.

Proof Let b^n, X^n, and T_n be as above. Let $g(x) = 1 + |x|^2$ and, to ease the notation, let $Y_t = X_t^n$. Applying Itô's formula gives

$$g(Y_t) - g(Y_0) = \sum_i \int_0^t 2Y_s^i \, dY_s^i + \frac{1}{2}\sum_i \int_0^t 2 \, d\langle Y^i \rangle_s$$

$$= \text{a local martingale} + \sum_i \int_0^t 2Y_s^i b_n^i(Y_s) \, ds + td,$$

and our hypothesis implies that

$$\sum_i \int_0^t 2Y_s^i b_n^i(Y_s) \, ds \le 2C \int_0^t g(Y_s) \, ds,$$

so applying Itô's formula to $\exp(-(2C + d)t)g(Y_t)$ gives

$$\exp(-(2C+d)t)g(Y_t) - g(Y_0) = \text{a local martingale}$$
$$+ \sum_i \int_0^t \exp(-(2C+d)s) 2Y_s^i b_n^i(Y_s) \, ds$$
$$+ \int_0^t -(2C+d)\exp(-(2C+d)s) g(Y_s) \, ds$$
$$+ \frac{1}{2}\sum_i \int_0^t \exp(-(2C+d)s) 2 \, d\langle Y^i \rangle_s.$$

The inequalities above show that the sum of the last three integrals is ≤ 0, so $S_t \equiv \exp(-(2C+d)t)g(Y_t)$ is a local supermartingale, and since $S_t \ge 0$, it is a supermartingale. Applying the optional stopping theorem at time $T_n \wedge t$ now shows that

$$E_x \exp(-(2C+d)t) g(Y_{T_n \wedge t}) \le g(x),$$

so

$$P_x(T_n < t) \le \frac{\exp((2C+d)t)}{1 + n^2} g(x),$$

and the desired result follows immediately. $\square$

Remark: (7) implies that if $b(x) = |x|^\delta v$ where $v \ne 0$ is some fixed vector, then the process does not explode if $\delta \le 1$. We will see in a minute that it does explode if $\delta > 1$, so the condition above is sharp.

Perhaps the best way to get a feel for the properties of solutions of

(∗) $dX_t = dB_t + b(X_t) \, dt$

8.4 The Cameron-Martin Transformation

is to consider the one-dimensional case. In this case, the analysis is simple because we can find a function φ that makes $\varphi(X_t)$ a local martingale (this is the "natural scale"). To see what φ to choose, use Itô's formula to conclude that

$$\varphi(X_t) - \varphi(X_0) = \int_0^t \varphi'(X_s)\, dX_s + \frac{1}{2}\int_0^t \varphi''(X_s)\, ds$$

$$= \text{a local martingale} + \int_0^t \varphi'(X_s)b(X_s) + \frac{1}{2}\varphi''(X_s)\, ds,$$

so if we want $\varphi(X_t)$ to be a local martingale, we must have

$$\varphi'(x)b(x) + \frac{1}{2}\varphi''(x) = 0,$$

that is,

$$\varphi''(x) = -2b(x)\varphi'(x)$$

$$\varphi'(y) = C \exp\left(\int_0^y -2b(x)\, dx\right)$$

$$\varphi(z) = B + \int_0^z C \exp\left(\int_0^y -2b(x)\, dx\right) dy.$$

Taking $B = 0$ and $C = 1$ in the last expression, we get a function that is very useful in studying the behavior of X. We have used φ to try to remind you of the results in Sections 1.7 and 3.1. If it did, you should have no trouble with the following exercises.

Exercise 2 Let $T_c = \inf\{t : X_t = c\}$. Then if $a < x < b$,

$$P_x(T_b < T_a) = \frac{\varphi(x) - \varphi(a)}{\varphi(b) - \varphi(a)}.$$

Exercise 3 Since φ is increasing, $\varphi(\infty) = \lim_{x \uparrow \infty} \varphi(x)$ and $\varphi(-\infty) = \lim_{x \downarrow -\infty} \varphi(x)$ exist. Show that X is recurrent (i.e., $P_x(T_y < \infty) = 1$ for all x and y) if and only if $\varphi(\infty) = \infty$ and $\varphi(-\infty) = -\infty$. To see what this means in a concrete case, consider $b(x) = C|x|^\delta$ for $|x| \geq 1$, and $b(x) = 0$ otherwise. In this case,

$$\varphi(z) = \int_0^z \exp\left(-\int_1^y 2C|x|^\delta\, dx\right) dy,$$

so when $\delta > -1$, $\varphi(\infty) < \infty$, and when $\delta < -1$,

$$\varphi(z) \sim z \exp\left(-\int_1^\infty 2C|x|^\delta\, dx\right)$$

as $z \to \infty$. In the critical case $\delta = -1$,

$$\varphi(z) = 1 + \int_1^z \exp(-2C \log y) \, dy,$$

so $\varphi(\infty) = \infty$ if and only if $2C \le 1$.

The last exercise shows that X is recurrent if and only if $\varphi(-\infty, \infty) = (-\infty, \infty)$. If we think about the results of Section 2.11, this conclusion is obvious. $\varphi(X_t)$ is a time change of Brownian motion run for a random amount of time τ. If $\varphi(-\infty, \infty) \ne (-\infty, \infty)$, then this time must be finite, whereas if $\varphi(-\infty, \infty) = (-\infty, \infty)$, it must be infinite. By looking at the scale function, we can also tell exactly when the process will explode.

Exercise 4 *A Special Case of Feller's Test.* Let T_∞ be the explosion time defined above. Either $P_x(T_\infty = \infty) \equiv 1$ or $P_x(T_\infty < \infty) \equiv 1$, depending on whether or not

$$\int_{-\infty}^0 (\varphi(x) - \varphi(-\infty)) \, d\varphi(x) = \infty = \int_0^\infty (\varphi(\infty) - \varphi(x)) \, d\varphi(x).$$

Solution: The key to the proof of (7) was the fact that

$$S_t = (1 + X_t^2) \exp(-\lambda t) \ge 0$$

is a supermartingale if λ is sufficiently large. To get the optimal result on explosions, we have to replace $1 + x^2$ by a function that is tailor-made for the process. A natural choice that "gives up nothing" is a function $g \ge 0$ that makes $e^{-t}g(X_t)$ a local martingale. To solve this problem, it is convenient to look at things on the natural scale, that is, let $Y_t = \varphi(X_t)$ and find $f = g \cdot \varphi^{-1}$, for then when we apply Itô's formula, the second term on the right is a local martingale:

$$e^{-t}f(Y_t) - f(Y_0) = -\int_0^t e^{-s} f(Y_s) \, ds + \int_0^t e^{-s} f'(Y_s) \, dY_s$$
$$+ \frac{1}{2} \int_0^t e^{-s} f''(Y_s) \, d\langle Y \rangle_s.$$

To evaluate the third term on the right, we observe that

$$Y_t - Y_0 = \int_0^t \varphi'(X_s) \, dX_s + \text{bounded variation},$$

so

$$\langle Y \rangle_t = \int_0^t \varphi'(X_s)^2 \, ds.$$

Combining this result with other observations above, we see that $e^{-t}f(Y_t) = e^{-t}g(X_t)$ is a local martingale if and only if

$$\tfrac{1}{2} f''(x) \varphi'(x)^2 - f(x) = 0.$$

8.4 The Cameron-Martin Transformation

To solve this equation, we iterate. Let

$$f_0 \equiv 1$$

$$f_n(x) = \int_0^x d\varphi(x) \int_0^y d\varphi(y) f_{n-1}(y)$$

and let

$$f = \sum_{n=0}^\infty f_n.$$

It is easy to show that $f_n \leq (f_1)^n/n!$, so the series converges and has $1 + f_1 \leq f \leq \exp(f_1)$. If the solution f that we construct $\to \infty$ as we approach either end of $\varphi(-\infty, \infty)$, then we are done, because then we can let $[a_n, b_n] \uparrow \varphi(-\infty, \infty)$, let $T_n = \inf\{t : X_t \notin [a_n, b_n]\}$, and apply the optional stopping theorem at time $T_n \wedge t$ to conclude that if $x \in [a_n, b_n]$, then

$$P_x(T_n < t) \leq \frac{e^t f(x)}{f(a_n) \wedge f(b_n)} \to 0$$

as $n \to \infty$ (generalizing the inequality in the proof of (7)), so there is no explosion.

On the other hand, if $f(x)$ stays bounded as, say, $x \uparrow \varphi(\infty)$, we are also done, for then we let $\tau = \inf\{t : X_t = 0\}$, let $0 < x < \infty$, and apply the optional stopping theorem at time $\tau \wedge T_n$ to conclude that

$$1 < g(x) = E_x(e^{-\tau \wedge T_n} g(X_{\tau \wedge T_n}))$$
$$\leq 1 + g(\infty) E_x(e^{-\tau \wedge T_n}; T_n < \tau)$$
$$= 1 + g(\infty) E_x(e^{-T_n}; T_n < \tau).$$

Letting $n \to \infty$, now have

$$E_x(e^{-T_n}; T_n < \tau) \downarrow E_x(e^{-T_\infty}; T_\infty < \tau),$$

which is a contradiction, unless $P_x(T_\infty < \tau) > 0$, so X explodes in this case.

The last detail now is to relate the behavior of g to the behavior of the integrals above. Going back to natural scale and using the inequality $1 + f_1 \leq f \leq \exp(f_1)$, we see that X does not explode if and only if $f_1 \to \infty$ at both ends of $\varphi(-\infty, \infty)$, which is the condition given in the theorem.*

Note: This exercise is from McKean (1969). The result is due to Feller.

Now that the one-dimensional case has been discussed in general, the last step is to consider two concrete examples.

Example 1 *The d-Dimensional Bessel Process.* In Section 2.10 (see (8) and (9)), we showed that if $R_t = |B_t|$, where B_t is a d-dimensional Brownian motion, then

$$R_t - \frac{1}{2} \int_0^t \frac{d-1}{2} R_s^{-1} \, ds = \int_0^t R_s^{-1} B_s \cdot dB_s$$

and $\langle R \rangle_t = t$, so R_t is a solution of (*) with $b(x) = (d-2)/2x$ and underlying Brownian motion

$$B'_t = \int_0^t R_s^{-1} B_s \cdot dB_s.$$

Since the drift coefficient here is of the form $b(x) = Cx^{-1}$, by applying Exercise 3 we see that R_t is recurrent if and only if

$$2C = d - 1 \leq 1, \text{ that is, } d \leq 2.$$

Example 2 *The Ornstein-Uhlenbeck Process.*

$$b(x) = -\alpha x.$$

In this case there is an amusing way of solving

(*) $dX = -\alpha X \, dt + dB$

by purely formal calculations:

$$\frac{dX}{dt} = -\alpha X + \frac{dB}{dt}$$

$$e^{\alpha t} \frac{dX}{dt} = (-\alpha e^{\alpha t}) X + e^{\alpha t} \frac{dB}{dt}$$

$$\frac{d}{dt}(e^{\alpha t} X) = e^{\alpha t} \frac{dB}{dt}$$

$$e^{\alpha t} X_t - X_0 = \int_0^t e^{\alpha s} dB_s$$

$$X_t = e^{-\alpha t} X_0 + e^{-\alpha t} \int_0^t e^{\alpha s} dB_s.$$

All the calculations above are formal, but it doesn't really matter. Once we know what to guess, it is easy to check that the formula defines an Ornstein-Uhlenbeck process

$$E_x(X_t - X_0) = (e^{-\alpha t} - 1)x \sim -\alpha x t \text{ as } t \to 0$$

$$E_x(X_t - X_0)^2 = ((1 - e^{-\alpha t})x)^2 + E_x\left(e^{-\alpha t}\int_0^t e^{\alpha s} dB_s\right)^2$$

$$= 0(t^2) + t + o(t) \text{ as } t \to 0.$$

so it follows that

$$X_t + \int_0^t \alpha X_s \, ds \text{ is a local martingale and}$$

$$\langle X \rangle_t \equiv t.$$

The representation given above for the Ornstein-Uhlenbeck process is nice because it makes certain facts about the process obvious.

(a) If $X_0 = x$, then X_t has a normal distribution with mean $e^{-\alpha t}x$ and variance $(1 - e^{-2\alpha t})/2\alpha$, and hence
(b) As $t \to \infty$, X_t converges in distribution to normal with mean 0 and variance $1/2\alpha$.

Note: The discussion above of Example 2 is based on Section 16.1 of Breiman (1968). Although the first steps in developing the theory in this section were taken by Cameron and Martin (1944b), the formulation given above is due to Girsanov (1960). In our development above, we have basically followed Meyer (1976), but we have also incorporated some material from Friedman (1975) and Stroock and Varadhan (1979).

B Elliptic Equations

In the next three sections of this chapter, we show how Brownian motion can be used to construct (classical) solutions of the following equations:

$$0 = \tfrac{1}{2}\Delta u$$
$$0 = \tfrac{1}{2}\Delta u + g$$
$$0 = \tfrac{1}{2}\Delta u + cu$$

in an open set G subject to the following boundary condition: u is continuous in $\bar{G}$ and $u = f$ on ∂G.

The solutions to these equations are (under suitable assumptions) given by

$$E_x f(B_\tau)$$

$$E_x\left(f(B_\tau) + \int_0^\tau g(B_s)\,ds\right)$$

$$E_x\left(f(B_\tau)\exp\left(\int_0^\tau c(B_s)\,ds\right)\right)$$

where $\tau = \inf\{t : B_t \notin G\}$. Comparing the solutions above to the solutions given for the equations in Part A of this chapter shows that (except for a minor modification of the second solution) all we have done is replace t by τ. From this viewpoint, the changes made above may seem ad hoc, but if we reverse our perspective and rewrite the first solutions in terms of space-time Brownian motions $\hat{B}_s = (t - s, B_s)$ run until time $\tau = \inf\{s : \hat{B}_s \notin (0, \infty) \times R^d\}$, we see that the recipes are exactly the same.

8.5 The Dirichlet Problem

In this section, we will consider the most classical form of the Dirichlet problem, that is, we will study:

(1)   (a) $\Delta u = 0$ in G
       (b) u is continuous on $\bar{G}$ and $u = f$ on ∂G.

If we let $h(t, x) = u(x)$, then h satisfies the heat equation with $h(0, x) = u(x)$, that is, u is an equilibrium distribution for the heat equation in which the ∂G is held at a fixed temperature f (which may vary from point to point).

As in the first half of this chapter, the first step in solving (1) is to show:

(2)   Let $\tau = \inf\{t : B_t \notin G\}$. If u satisfies (a), then $M_t = u(B_t)$ is a local martingale on $[0, \tau)$.

Proof Applying Itô's formula gives

$$u(B_t) - u(B_0) = \int_0^t \nabla u(B_s) \cdot dB_s + \frac{1}{2} \int_0^t \Delta u(B_s) \, ds,$$

which proves (2), since $\Delta u \equiv 0$ and the first term is a local martingale on $[0, \tau)$.

If u is bounded and satisfies (a), then M_s, $0 \leq s < \tau$, is a bounded local martingale, so M_s converges to a limit as $s \uparrow \tau$. If G is bounded, $\tau < \infty$ a.s. (see Exercise 1 in Section 1.7), so if u satisfies (ii), the limit must be $f(B_\tau)$ and since M_s, $s < \tau$, is bounded, it follows that

$$M_s = E_x(f(B_\tau)|\mathcal{F}_s).$$

Taking $s = 0$ in the last equation gives

(3)   Suppose that G is bounded. If there is a solution of (1) that is bounded, it must be

$$v(x) = E_x f(B_\tau).$$

As in the first half of the chapter, it is easy to show:

(4)   Suppose that G and f are bounded. If v is smooth, then it satisfies (i).

Proof The Markov property implies that on $\{\tau > s\}$

$$E_x(f(B_\tau)|\mathcal{F}_s) = v(B_s).$$

Since the left-hand side is a local martingale on $[0, \tau)$, it follows that $v(B_s)$ is also. If v is smooth, then repeating the calculation in the proof of (2) shows that, for $s \in [0, \tau)$,

$$v(B_s) - v(B_0) = \frac{1}{2} \int_0^s \Delta v(B_r) \, dr + \text{a local martingale},$$

so it follows that the integral on the right-hand side is a local martingale and, hence, must be $\equiv 0$, so we have $\Delta v = 0$ in D.

Up to this point, everything has been the same as that in Section 8.1.

8.5 The Dirichlet Problem

Differences appear when we consider the boundary condition (b), since it is no longer sufficient for f to be bounded and continuous. The open set G must satisfy a regularity condition.

A point $y \in \partial G$ is said to be a regular point if $P_y(\tau = 0) = 1$.

(5) Let G be any open set. Suppose that f is bounded and continuous and y is a regular point of ∂G. If $x_n \in G$ and $x_n \to y$, then $v(x_n) \to v(y)$.

Proof The first step is to show

(5a) If $t > 0$, then $x \to P_x(\tau \le t)$ is lower semicontinuous.

Proof $P_x(X_s \in G^c \text{ for some } s \in (\varepsilon, t]) = \int p_\varepsilon(x, y) P_y(\tau \le t - s)$. Since $y \to (\tau \le t - s)$ is bounded and measurable and

$$p_\varepsilon(x, y) = (2\pi\varepsilon)^{-d/2} e^{-|x-y|^2/2\varepsilon},$$

it follows from the dominated convergence theorem that

$$x \to P_x(X_s \in G^c \text{ for some } s \in (\varepsilon, t])$$

is continuous for each $\varepsilon > 0$. Letting $\varepsilon \downarrow 0$ shows that $x \to P_x(\tau \le t)$ is an increasing limit of continuous functions and, hence, by a standard argument, that if $x_n \to y$, then

$$\liminf_{n \to \infty} P_{x_n}(\tau \le t) \ge P_y(\tau \le t).$$

If y is regular for G and $t > 0$, then $P_y(\tau \le t) = 1$, so it follows from (5a) that if $x_n \to y$, then

$$\liminf_{n \to \infty} P_{x_n}(\tau \le t) \ge 1.$$

With this established, it is easy to complete the proof. Since f is bounded and continuous, it suffices to show

(5b) If y is regular for G and $x_n \to y$, then for all $\delta > 0$

$$P_{x_n}(B_\tau \in D(y, \delta)) \to 1.$$

Proof Let $\varepsilon > 0$ and pick t so small that

$$P_0\left(\sup_{0 \le s \le t} |B_s| > \frac{\delta}{2}\right) < \varepsilon.$$

Since $P_{x_n}(\tau \le t) \to 1$ as $x_n \to y$, it follows from the choices above that

$$\liminf_{n \to \infty} P_{x_n}(B_\tau \in D(y, \delta)) \ge \liminf_{n \to \infty} P_{x_n}\left(\tau \le t, \sup_{0 \le s \le t} |B_s - x_n| \le \frac{\delta}{2}\right)$$

$$\ge \liminf_{n \to \infty} P_{x_n}(\tau \le t) - P_0\left(\sup_{0 \le s \le t} |B_s| > \frac{\delta}{2}\right)$$

$$> 1 - \varepsilon.$$

Since ε was arbitrary, this result proves (5b) and, hence, (5).

(5) shows that if G is regular (i.e., every point of ∂G is regular), then $v(x) = E_x f(B_\tau)$ will satisfy the boundary condition (b) for any bounded continuous f. It is easy to see that there is a converse to this result.

Exercise 1 Let G be an open set and let $y \in \partial G$ have $P_y(\tau = 0) < 1$ (and, hence, by Blumenthal's zero-one law, $P_y(\tau = 0) = 0$). Let f be a continuous function on ∂G with $f(y) = 1$ and $f(z) < 1$ for all other $z \in \partial G$. Show that there is a sequence of points $x_n \to y$ such that $\liminf_{n \to \infty} v(x_n) < 1$.

Hint: To find these points, start a Brownian motion at y and run it until it exits $D(y, 1/n)$.

From the discussion above, we see that for v to satisfy (b) it is sufficient (and almost necessary) that each point of ∂G be a regular point. This situation raises two questions:

Do irregular points exist?
What are sufficient conditions for a point to be regular?

In order to answer the first question we will give two examples.

Example 1 (trivial) Let $d \geq 2$ and let $G = D - \{0\}$, where $D = \{x : |x| < 1\}$. If we let $T_0 = \inf\{t > 0 : B_t = 0\}$, then $P_0(T_0 = \infty) = 1$, so 0 is not a regular point of ∂D.

Example 2 *Lebesgue's Thorn.* Let $d \geq 3$ and let $G = (-1, 1)^d - \bigcup_{n=1}^{\infty} [2^{-n}, 2^{-n-1}] \times [-a_n, a_n]^{d-1}$. (See Figure 8.1 for a look at $G \cap \{x : x_3 = \cdots = x_d = 0\}$. Younger readers will notice that G is a cubistic cousin of Pac-Man with infinitely many very small teeth.) I claim that if $a_n \downarrow 0$ sufficiently fast, then 0 is not a regular point of ∂G. To prove this result, we observe that since three-dimensional Brownian motion is transient and $P_0((B_t^2, B_t^3) = (0, 0)$ for some $t > 0) = 1$, then with probability 1, a Brownian motion B_t starting at 0 will not hit $I_n = \{x : x_1 \in [2^{-n}, 2^{-n-1}], x_2 = x_3 = \cdots = x_d = 0\}$, and furthermore, for a.e. ω the distance between $\{B_s : 0 \leq s < \infty\}$ and I_n is positive. From the last observation, it follows immediately that if we let $T_n = \inf\{t : B_t \in [2^{-n}, 2^{-n-1}] \times [a_n, a_n]^{d-1}\}$ and pick a_n small enough, then $P_0(T_n < \infty) \leq 3^{-n}$. Now $\sum_{n=1}^{\infty} 3^{-n} = 3^{-1}(3/2) = 1/2$, so if we let $\tau = \inf\{t > 0 : B_t \notin G\}$ and $\sigma = \inf\{t > 0 : B_t \notin (-1, 1)^d\}$, then we have

$$P_0(\tau < \sigma) \leq \sum_{n=1}^{\infty} P_0(T_n < \infty) < \frac{1}{2},$$

so

$$P_0(\tau > 0) \geq P_0(\tau = \sigma) > \frac{1}{2}$$

and 0 is an irregular point.

The last two examples show that if G^c is too small near y, then y may be

8.5 The Dirichlet Problem

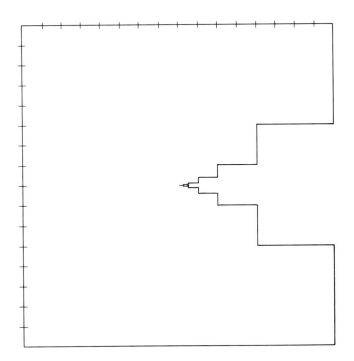

Figure 8.1

irregular. The next result shows that if G^c is not too small near y, then y is regular.

(5c) *Poincaré's Cone Condition.* If there is a cone V having vertex y such that $V \cap D(y,r) \subset G^c$, then y is a regular point.

Proof The first thing we have to do is explain what we mean by a cone. In Section 4.1, we defined

$$V_a^\theta = \{(x,y) \in H : |x - \theta| < ay, y < 1\},$$

an object that might be called a cone with opening a, vertex $(\theta, 0)$, and direction $e_d = (0, \ldots, 0, 1)$. Generalizing this definition, we define a cone with opening a, vertex z_1, and direction z_2 as follows:

$$V_a(z_1, z_2) = \{z : z = z_1 + y(z_2 + w) \text{ where } w \perp z_2 \text{ and } \|w\| < a\}.$$

Now that we have defined *cone*, the rest is easy. Since the normal distribution is spherically symmetric,

$$P_{z_1}(B_t \in V_a(z_1, z_2)) = \varepsilon > 0,$$

where ε is a constant that depends only on a, so an easy argument shows that if $V_a(z,z') \cap D(z,r) \subset G^c$ for some $r > 0$, then

$$\liminf_{t \downarrow 0} P_z(B_t \in G^c) \geq \varepsilon.$$

Combining the last conclusion with the trivial inequality $P_z(\tau \leq t) \geq P_z(B_t \in G^c)$ shows that

$$P_z(\tau = 0) = \lim_{t \downarrow 0} P_z(\tau \leq t) \geq \varepsilon,$$

and it follows from Blumenthal's zero-one law that $P_z(\tau = 0) = 1$.

The last result, called Poincaré's cone condition, is sufficient for most examples (e.g., if G is a region with a smooth boundary). The ultimate result on regularity is a necessary and sufficient condition due to Wiener (1924). To describe Wiener's test, we would have to define and explain the notion of capacity, so we will content ourselves to state what Wiener's test says about Example 2 above and refer the reader to Itô and McKean (1964), page 259, or Port and Stone (1978), page 68, for details.

(5d) In $d = 3$, $P_0(\tau = 0) = 0$ if and only if

$$-\infty \leq \sum_{n=1}^{\infty} \log(2^n a_n) < \infty.$$

In $d \geq 4$, $P_0(\tau = 0) = 0$ if and only if

$$\sum_{n=1}^{\infty} (2^n a_n)^{d-3} < \infty.$$

In contrast, Poincaré's cone condition implies that $P_0(\tau = 0) = 1$ if

$$\liminf_{n \to \infty} 2^n a_n > 0.$$

The last result completes our discussion of the boundary condition, so we now turn our attention to determining when v is smooth. As in Section 8.1, this is true under minimal assumptions on f.

(6) Let G be any open set. If f is bounded, then v is smooth and, hence, satisfies (a).

Proof Let $x \in G$ and pick $\delta > 0$ so that $D(x, \delta) \subset G$. If we let $\sigma = \inf\{t : B_t \notin D(x, \delta)\}$, then the strong Markov property implies that

$$(*) \quad v(x) = E_x f(B_\tau) = E_x[E_{B(\sigma)} f(B_\tau)]$$

$$= \int_{D(x,\delta)} v(y) \, d\pi(y),$$

where π is surface measure on $D(x, \delta)$ normalized to be a probability measure, so it follows from (7) in Section 2.10 that $v \in C^\infty$.

As in the last four sections, our last topic is to discuss what happens when something becomes unbounded. This time we will focus on G and ignore f. By repeating the arguments above, we can easily show the following:

(7a) Suppose that f is bounded and continuous and that each point of ∂G is regular. If for all $x \in G$, $P_x(\tau < \infty) = 1$, then v satisfies (1) and is the unique solution.

Conversely, we have

(7b) Suppose that f is bounded and continuous and that each point of ∂G is regular. If for some $x \in G$, $P_x(\tau < \infty) < 1$, then the solution of (1) is not unique.

Proof of (7b) Since $h(x) = P_x(\tau = \infty)$ has the averaging property given in (∗), it is C^∞ and has $\Delta h = 0$. Since each point of ∂G is regular, $P_x(\tau = \infty) \leq P_x(\tau > 1) \to 0$ as $x \to y \in \partial G$. The last two observations show that h is a solution of (1) with $f \equiv 0$, proving (7b).

By working a little harder, one can show that adding $\alpha P_x(\tau = \infty)$ is the only way to produce new bounded solutions.

(7c) Suppose that f is bounded and continuous. If u is bounded and satisfies (1), then there is a constant C such that

$$u(x) = E_x f(B_\tau) + C P_x(\tau = \infty).$$

Proof See Port and Stone (1978), Theorem 4.2.12.

8.6 Poisson's Equation

In this section, we will see what happens when we add a function of x to the equation considered in the last section, that is, we will study

(1) (a) $\frac{1}{2}\Delta u = -g$ in G
(b) u is continuous in $\bar{G}$ and $u = 0$ on ∂G.

If $G = R^d$ and the boundary condition is ignored, then any solution of (1) is called a potential of the charge distribution, because (if the units are chosen correctly) the gradient of the solution gives the force field associated with electrical charges distributed according to g.

As in the first five sections of this chapter, the first step in solving (1) is to show

(2) Let $\tau = \inf\{t : B_t \notin G\}$. If u satisfies (a), then

$$M_t = u(B_t) + \int_0^t g(B_s)\,ds$$

is a local martingale on $[0, \tau)$.

Proof Applying Itô's formula as we did in the last section gives

$$u(B_t) - u(B_0) = \int_0^t \nabla u(B_s) \cdot dB_s + \frac{1}{2}\int_0^t \Delta u(B_s)\,ds,$$

which proves (2), since $\frac{1}{2}\Delta u = -g$ and the first term on the right-hand side is a local martingale on $[0, \tau)$.

If G is bounded, then $E_x \tau < \infty$ for all $x \in G$ (see Exercise 1 in Section 1.7), so if g is bounded and u is bounded and satisfies (a), then for $s < \tau$

$$|M_s| \le \|u\|_\infty + \tau \|g\|_\infty,$$

so M_s, $s < \tau$, is a uniformly integrable martingale, and if u satisfies (b), then

$$\lim_{s \uparrow \tau} M_s = \int_0^\tau g(B_t)\, dt$$

and

$$M_s = E_x\left(\int_0^\tau g(B_t)\, dt \,\Big|\, \mathscr{F}_s\right).$$

Taking $s = 0$ now gives

(3) Suppose that G and g are bounded. If there is a solution of (1) that is bounded, it must be

$$v(x) = E_x\left(\int_0^\tau g(B_t)\, dt\right).$$

Again, it is easy to show

(4) Suppose that G and g are bounded. If v is smooth, then it satisfies (a) a.e. in G.

Proof The Markov property implies that on $\{\tau > s\}$,

$$E_x\left(\int_0^\tau g(B_t)\, dt \,\Big|\, \mathscr{F}_s\right) = \int_0^s g(B_t)\, dt + E_{B(s)}\left(\int_0^\tau g(B_t)\, dt\right).$$

Since the left-hand side is a local martingale on $[0, \tau)$, it follows that

$$\int_0^s g(B_t)\, dt + v(B_s)$$

is also. If v is smooth, then repeating the calculation in the proof of (2) shows that for $s \in [0, \tau)$,

$$v(B_s) - v(B_0) + \int_0^s g(B_r)\, dr = \int_0^s (\tfrac{1}{2}\Delta u + g)(B_r)\, dr + \text{a local martingale},$$

so again, we conclude that the integral on the right-hand side is a local martingale and, hence, must be $\equiv 0$, so we have $\tfrac{1}{2}\Delta u + g = 0$ a.e. in D.

After the discussion in the last section, the conditions needed to guarantee that the boundary conditions hold should come as no surprise.

(5) Suppose that G and g are bounded. Let y be a regular point of ∂G. If $x_n \in G$ and $x_n \to y$, then $v(x_n) \to 0$.

8.6 Poisson's Equation

Proof We begin by observing: (i) In the last section we showed that if $\varepsilon > 0$, $P_{x_n}(\tau > \varepsilon) \to 0$, and (ii) if G is bounded, then we have (see Exercise 1 of Section 1.7)

$$C = \sup_x E_x \tau < \infty.$$

Combining the last two observations with the Markov property shows that for any $\varepsilon > 0$,

$$|v(x_n)| \leq \varepsilon \|g\|_\infty + C\|g\|_\infty P_{x_n}(\tau > \varepsilon),$$

which proves (5).

Last, but not least, we come to the question of smoothness. We begin with the case $d \geq 3$, because in this case,

$$w(x) = E_x \int_0^\infty g(B_t)\,dt < \infty,$$

so the strong Markov property implies that

$$w(x) = E_x \int_0^\tau g(B_t)\,dt + E_x w(B_\tau),$$

and we have

(*) $\quad v(x) = w(x) - E_x w(B_\tau).$

The last equation allows us to verify that v is smooth by proving that w is, a task that is made simple by the fact that

$$w(x) = c \int |x-y|^{2-d} g(y)\,dy.$$

The first derivative is easy:

(6a)   If g is bounded and has compact support, then w is C^1.

Proof As before, we will content ourselves to show that the expression we get by differentiating under the integral sign converges and leave it to the reader to apply Exercise 1 of Section 1.10 to make the argument rigorous.

$$|x-y|^{2-d} = \left(\sum_i (x_i - y_i)^2\right)^{(2-d)/2},$$

so

$$D_i |x-y|^{2-d} = \left(\frac{2-d}{2}\right)\left(\sum_i (x_i - y_i)^2\right)^{-d/2} 2(x_i - y_i),$$

and we have

$$D_i w(x) = c \int (2-d) \frac{(x_i - y_i)}{|x-y|^d} g(y)\,dy,$$

the integral on the right-hand side being convergent, since

$$\int \left| \frac{(x_i - y_i)}{|x - y|^d} g(y) \right| dy \le \|g\|_\infty \int_{\{g>0\}} \frac{dy}{|x - y|^{d-1}} < \infty.$$

As in Section 8.2, trouble starts when we consider second derivatives. If $i \ne j$, then

$$D_{ij}|x - y|^{2-d} = (2-d)(-d)|x - y|^{-d-2}(x_i - y_i)(x_j - y_j).$$

In this case, the estimate used above leads to

$$|D_{ij}|x - y|^{2-d}| \le |x - y|^{-d},$$

which is (just barely) not locally integrable. As in Section 8.2, if g is Hölder continuous of order α, we can get an extra $|x - y|^\alpha$ to save the day. The details are tedious, so we will content ourselves to state the result:

(6b) If g is Hölder continuous, then w is C^2.

The reader can find a proof either in Port and Stone (1978), pages 116–117, or in Gilbarg and Trudinger (1977), pages 53–55.

Combining (∗) with (6a) and (6b) gives

(6) Suppose that G is bounded. If g is Hölder continuous, then v is smooth and hence satisfies (a).

Proof (6b) implies that w is C^2. Since w is bounded, it follows from (6) in the last section that $x \to E_x w(B_t)$ is C^∞.

The last result settles the question of smoothness in $d \ge 3$. To settle the question in $d = 1$ and $d = 2$, we need to find a substitute for (∗). To do this, we let

$$w(x) = \int G(x, y) g(y) \, dy$$

where G is the potential kernel defined in Section 1.8, that is,

$$G(x, y) = \begin{cases} -\frac{1}{\pi} \log(|x - y|) & d = 2 \\ -1|x - y| & d = 1 \end{cases}$$

G was defined as

$$\int_0^\infty p_t(x, y) - a_t \, dt$$

where the a_t were chosen to make the integral converge, so if $\int g\,dx = 0$, we see that

$$\int G(x, y) g(y) \, dy = \lim_{T \to \infty} E_x \int_0^T g(B_t) \, dt.$$

Using this interpretation of w, we can easily show that (*) holds, so again our problem is reduced to proving that w is C^2, which is a problem in calculus. Once all the computations are done, we find that (6) holds in $d \le 2$ and that in $d = 1$, it is sufficient to assume that g is continuous. The reader can find details in either of the sources given above.

On the basis of what we have done in the last five sections, our next step should be to consider what happens when something becomes unbounded. For the sake of variety, however, we will not do this, but instead, we will show how (1) can be used to study Brownian motion.

Example 1 Let $d = 1$, $G = (-1, 1)$, and $g \equiv 1$. In this case, formulas (3) through (6) imply that $v(x) = E_x \tau$ is the unique solution of

(1)   (a) $\frac{1}{2} u''(x) = -1$ in $(-1, 1)$
    (b) u is continuous on $[-1, 1]$ and $u(-1) = u(1) = 0$,

so $u(x) = 1 - x^2$. Once you see the one-dimensional case, it is easy to do the general case.

Exercise 2 Let $d \ge 2$, $D = \{x : |x| < 1\}$, $\tau = \inf\{t : B_t \notin D\}$. Then

$$E_x \tau = \frac{1}{d}(1 - |x|^2).$$

Remark: This result can also be proved by observing that $|B_t|^2 - dt$ is a martingale, so $|x|^2 = 1 - dE_x\tau$.

8.7 The Schrödinger Equation

In this section, we will consider what happens when we add cu to the left-hand side of the equation considered in Section 8.5, that is, we will study:

(1)   (a) $\frac{1}{2}\Delta u + cu = 0$ in G
    (b) u is continuous in $\overline{G}$ and $u = f$ on ∂G.

We will explain the physical significance of this equation in the next section. For the moment, you should consider it simply as the inevitable next step in the progression established in Sections 8.1, 8.2, 8.3, 8.5, and 8.6.

As in the first six sections of this chapter, the first step in solving (1) is to show

(2)   Let $\tau = \inf\{t : B_t \notin G\}$. If u satisfies (a) then

$$M_t = u(B_t) \exp\left(\int_0^t c(B_s)\, ds\right)$$

is a local martingale on $[0, \tau)$.

Proof Let $c_t = \int_0^t c(B_s)\, ds$. Applying Itô's formula gives

$$u(B_t)\exp(c_t) - u(B_0) = \int_0^t \exp(c_s)\nabla u(B_s) \cdot dB_s + \int_0^t u(B_s)\exp(c_s)\, dc_s$$

$$+ \frac{1}{2}\int_0^t \Delta u(B_s)\exp(c_s)\, ds,$$

which proves (2), since $dc_s = c(B_s)\, ds$, $\frac{1}{2}\Delta u + cu = 0$, and the first term on the right-hand side is a local martingale on $[0, \tau)$.

At this point, the reader might expect that the next step, as it has been six times before, is to assume that everything is bounded and conclude that if there is a solution of (1) that is bounded, it must be

$$v(x) = E_x(f(B_\tau)\exp(c_\tau)).$$

We will not do this, however, because the following simple example shows that this result is false.

Example 1 Let $d = 1$, $G = (-\pi/2, \pi/2)$, and $c \equiv 1/2$. If $u(x) = \cos x$, then $u'(x) = -\sin x$ and $u''(x) = -\cos x$, so $\frac{1}{2}u'' + cu = 0$ and $u = 0$ on ∂G. But there is obviously another solution: $u \equiv 0$.

We will see below that the trouble with the last example is that $c \equiv 1/2$ is too large or, to be precise, if we let

$$w(x) = E_x \exp\left(\int_0^\tau c(B_s)\, ds\right),$$

then $w \equiv \infty$. The rest of this section is devoted to showing that if $w \not\equiv \infty$, then "everything is fine." The development will require several stages. The first step is to show

(3a)   If $w \not\equiv \infty$ and G is connected, then $w(x) < \infty$ for all $x \in G$.

Proof Let $c^* = \sup|c(x)|$. By Exercise 1 in Section 1.7, we can pick r_0 so small that if $T_r = \inf\{t : |B_t - B_0| > r\}$ and $r \leq r_0$, then $E_x \exp(c^* T_r) \leq 2$ for all $x \in G$. If $D(x, r) \subset G$, then the strong Markov property implies that

$$w(x) = E_x[\exp(c(T_r))w(B(T_r))]$$
$$\leq E_x[\exp(c^* T_r)w(B(T_r))]$$
$$= E_x[\exp(c^* T_r)]\int_{\partial D(x,r)} w(y)\, d\pi(y),$$

since the exit time T_r and location are independent (here π is surface measure on $D(x, r)$ normalized to be a probability measure). If $\delta < r_0$ and $D(x, \delta) \subset G$, multiplying the last equality by r^{d-1} and integrating from 0 to δ gives

$$w(x) \leq 2 \cdot \frac{C}{\delta^d}\int_{D(x,\delta)} w(z)\, dz$$

where C is a constant (that depends only on d).

8.7 The Schrödinger Equation

Repeating the argument above and using $c(T_r) \geq -c^* T_r$, gives a lower bound of

$$w(x) \geq 2^{-1} \frac{C}{\delta^d} \int_{D(x,\delta)} w(y)\, dy.$$

Combining the last two bounds gives:

(3b) Let $\delta \leq r_0$. If $D(x, 2\delta) \subset G$ and $y \in D(x, \delta)$, then
$$w(x) \geq 2^{-(d+2)} w(y).$$

Proof

$$w(x) \geq 2^{-1} \frac{C}{(2\delta)^d} \int_{D(x,2\delta)} w(z)\, dz$$

$$\geq 2^{-1} \frac{C}{(2\delta)^d} \int_{D(y,\delta)} w(z)\, dz$$

$$\geq 2^{-1} \frac{C}{(2\delta)^d} \cdot 2^{-1} \frac{\delta^d}{C} w(y) = 2^{-(d+2)} w(y).$$

From (3b), we see that if $w(x) < \infty$, $2\delta < r_0$, and $D(x, 2\delta) < G$, then $w < \infty$ on $D(x, \delta)$. From this result, it follows that $G_0 = \{x : w(x) < \infty\}$ is an open subset of G. It is easy to see that G_0 is also closed (if $x_n \to y \in G$ and $D(y, 3\delta) \subset G$, then for n sufficiently large, $D(x_n, 2\delta) \subset G$ and $y \in D(x_n, \delta)$, so $w(y) < \infty$). From the last results, it follows (if G is connected) that $G_0 = G$, so (3a) holds.

With (3a) established, we are ready to prove our uniqueness result.

(3) Suppose that f and c are bounded and that $w \not\equiv \infty$. If there is a solution of (1) that is bounded, it must be
$$v(x) = E_x(f(B_\tau)\exp(c_\tau)).$$

Proof If u satisfies (a), then (2) implies that $X_s = u(B_{s\wedge t})\exp(c_{s\wedge t})$ is a local martingale on $[0, \tau)$. If u, f, and c are bounded and u satisfies (b), then letting $s \uparrow \tau$ gives

$$u(x) = E_x(f(B_\tau)\exp(c_\tau); \tau \leq t) + E_x(u(B_t)\exp(c_t); \tau > t).$$

Since f is bounded and $w(x) = E_x \exp(c_\tau) < \infty$, the dominated convergence theorem implies that as $t \to \infty$, the first term converges to

$$E_x(f(B_\tau)\exp(c_\tau)).$$

To show that the second term $\to 0$, we observe that

$$E_x[u(B_t)\exp(c_t); \tau > t] = E_x[E_x(u(B_\tau)\exp(c_\tau)|\mathscr{F}_t); \tau > t]$$
$$= E_x[u(B_t)\exp(c_t)w(B_t); \tau > t]$$

and use the trivial inequality

$$w(x) \geq \exp(-c^*) P_x(\tau \leq 1)$$

to conclude that

$$\inf_{x \in G} w(x) = \varepsilon > 0.$$

Replacing $w(B_t)$ by ε,

$$E_x[u(B_t)\exp(c_t); \tau > t] \leq \varepsilon^{-1} E_x[u(B_t)\exp(c_\tau); \tau > t]$$
$$\leq \varepsilon^{-1}\|u\|_\infty E_x[\exp(c_\tau); \tau > t] \to 0$$

as $t \to \infty$, since $w(x) = E_x \exp(c_\tau) < \infty$.

This completes our consideration of uniqueness. The next stage in our program, fortunately, is as easy as it always has been.

(4) Suppose that f and c are bounded and that $w \not\equiv \infty$. If v is smooth, then it satisfies (a) a.e. in G.

Proof The Markov property implies that on $\{\tau > s\}$,

$$E_x(\exp(c_\tau)f(B_\tau)|\mathscr{F}_s) = \exp(c_s)E_{B(s)}(\exp(c_\tau)f(B_\tau)).$$

Since the left-hand side is a local martingale on $[0, \tau)$, it follows that $\exp(c_s)v(B_s)$ is also. If v is smooth, then repeating the calculation in the proof of (2) shows that for $s \in [0, \tau)$,

$$v(B_s)\exp(c_s) - v(B_0) = \int_0^s (\tfrac{1}{2}\Delta v + cv)(B_r)\exp(c_r)\,dr + \text{a local martingale},$$

so it follows that the integral on the right-hand side is a local martingale and, hence, must be $\equiv 0$, so we have $\tfrac{1}{2}\Delta v + cv = 0$ a.e. in G.

Having proved (4), the next step is to consider the boundary condition. As in the last two sections, we need the boundary to be regular.

(5) Suppose that f and c are bounded and that $w \not\equiv \infty$. If f is continuous, then v satisfies (b) at each regular point of ∂G.

Proof Let y be a regular point of ∂G. We showed in (5a) and (5b) of Section 8.5 that if $\delta > 0$ and $x_n \to y$, then $P_{x_n}(\tau < \delta) \to 1$ and $P_{x_n}(B_\tau \in D(y, \delta)) \to 1$, so if $|c| \leq M$, then

$$P_{x_n}(e^{-M\delta} \leq \exp(c_\tau) \leq e^{M\delta}) \to 1$$

and, since f is bounded and continuous,

$$E_{x_n}(\exp(c_\tau)f(B_\tau); \tau < \delta) \to f(x).$$

To control the contribution from the rest of the space, we observe that

$$E_{x_n}(\exp(c_\tau)f(B_\tau); \tau \geq \delta) \leq e^{M\delta}\|f\|_\infty E_{x_n}(w(B_\tau); \tau \geq \delta)$$
$$\leq e^{M\delta}\|f\|_\infty \|w\|_\infty P_{x_n}(\tau \geq \delta) \to 0.$$

This brings us finally to the problem of determining when v is smooth enough to be a solution. To solve the problem in this case, we use the same

8.7 The Schrödinger Equation

trick used in Section 8.3 to reduce our result to the previous case. We observe that

$$\exp\left(\int_0^\tau c(B_s)\,ds\right) = 1 + \int_0^\tau c(B_s)\exp\left(\int_s^\tau c(B_r)\,dr\right)ds,$$

so multiplying by $f(B_\tau)$ and taking expected values gives

$$v(x) = 1 + \int_0^\infty E_x\left(c(B_s)\exp\left(\int_s^\tau c(B_r)\,dr\right)1_{(s<\tau)}\right)ds.$$

Conditioning on $\mathscr{F}_s$ and using the Markov property, we can write the above as

$$v(x) = 1 + \int_0^\infty E_x(c(B_s)w(B_s); \tau > s)\,ds.$$

The second term on the right-hand side is of the form considered in the last section, so if c and f are bounded and $w \not\equiv \infty$, then v is bounded, so it follows from results in the last section that v is C^1. If c is Hölder continuous, then the right-hand side is Hölder continuous, and we can use (6) from the last section to conclude

(6) Suppose that f and c are bounded and that $w \not\equiv \infty$. If c is Hölder continuous, then v is smooth and, hence, satisfies (a).

Combining (4) through (6), we see that if, in addition to the conditions in (6), we assume that f is continuous, then v satisfies (1). Just as in the last section, this fact can be used to study Brownian motion.

Example 2 Let $d = 1$, $G = (-1, 1)$, $c \equiv -\beta < 0$, and $f \equiv 1$. In this case, $v(x) = E_x \exp(-\beta\tau) \leq 1$, so v is the unique solution of (1), and we can find v by "guessing and verifying." Let

$$u(x) = a\cosh(bx)$$

and recall that

$$\cosh x = \frac{e^x + e^{-x}}{2} \qquad \sinh x = \frac{e^x - e^{-x}}{2}$$

$$\cosh' x = \sinh x \qquad \sinh' x = \cosh x.$$

so we have

$$\frac{1}{2}u'' - \beta u = \left(\frac{ab^2}{2} - \beta a\right)\cosh(bx).$$

From the last equation, we see that $u(x) = a\cosh(bx)$ satisfies (a) if and only if $b^2/2 = \beta$, that is, $b = \sqrt{2\beta}$, so picking a to satisfy the boundary condition, we find that

$$E_x \exp(-\beta\tau) = \frac{\cosh(x\sqrt{2\beta})}{\cosh(\sqrt{2\beta})}.$$

When $x = 0$, the expression reduces to
$$E_0 \exp(-\beta\tau) = \cosh(\sqrt{2\beta})^{-1}.$$
The formula above is valid only for $\beta > 0$, but if we let $\beta = -\alpha$, $\alpha > 0$, then
$$\cosh(\sqrt{-2\alpha}) = \frac{e^{i\sqrt{2\alpha}} + e^{-i\sqrt{2\alpha}}}{2} = \cos(\sqrt{2\alpha}).$$
The resulting expression,
$$E_0 \exp(\alpha\tau) = \cos(\sqrt{2\alpha})^{-1},$$
makes sense for $\alpha < \pi^2/8$ and $\uparrow \infty$ as $\alpha \uparrow \pi^2/8$. We leave it to the reader to prove that the formula above is correct. We will return to this example in the next section.

If you know something about Bessel functions, you can extend the last result to higher dimensions.

Example 3 Let $d \geq 2$, $G = \{z : |z| < 1\}$, and $c \equiv -\beta < 0$. Again, $v(x) = E_x \exp(-\beta\tau) \leq 1$, so v is the unique solution of (1). This time, however, it is not so trivial to guess the answer, so we just state the result:
$$v(x) = C|x|^{1-d/2} I_{d/2-1}(\sqrt{2\beta}|x|)$$
where I_v is the modified Bessel function
$$I_v(z) = \sum_{m=0}^{\infty} \left(\frac{z}{2}\right)^{v+2m} /(m!\Gamma(v+m+1))$$
(see Ciesielski and Taylor (1962) or Knight (1981), pages 88–89, for details). It is one of the great mysteries of life that the distribution of τ starting from 0 is the same as the total amount of time a $(d+2)$-dimensional Brownian motion spends in $\{z : |z| > 1\}$ (which can also be computed using the methods of this section). For more on this phenomenon, see Getoor and Sharpe (1979).

Example 4 $G = R^d$, $c(x) = -\alpha - \beta k(x)$, where α, $\beta \in (0, \infty)$, and $k(x) \geq 0$. Since G is unbounded, this example is, strictly speaking, not covered by the results above. However,
$$E_x\left(\exp\left(-\beta \int_0^t k(B_s)\,ds\right)\right) \leq 1$$
and we have supposed that $\alpha > 0$, so
$$v(x) = \int_0^\infty dt\, e^{-\alpha t} E_x\left(\exp\left(-\beta \int_0^t k(B_s)\,ds\right)\right)$$
is nicely convergent, and it is not hard to show (details are left to the reader) that v is the bounded solution of (1), the boundary condition (b) being regarded as vacuous.

8.7 The Schrödinger Equation

The most famous instance of this solution occurs when $d=1$ and

$$k(x) = \begin{cases} 1 & x > 0 \\ 0 & x \leq 0 \end{cases}$$

(which again does not satisfy our hypothesis). In this case, Kac (1951) showed that

$$v(0) = 1/\sqrt{\alpha(\alpha+\beta)},$$

so inverting the Laplace transform,

$$\frac{1}{\sqrt{\alpha(\alpha+\beta)}} = \int_0^\infty e^{-\alpha t} \frac{1}{\pi} \int_0^t \frac{e^{-\beta s}}{\sqrt{s(t-s)}} ds\, dt,$$

and observing that under P_0 the distribution of $t^{-1}|\{s \in [0,t] : B_s \geq 0\}|$ is independent of t, we get Lévy's arcsine law:

$$P_0(|\{s \in [0,t] : B_s \geq 0\}| \leq \theta t) = \frac{1}{\pi} \int_0^\theta \frac{dr}{\sqrt{r(1-r)}} = \frac{2}{\pi} \arcsin(\theta).$$

The reader should note that

$$t^{-1}|\{s \in [0,t] : B_s \geq 0\}| \not\to 1/2$$

as $t \to \infty$. In fact, the distribution of this quantity is independent of t, and its density has a minimum at $t = 1/2$!

Examples where (1) can be solved explicitly are rare. A second famous example, due to Cameron and Martin (1944b), is $k(x) = x^2$. In this case,

$$E_0\left(\exp\left(-\beta \int_0^t (B_s)^2\, ds\right)\right) = (\text{sech}((2\beta)^{1/2}t))^{1/2}.$$

The reader is invited to try to derive this equation. The proof given on pages 10–11 of Kac (1949) is a beautiful example of Kac's computational ability.

Up to now, we have focused our attention on the question of the existence of solutions to (1). The probabilistic formula for the solution can also be used to study properties of the solution. Perhaps the most basic result is

(7) **Harnack's Inequality.** Let $u \geq 0$ and satisfy $\frac{1}{2}\Delta u + cu = 0$ in $D = \{x : |x| < 1\}$. Then for any $r < 1$, there is a constant C (depending only on r, c) such that if $x, y \in D(0, r)$, then

$$u(x) \leq Cu(y).$$

Proof Pick r_0 so small that if $T_r = \inf\{t : |B_t - B_0| > r\}$ and $r \leq r_0$, then $E_x \exp(c^* T_r) \leq 2$, where $c^* = \sup|c(x)|$. Repeating the first computation in the proof of (3a), we see that if $\delta \leq r_0$ and $D(x, \delta) \subset G$, then

$$u(x) \leq 2 \cdot \frac{C}{\delta^d} \int_{D(x,\delta)} u(z)\, dz$$

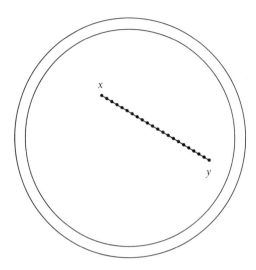

Figure 8.2

and

$$u(x) \geq 2^{-1} \frac{C}{\delta^d} \int_{D(x,\delta)} u(y) \, dy.$$

Therefore, since we have assumed that $u \geq 0$, we can repeat the proof of (3b) to conclude that if $\delta \leq r_0$, $D(x, 2\delta) \subset G$, and $y \in D(x, \delta)$, then

(*) $u(x) \geq 2^{-(d+2)} u(y)$.

The desired result now follows from a simple covering argument. Fix $r < 1$ and pick $\delta < r_0 \wedge (1-r)/2$. Given $x, y \in D(0, 1-r)$, there is a sequence of points $x_0 = x, x_1, \ldots, x_m = y$ (with $m \leq 3/\delta$), such that for all i, we have $|x_i - x_{i-1}| \leq 2\delta/3$ and $D(x_i, 2\delta) \subset D$ (see Figure 8.2). It follows from (*) that the inequality holds with $C = 2^{(d+2)3/\delta}$.

Remarks:

(a) The constant in the proof above grows like $\exp(C(1-r)^{-1})$; by working harder, you can get $(1-r)^{-\beta}$ (exercise).

(b) The result in (7) is true if D is replaced by any connected open set G and $D(0, r)$ is replaced by K, a compact subset of G. The details of the covering argument are much more complicated in this case. See Chung (1982), Exercise 3 on page 205, for a proof, or spend an hour or two and find your own.

(c) The result above is also true if we assume only that $c \in K_d^{\text{loc}}$ (the space defined in Section 8.2). The original proof, due to Aizenman and Simon (1982), was given in this generality. The fact that c can be unbounded causes quite a bit of trouble, but by following the outline of the proof of Theorem B.1.1 given in Section 8.3 and using a clever time reversal argument as a substitute for the self-adjointness used in Step 3, they succeed in proving the key estimate: If $\delta < \delta_0$, then

$$u(x) \le C \int_{\partial D(x,\delta)} u(y) \, d\pi(y)$$

(their (1.13)), and once this result is established, things follow pretty much as before. The reader can find the details clearly explained in their paper.

Exercise 1 Use the Poisson integral representation ((2) of Section 3.3) to show that if $u \ge 0$ and $\Delta u = 0$ in D, then for any $r < 1$, we have for all $x, y \in D(0, r)$ that

$$u(x) \le \left(\frac{1+r}{1-r}\right)^d u(y).$$

Hint: The worst case is $u(z) = (1 - |z|^2)/|z - 1|^d$.

Problem Let $\tau = \inf\{t : B_t \notin D\}$ and suppose that $E_z \exp(c_\tau) < \infty$. If we let

$$g_y(z) = E_z(\exp(c_\tau)|B_\tau = y),$$

then any solution of (1) satisfies

$$u(z) = \int_{\partial D} g_y(z) u(y) k_y(z) \, d\pi(y).$$

If we had good estimates on $g_y(z)$, then we could give a proof of (7) that is similar to the proof of Harnack's inequality which is given in Exercise 1, but I do not know how to do this.

Note: The applications to Brownian motion are based in part on Section 2.6 of Itô and McKean (1964), where the reader can find more details and other applications. The proofs of (3a), (3b), and Harnack's inequality follow Chung (1982), which is, in turn, based on his previous work with several coauthors. These results were also discovered independently by Aizenman and Simon (1982), who proved their results for $c \in K_d^{loc}$. Harnack's inequality is just one of several properties of solutions of Schrödinger's equation that can be studied using probabilistic methods. A good place to start learning about these results is Carmona and Simon (1981). They use probabilistic methods to study the exponential decay of Schrödinger eigenfunctions, and they give numerous references to earlier work on related topics.

8.8 Eigenvalues of $\Delta + c$

In this section, we will break the pattern set down in the first seven sections of this chapter and study a new type of problem:

(1) (a) $\frac{1}{2}\Delta u + cu = \lambda u$ in G
     (b) u is continuous in $\bar{G}$ and $u = 0$ on ∂G.

A function u that satisfies (1) is said to be an eigenfunction of $\frac{1}{2}\Delta + c$ (with Dirichlet boundary conditions), and λ is the corresponding eigenvalue. These functions are of interest because they correspond to the pure tones of a drum made in the shape of G. More generally, they are stationary states of the Schrödinger equation

$$iu_t = \tfrac{1}{2}\Delta u + cu.$$

A good way of getting a feel for (1) is to consider a simple example.

Example 1 Let $D = (0, 1)$, $c \equiv 0$. In this case, if we let $u(x) = \sin(n\pi x)$, then

$$\frac{1}{2}u''(x) = -\frac{(n\pi)^2}{2}\sin(n\pi x) = -\frac{(n\pi)^2}{2}u(x),$$

so (a) has solutions for $\lambda = -\frac{\pi^2}{2}, -2\pi^2, -\frac{9\pi^2}{2}, \ldots$. The next result, which is well known (see Courant and Hilbert (1953), vol. I), shows that this example is typical.

(A) If c is bounded and smooth enough (e.g., Hölder continuous), then there is an infinite sequence of (real) eigenvalues

$$\infty > \lambda_0 > \lambda_1 \geq \lambda_2 \geq \lambda_3 \ldots$$

such that $\lambda_n \to -\infty$ as $n \to \infty$. Some of the eigenvalues may be multiple, but the first is simple, and the corresponding eigenfunction can be chosen to be > 0.

In this section, we will investigate the probabilistic meaning of λ_0 and the corresponding eigenfunction u_0, and derive some characterizations of λ_0, ending with a "variational formula" for λ_0. Symbolically, we will show that

$$\lambda_0 \leq \varphi_0 \leq \varphi_1 \leq \psi_2 = \psi_1 \leq \varphi_1 \leq \lambda_0.$$

We will define the symbols above as we need them. The first inequality involves the function w used in the last section (and is from Chung and Li (1983)). Let

$$w_f(x) = E_x\left(\exp \int_0^\tau f(B_t)\, dt\right)$$

$$\varphi_0 = -\sup\{\theta : w_{c+\theta} \not\equiv \infty \text{ in } G\}.$$

(2) $\lambda_0 \leq \varphi_0.$

Proof If $E_x \exp(\int_0^\tau \theta + c(B_s)\, ds) \not\equiv \infty$, then it follows from results in Section 8.7 that

8.8 Eigenvalues of $\Delta + c$

$$\tfrac{1}{2}\Delta u + cu + \theta u = 0 \quad \text{in } G$$
$$u = 0 \quad \text{in } \partial G$$

has a unique solution $u \equiv 0$, so $-\theta$ is not an eigenvalue.

To work with φ_0, it is convenient to recast it in a more computational form. Let

$$a_t = \sup_x E_x(\exp(c_t); \tau > t)$$

where

$$c_t = \int_0^t c(B_s)\, ds.$$

The strong Markov property implies that

$$\begin{aligned}
a_{s+t} &= E_x(\exp(c_{s+t}); \tau > s+t) \\
&= E_x(\exp(c_s) E_{B(s)}(\exp(c_t) 1_{(\tau > t)}); \tau > s) \\
&\le a_t E_x(\exp(c_s); \tau > s) \\
&= a_t a_s.
\end{aligned}$$

Taking logarithms gives

$$\log a_{s+t} \le \log a_s + \log a_t,$$

that is, $b_t = \log a_t$ is subadditive.

From the last observation, it follows easily that

(*) $\quad \lim_{t \to \infty} b_t/t = \inf_{s>0} b_s/s.$

Proof It is clear that

$$\liminf_{t \to \infty} b_t/t \ge \inf_{s>0} b_s/s.$$

To prove the other inequality, observe that if $s > 0$, $ns \le t < (n+1)s$, and $r = t - ns$, then subadditivity implies

$$b_t \le nb_s + b_r,$$

so dividing by t and letting $t \to \infty$ gives

$$\limsup_{t \to \infty} b_t/t \le b_s/s,$$

which proves (*) since s is arbitrary.

Let $\varphi_1 = \lim_{t \to \infty} (1/t) \log a_t$. It is easy to see that

(3) $\quad \varphi_0 \le \varphi_1.$

Proof Let $l > \varphi_1$. Then if $t \ge t_0$, $\log a_t \le lt$ (i.e., $a_t \le e^{lt}$), so if $\theta < -l$, then

$$E_x\left(\exp\left(\int_0^t (\theta + c)(B_s)\, ds\right); \tau > t\right) = e^{\theta t} a_t \le e^{(\theta + l)t},$$

from which it follows easily that

$$E_x\left(\exp\left(\int_0^\tau (\theta + c)(B_s)\,ds\right)\right) < \infty.$$

Since this holds whenever $-\theta > \varphi_1$, it follows that

$$\varphi_0 = \inf\{-\theta : w_{c+\theta} \not\equiv \infty \text{ in } G\} \le \varphi_1.$$

Having proved that $\lambda_0 \le \varphi_1$, we are now in a situation where we can use results of Donsker and Varadhan (1974–1976). Our next step (following their (1976) paper) is to show that

(4) $$\varphi_1 \le \psi_2 = \inf_u \sup_x c + \frac{\Delta u}{2u}$$

where the infimum is taken over all $u \in C_+^\infty = \{u \in C^\infty : u > 0 \text{ on } \bar G\}$.

Proof If $l > \psi_2$, there is a $u \in C^\infty$ that is > 0 on $\bar G$ and has

$$c + \frac{\Delta u}{2u} \le l \quad \text{for all } x \in G.$$

Let $E = \inf\{u(x) : x \in G\}$ and $V(t,x) = u(x)e^{lt}/E$. I claim that

$$E_x\left(\exp\left(\int_0^t c(B_s)\,ds\right); \tau > t\right) \le v(x,t).$$

To prove this inequality, we observe that $\Delta u/2 + cu \le lu$, so

$$v_t - \frac{\Delta}{2}v - cv = lue^{lt} - \left(\frac{\Delta}{2}u + cu\right)e^{lt} \ge 0 \quad \text{in } G \times (0, \infty)$$

$$v(t,x) \ge 0 \qquad\qquad\qquad\qquad\qquad \text{on } \partial G \times (0, \infty)$$

$$v(0,x) \ge 1 \qquad\qquad\qquad\qquad\qquad \text{on } G.$$

and using Itô's formula gives

$$v(t-s, B_s)\exp(c_s) - v(t,x) = \int_0^s -v_t(t-r, B_r)\exp(c_r)\,dr$$

$$+ \int_0^s \exp(c_r)\nabla v(t-r, B_r)\cdot dB_r$$

$$+ \int_0^s v(t-r, B_r)\exp(c_r)c(B_r)\,dr$$

$$+ \frac{1}{2}\int_0^s \Delta v(t-r, B_r)\exp(c_r)\,dr.$$

Since $\left(v_t - \frac{\Delta}{2}v - cv\right) \ge 0$, we see that

$$v(t-s, B_s)\exp(c_s)$$

8.8 Eigenvalues of $\Delta + c$

is a nonnegative local supermartingale on $[0, \tau)$, so

$$v(t, x) \geq Ev((t - \tau)^+, B_{t\wedge\tau})\exp(c_{t\wedge\tau})$$
$$\geq E_x(\exp(c_t); \tau > t),$$

proving the desired inequality and completing the proof of (4).

Let $\psi_1 = \sup_\mu \inf_u \int \left(c + \dfrac{\Delta u}{2u}\right) d\mu$, where the supremum is taken over all probability measures on G and the infimum over all $u \in C_+^\infty$. In this notation,

$$\psi_2 = \inf_u \sup_\mu \left(\int c + \frac{\Delta u}{2u} d\mu \right).$$

The next step is to show that, in our situation, the infimum and supremum can be interchanged.

(5) $\quad\quad \psi_2 = \psi_1.$

Proof We only need to know that $\psi_2 \leq \psi_1$, but since inequality in the other direction is trivial, we begin by proving it. Let $F(x, y)$ be a function defined on some product space $A \times B$. If $(x_n, y_n) \in A \times B$, then

$$\sup_y F(x_n, y) \geq F(x_n, y_n) \geq \inf_x F(x, y_n),$$

so if we pick x_n and y_n such that

$$\sup_y F(x_n, y) \downarrow \inf_x \sup_y F(x, y)$$
$$\inf_x F(x, y_n) \uparrow \sup_y \inf_x F(x, y),$$

it follows that

$$\inf_x \sup_y F(x, y) \geq \sup_y \inf_x F(x, y),$$

proving $\psi_2 \geq \psi_1$.

The last inequality is valid for any function F, but as anyone who has heard of game theory can tell you,

(i) simple examples, for example, $F(x, y) = |x - y|$, show that we may have (here $x \sim u, y \sim \mu$)

$$\inf_x \sup_y F(x, y) > 0 = \sup_y \inf_x F(x, y)$$

(ii) but if $x \to F(x, y)$ is convex for fixed y, $y \to F(x, y)$ is concave for fixed x, and F is continuous in a suitable sense, then

$$\inf_x \sup_y F(x, y) = \sup_y \inf_x F(x, y).$$

Theorems of the type described in (ii) are called min-max theorems. The one we use is due to Sion (1958).

To apply this result, we write, with $u = e^h$,

$$D_i e^h = e^h D_i h$$
$$D_{ii} e^h = e^h D_{ii} h + e^h (D_i h)^2.$$

This change of variables makes

$$\int \left(c + \frac{\Delta u}{2u} \right) d\mu = \int \left(c + \frac{\Delta h}{2} + \frac{|\nabla h|^2}{2} \right) d\mu.$$

Let $G(\mu, h)$ denote the right-hand side of the last equation. If we use the usual weak topology on μ, then for fixed $h \in bC^\infty$, $\mu \to G(\mu, h)$ is continuous and linear. On the other hand, if we use the C^2 topology on C^∞ (i.e., $h_n \to h$ if and only if $D_\alpha h_n \to D_\alpha h$ uniformly on G for all α with $|\alpha| \le 2$), then for a fixed probability measure, $h \to G(\mu, h)$ is continuous and

$$G(\mu, \theta h_1 + (1-\theta) h_2) = \int c \, d\mu + \theta \int \frac{\Delta h_1}{2} d\mu + (1-\theta) \int \frac{\Delta h_2}{2} d\mu$$
$$+ \frac{1}{2} \int |\theta \nabla h_1 + (1-\theta) \nabla h_2|^2 d\mu.$$

To conclude that $h \to G(\mu, h)$ is convex, it suffices to show that for all a, b, and $\theta \in [0,1]$

$$(\theta a + (1-\theta) b)^2 \le \theta a^2 + (1-\theta) b^2.$$

Proof Since $(\theta a + (1-\theta) b)^2 \le (\theta |a| + (1-\theta)|b|)^2$ and the result is trivial when $a = 0$, letting $c = |b|/|a|$ it suffices to show that for all $c > 0$, $(\theta + (1-\theta)c)^2 \le \theta + (1-\theta)c^2$. This is true when $c = 0$. Differentiating the difference reveals that

$$\frac{\partial}{\partial c} = 2(1-\theta)c - 2(1-\theta)(\theta + (1-\theta)c)$$
$$> 0 \quad \text{if } c > 1$$
$$< 0 \quad \text{if } c < 1.$$

Checking the value at $c = 1$ reveals that $\theta + (1-\theta)c^2 - (\theta + (1-\theta)c)^2 = 0$, so the inequality holds for all c.

With the inequality above verified, we have shown that the hypotheses of Sion's theorem are satisfied and, hence, that (5) holds. This brings us to the fifth stage in the cycle:

(6)
$$\psi_1 \le \varphi_1 = \lim_{t \to \infty} \frac{1}{t} \log \sup_x E_x(\exp(c_t); \tau > t).$$

Proof We begin by making the simple observation that $\mu \to \int c \, d\mu$ is continuous and $J(\mu) = \inf_u \int \Delta u / 2u \, d\mu$ is upper semicontinuous, so there is a measure μ_0 with

$$\psi_1 = \int c \, d\mu_0 + J(\mu_0).$$

8.8 Eigenvalues of $\Delta + c$

This result frees us from having to deal with supremum over μ in the definition of ψ_1 and prepares us for the main part of the proof. In order to focus attention on the main steps of the proof, we will not prove first equality in detail. A complete proof (which involves discretizing time and passing to the limit) can be found on pages 601–602 of Donsker and Varadhan (1976).

Let $v(t, x) = E_x(\exp(c_t); \tau > t)$.

$$\frac{\partial}{\partial t} \int \log(v(t, x)) \, d\mu_0(x) = \int \frac{1}{v(t, x)} \frac{\partial v}{\partial t}(t, x) \, d\mu_0(x)$$

$$= \int \frac{1}{v(t, x)} \left(\frac{\Delta}{2} + c\right) v(t, x) \, d\mu_0(x)$$

$$= \int c \, d\mu_0 + \int \frac{\Delta v}{2v}(t, x) \, d\mu_0(x)$$

$$\geq \int c \, d\mu_0 + J(\mu_0) = \psi_1,$$

so

$$\int \log(v(t, x)) \, d\mu_0 \geq \psi_1 t.$$

Jensen's inequality implies that

$$\log \int d\mu_0(x) v(t, x) \geq \psi_1 t,$$

and we have

$$\int d\mu_0(x) v(t, x) \geq \exp(\psi_1 t),$$

which proves (6).

The last link in the chain is:

(7) $\quad \varphi_1 \leq \lambda_0.$

Proof Let $l = \varphi_1$. We want to show that l is in the spectrum, that is, if we let

$$R_l f(x) = \int_0^\infty e^{-lt} E_x(f(B_t) \exp(c_t)) \, dt,$$

then R_l is not a bounded operator on $C(\bar{G})$. If $R_l 1$ is bounded, then we have

$$\infty > \int d\mu_0(x) \int_0^\infty e^{-lt} v(t, x) \, dt$$

$$= \int_0^\infty dt \, e^{-lt} \int d\mu_0(x) v(t, x),$$

but it follows from (4), (5), and the proof of (6), that the last expression is

$$\geq \int_0^\infty dt\, e^{-lt} \exp(\psi_1 t) = \infty.$$

This is a contradiction, which proves (7) and completes the chain

$$\lambda_0 \leq \varphi_0 \leq \varphi_1 \leq \psi_2 = \psi_1 \leq \varphi_1 \leq \lambda_0.$$

At this point, we have just completed a rapid trip around the world, so we will try to fill in the picture with a number of remarks.

Remarks:

1. The equality of λ_0 and φ_0 in the case $c \geq 0$ is due to Khas'minskii (1959), who showed that $\lambda_0 < 0$ was equivalent to the existence of a solution of $\frac{1}{2}\Delta u + cu = 0$ that is > 0 on $\bar{G}$.

2. The function u that minimizes

$$\sup_x c + \frac{\Delta u}{2u}$$

is the eigenfunction g_0 associated with λ_0. The equality of ψ_2 and λ_0 was first proved by Protter and Weinberger (1966).

3. The function

$$J(\mu) = \inf_u \int \frac{\Delta u}{2u}\, d\mu$$

defined in the proof of (6) is -1 times what Donsker and Varadhan call $I(\mu)$. If $d\mu = f\, dx$ where f is smooth, then

$$J(\mu) = -\int |\nabla g|^2\, dx$$

where $g = f^{1/2}$ (see Donsker and Varadhan (1975–1976), I, Section 4), so substituting this equality in the definition of ψ_1 gives

$$\lambda_0 = -\inf_{\substack{g \in L^2 \\ \|g\|_2 = 1}} \int cg^2 + |\nabla g|^2\, dx,$$

the classical Rayleigh-Ritz variational formula for the first eigenvalue. The approach we have taken is certainly not the most direct way of proving this result; see Courant and Hilbert (1953), vol. I, Chapter 6.

4. The results we have proved above for Brownian motion are only a small part (and a relatively trivial one) of the theory of large deviations for Markov processes developed by Donsker and Varadhan. Their theory is one of the most important developments in probability theory in the last ten years, but you will have to learn about this from somebody else.

9 Stochastic Differential Equations

9.1 PDE's That Can Be Solved by Running an SDE

Let $Lf(x) = \frac{1}{2}\sum_{ij} A_{ij}(x)D_{ij}f(x) + \sum_i b_i(x)D_i f(x)$, where the $A_{ij}(x)$ and $b_i(x)$ are (for the moment) arbitrary. In this chapter, we will consider the following equation:

(1)   (a) $u_t = Lu$ in $(0, \infty) \times R^d$
         (b) u is continuous in $[0, \infty) \times R^d$ and $u(0, x) = f(x)$.

In Section 8.4, we solved (1) in the special case $A(x) \equiv I$ by first solving the stochastic differential equations

$$dX_t^x = dB_t + b(X_t^x)\,dt \quad X_0^x = x$$

and then running the resulting processes to solve (1):

$$u(t, x) = Ef(X_t^x).$$

On the basis of the results for $A \equiv I$, it seems reasonable to try to solve the general case by solving

$(*)\ dX_t^x = \sigma(X_t^x)\,dB_t + b(X_t^x)\,dt \quad X_0^x = x$

(where σ is a $d \times d$ matrix) and letting $u(t, x) = Ef(X_t^x)$. To see which σ to pick, we apply Itô's formula to get

$$u(t - s, X_s) - u(t, X_0) = \int_0^s -u_t(t - r, X_r)\,dr$$
$$+ \sum_i \int_0^s D_i u(t - r, X_r)\,dX_r^i$$
$$+ \frac{1}{2}\sum_{ij} \int_0^s D_{ij}u(t - r, X_r)\,d\langle X^i, X^j\rangle_r$$

(here we have taken the liberty of dropping the superscript x to simplify the formulas). Now

$$dX_s^i = b_i(X_s)\,ds + \sum_j \sigma_{ij}(X_s)\,dB_s^j,$$

so it follows from the formula for the covariance of two stochastic integrals that

$$\langle X^i, X^j \rangle_s = \sum_k \int_0^s \sigma_{ik}(X_s)\sigma_{jk}(X_s)\,ds$$

and we have

$$u(t - s, X_s) - u(t, X_0) = \int_0^s -u_t(t - r, X_r)\,dr$$
$$+ \sum_i \int_0^s D_i u(t - r, X_r) b_i(X_r)\,dr$$
$$+ \text{a local martingale}$$
$$+ \frac{1}{2}\sum_{ij} \int_0^s D_{ij}u(t - r, X_r)(\sigma\sigma^T)_{ij}(X_r)\,dr,$$

so if $\sigma\sigma^T = A$, then we have

$$u(t - s, X_s) - u(t, X_0) = \int_0^s (-u_t + Lu)(t - r, X_r)\,dr + \text{a local martingale}.$$

The condition $A = \sigma\sigma^T$ obviously restricts the set of A's we can consider, for if $A = \sigma\sigma^T$, then for each $x \in R^{d \times 1}$ (i.e., R^d viewed as $d \times 1$ matrices),

$$x^T A x = x^T \sigma\sigma^T x = |\sigma^T x|^2 \geq 0,$$

that is, A is nonnegative definite. If we assume (as we can without loss of generality) that A is symmetric, then this condition is also sufficient, because results in linear algebra tell us that any nonnegative definite symmetric matrix can be written as $U^T D U$, where U is an orthogonal matrix (i.e., $U^T U = I$) and D is a diagonal matrix. This observation allows us to define σ by setting $\sigma = U^T C U$, where $C \geq 0$ is the diagonal matrix that has $C^2 = D$.

With this choice of σ, we have:

(2) If u satisfies (i) and X satisfies ($*$), then $M_s = u(t - s, X_s)$ is a local martingale on $[0, t)$.

Proof By computations above,

$$u(t - s, X_s) - u(t, X_0) = \int_0^s (-u_t + Lu)(t - r, X_r)\,dr + \text{a local martingale}.$$

If u satisfies (i), the first term is $\equiv 0$.

Remark: Looking back at the proof of (2), we see that after we used Itô's formula to conclude that

9.1 PDE's That Can Be Solved by Running an SDE

$$u(t-s, X_s) - u(t, X_0) = \int_0^s -u_t(t-r, X_r)\,dr$$
$$+ \sum_i \int_0^s D_i u(t-r, X_r)\,dX_r^i$$
$$+ \frac{1}{2}\sum_{ij} \int_0^s D_{ij} u(t-r, X_r)\,d\langle X^i, X^j\rangle_r,$$

all we did was work out what dX_r^i and $d\langle X^i, X^j\rangle_r$ were and plug in their values, so we have the same conclusion if X satisfies:

(**)   (i) For each i, $X_t^i - \int_0^t b_i(X_s)\,ds$ is a local martingale
(ii) For each i, j,

$$\langle X^i, X^j\rangle_t = \int_0^t A_{ij}(X_s)\,ds.$$

We will see in the next section that there is essentially no difference between (*) and (**).

As has been the case many times before in Chapter 8, the last result leads immediately to a uniqueness theorem.

(3)   Let X_t be a solution of (*) (or (**)) with $X_0 = x$. If there is a solution of (1) that is bounded, it must be

$$v(t, x) = Ef(X_t).$$

Proof   If u satisfies (1), then $M_s = u(t-s, X_s)$ is a bounded local martingale on $[0, t)$ that converges to $f(X_t)$ as $s \uparrow t$ and satisfies

$$M_s = E(f(X_t)|\mathcal{F}_s),$$

so taking $s = 0$ proves (3).

Remark: The reader should note that (3) is also a uniqueness result for (*):

(3')   Suppose there is a solution of (1) that is bounded. If X_t and X_t' are two solutions of (*) with $X_0 = X_0' = x$ (constructed, perhaps, on different probability spaces using different Brownian motions), then

$$Ef(X_t) = Ef(X_t').$$

(2) and (3) may look the same as the corresponding steps in Sections 8.1 through 8.7, but when we start to consider the existence of solutions, things become very different. We first have to construct solutions of (*) and then run them to produce solutions of (1). The first task will be accomplished in Sections 9.2 through 9.5 (since I am a probabilist, a neophyte, and a pedagogue, we will spend some time investigating the countryside on the way to our destination—constructing "weak" solutions of (*)). In Sections 9.6 and 9.7, we turn our attention to $v(t, x) = Ef(X_t^x)$ and prove the analogues of the results

that we called (4), (5), and (6) in Sections 8.1 through 8.7. The first two results, which are easy consequences of the Markov and Feller properties, are dispensed with in Section 9.6. In Section 9.7, we confront (but do not conquer) the problem of proving (6).

9.2 Existence of Solutions to SDE's with Continuous Coefficients

In this section, we will describe Skorohod's approach to constructing solutions of stochastic differential equations. In order to focus our attention on σ and not on b (which we have already considered in Section 8.4), we assume that $b \equiv 0$. The same method works when there is a (bounded) continuous $b \not\equiv 0$, but there are more estimates to do and, as parenthetical qualification suggests, many of the complications involve issues (e.g., explosions if b is too big) that we have considered earlier.

Skorohod's idea for solving stochastic differential equations was to discretize time to get an equation that can be solved by induction, and then pass to the limit and extract subsequential limits to solve the original equation. Given this approach, it is natural (and almost necessary) to assume that each A_{ij} is a continuous function of x, and, for the moment, we will suppose that A is bounded, that is, $|A_{ij}(x)| \leq M$ for all i, j, x.

For each n, define $X_n(t)$ by setting $X_n(0) = x$ and, for $m2^{-n} < t \leq (m+1)2^{-n}$,

$$X_n(t) = X_n(m2^{-n}) + \sigma(X_n(m2^{-n}))(B_t - B(m2^{-n})),$$

where the second term is the matrix $\sigma(X_n(m2^{-n}))$ times the Brownian increment $(B_t - B(m2^{-n}))$. Since X_n is a stochastic integral with respect to Brownian motion, the formula for covariance of stochastic integrals implies

$$\langle X_n^i, X_n^j \rangle_t = \sum_k \int_0^t (\sigma_{ik}\sigma_{jk})(X_n([2^n s]/2^n))\, ds$$

$$= \int_0^t A_{ij}(X_n([2^n s]/2^n))\, ds,$$

and it follows that if $s < t$, then

$$|\langle X_n^i, X_n^j \rangle_t - \langle X_n^i, X_n^j \rangle_s| \leq M(t - s),$$

so we have (see Section 6.3)

$$E \sup_{u \in [s,t]} |X_n^i(u) - X_n^i(s)|^p \leq CE |\langle X_n^i \rangle_t - \langle X_n^i \rangle_s|^{p/2}$$

$$\leq CM(t - s)^{p/2}.$$

Taking $s = 0$ and $p = 1$, we see that

$$E|X_n^i(t)| \leq CMt^{1/2},$$

so by taking subsequences, we can guarantee that for each i and rational t, $X_{n(k)}^i(t)$ converges weakly to a limit as $k \to \infty$. Taking $p = 4$, we see that

9.2 Existence of Solutions to SDE's with Continuous Coefficients

$$E \sup_{u\in[s,t]} |X_n^i(u) - X_n^i(s)|^4 \le CM(t-s)^2,$$

so the processes X_n^i satisfy Kolmogorov's continuity condition uniformly in n.

Combining the last observation with a standard result (Theorem 14.3 in Billingsley (1968)), we can conclude that the measures $P_{n(k)}$ induced on $(C, \mathscr{C})$ by the $X_{n(k)}$ converge weakly to a limit Q_x on $(C, \mathscr{C})$. I claim that under Q_x, the coordinate maps $X_t(\omega) = \omega_t$ satisfy (*). To prove this, we will first show that if $f \in C^2$ and we let

$$Lf = \frac{1}{2} \sum_{ij} A_{ij} D_{ij} f,$$

then we have

(1) $$f(X_t) - f(X_0) - \int_0^t Lf(X_s)\, ds$$

is a local martingale/Q_x.

Proof It suffices to show that if f, $D_i f$, and $D_{ij} f$ are bounded, then the process above is a martingale.

Itô's formula implies that

$$f(X_n(t)) - f(X_n(s)) = \sum_i \int_s^t D_i f(X_n(r))\, dX_n^i(r)$$
$$+ \frac{1}{2} \sum_{ij} \int_s^t D_{ij} f(X_n(r))\, d\langle X_n^i, X_n^j \rangle_r,$$

and it follows from the definition of X^n that

$$\langle X_n^i, X_n^j \rangle_r = \int_0^r A_{ij}(X_n([2^n u] 2^{-n}))\, du,$$

so if we let

$$L_n f(r) = \sum_{ij} A_{ij}(X_n([2^n r] 2^{-n})) D_{ij} f(X_r^n),$$

then $f(X_n(t)) - f(X_n(s)) - \int_s^t L_n f(r)\, dr$ is a local martingale.

The Skorohod (1956) representation theorem implies that we can construct processes $Y^k \stackrel{d}{=} X_{n(k)}$ on some probability space in such a way that with probability 1 as $k \to \infty$, Y_t^k converges to a limit Y_t uniformly on $[0, T]$ for any $T < \infty$. If $s < t$ and $g : C \to R$ is a bounded continuous function that is measurable with respect to $\mathscr{F}_s$, then

$$E\left(g(Y) \cdot \left\{ f(Y_t) - f(Y_s) - \int_s^t Lf(Y_r)\, dr \right\}\right)$$
$$= \lim E\left(g(Y^k) \cdot \left\{ f(Y_t^k) - f(Y_s^k) - \int_s^t L_n f(r)\, dr \right\}\right) = 0,$$

which proves (1).

Applying (1) to $f_i(x) = x_i$ and $f_{ij}(x) = x_i x_j$, we see that under Q_x, the coordinates X^i_t are local martingales with

$$\langle X^i, X^j \rangle_t = \int_0^t A_{ij}(X_s)\,ds,$$

that is, X solves the problem we called (∗∗) in Section 9.1. The final step is to construct a Brownian motion B such that

(∗) $\quad X_t - X_0 = \int_0^t \sigma(X_s)\,dB_s.$

If σ is invertible for each x, then the proof is trivial. We let $B_t = \int_0^t \sigma^{-1}(X_s)\,dX_s$. The associative law implies that this process satisfies (∗). To see that it is a Brownian motion, observe that each component B^i_t is a local martingale and that

$$\langle B^i, B^j \rangle_t = \sum_{kl} \int_0^t \sigma^{-1}_{ik}(X_s)\sigma^{-1}_{jl}(X_s)A_{kl}(X_s)\,ds$$

$$= \int_0^t (\sigma^{-1} A \sigma^{-1})_{ij}(X_s)\,ds = \delta_{ij} t,$$

since σ is symmetric and

$$\sigma^{-1} A \sigma^{-1} = \sigma^{-1} \sigma^2 \sigma^{-1} = I.$$

When σ is not invertible, for example, when $\sigma \equiv 0$, one has to first enlarge the space by adding independent Brownian motions and then use some linear algebra to get around the fact that σ^{-1} does not exist. Since the details get a little messy, we leave it to the reader either to figure out how to do this or to look up the answer in Ikeda and Watanabe (1981) on pages 89–91.

The discussion above shows how to solve (∗) when A is bounded and continuous. We now deal with a general continuous A. Let $0 < g(x) \le 1$ be a continuous function such that $\bar{A}(x) = g(x)A(x)$ is bounded. Let $\bar{\sigma}(x) = g(x)^{1/2}\sigma(x)$ and let Y_s be a solution of $dY_s = \bar{\sigma}(Y_s)\,dB_s$. Let

$$\sigma_t = \int_0^t g(Y_s)\,ds \quad (\le t)$$

$$T = \int_0^\infty g(Y_s)\,ds,$$

and for $s < T$, let

$\gamma_s = \inf\{t : \sigma_t > s\}$
$X_s = Y(\gamma_s).$

Since γ_s is an increasing family of stopping times, each component X^i_s is a local martingale. To compute $\langle X^i, X^j \rangle_s$, we observe that if $t < T$, then

$$X^i_t X^j_t - \int_0^t A_{ij}(X_s)\,ds = Y^i_{\gamma(t)} Y^j_{\gamma(t)} - \int_0^t A_{ij}(Y_{\gamma(s)})\,ds.$$

9.2 Existence of Solutions to SDE's with Continuous Coefficients

Changing variables $s = \sigma_r$ in the integral above and observing that $\gamma(\sigma_r) = r$ converts the right-hand side to

$$Y^i_{\gamma(t)} Y^j_{\gamma(t)} - \int_0^{\gamma(t)} A_{ij}(Y^j_r) g(Y^j_r) \, dr,$$

which is a local martingale, so we have

$$\langle X^i, X^j \rangle_t = \int_0^t A_{ij}(X_s) \, ds.$$

As before, we can construct a Brownian motion B such that for $t < T$,

$$X_t - X_0 = \int_0^t \sigma(X_s) \, dB_s.$$

If $P(T = \infty) = 1$, then we have solved (*) for all times and we are done. If $P(T < \infty) > 0$, then, as you might guess from the discussion in Section 8.4, we are also done, for the process has exploded, in other words, $\lim_{t \uparrow T} Y_t = \infty$ a.s. on $\{T < \infty\}$. It is easy to explain why this is true—for any R, there is a δ such that if $|x| < R$ and $S_{2R} = \inf\{t : |X_t| > 2R\}$, then $P_x(S_{2R} > \delta) > \delta$; therefore, with probability 1, X_t cannot leave $|x| \le 2R$ and return to $|x| \le R$ infinitely many times in a finite time interval. A full proof, however, requires the strong Markov property, which we have not established, and a number of unpleasant details, so we leave the rest to the reader. Again, a complete proof can be found in Ikeda and Watanabe (1981), this time on pages 160–162.

To get an idea of when explosions occur, consider the following:

Example 1 $A_{ij}(x) = (1 + |x|^\delta)\delta_{ij}$.
In these processes, if we let $g(x) = (1 + |x|^\delta)^{-1}$, then $\bar{A}(x) = I$, so only the second part of the construction is necessary:

$$T = \int_0^\infty (1 + |B_s|^\delta)^{-1} \, ds.$$

In $d = 1, 2$, Brownian motion is recurrent, so $T \equiv \infty$ (and the process never explodes). In $d \ge 3$, Fubini's theorem implies that

$$E_x T = E_x \int_0^\infty (1 + |B_s|^\delta)^{-1} \, ds$$

$$= C \int \frac{1}{|x - y|^{d-2}} \frac{1}{1 + |y|^\delta} \, dy.$$

The integrand $\sim |y|^{2-\delta-d}$ as $y \to \infty$, so if

$$\int_1^\infty r^{2-\delta-d} r^{d-1} \, dr < \infty$$

(i.e., $\delta > 2$), we have $E_x T < \infty$.
Conversely, we have

(2) If trace$(A) \le C(1 + |x|^2)$, then $Q_x(T = \infty) = 1$.

Proof This proof is the same as the proof of (7) in Section 8.4. Let $\varphi(x) = 1 + |x|^2$. By Itô's formula,

$$\varphi(X_t) - \varphi(X_0) = \sum_{i=1}^{d} \int_0^t 2X_s^i \, dX_s^i + \frac{1}{2} \sum_{i=1}^{d} \int_0^t 2 \, d\langle X^i \rangle_s$$

$$= \text{a local martingale} + \int_0^t \sum_{i=1}^{d} A_{ii}(X_s) \, ds.$$

The last integral $\leq C \int_0^t \varphi(X_s) \, ds$, so another application of Itô's formula shows that

$$e^{-Ct} \varphi(X_t) - \varphi(X_0) = \text{a local martingale} + \sum_i \int_0^t e^{-Cs} A_{ii}(X_s) \, ds$$

$$+ \int_0^t (-C) e^{-Cs} \varphi(X_s) \, ds$$

$$= \text{a local supermartingale}.$$

If we let $T_n = \inf\{t : |X_t| > n\}$, it follows from the optional stopping theorem that $Q_x(T_n < t) \to 0$ as $n \to \infty$.

Note: Our treatment of the existence of solutions follows Section 4.2 of Ikeda and Watanabe (1981), who in turn got their proof from Skorohod (1965). The treatment of explosions here and in Section 8.4 is from Section 10.2 in Stroock and Varadhan (1979). They also give a more refined result due to Hasminskii.

9.3 Uniqueness of Solutions to SDE's with Lipschitz Coefficients

The first and simplest existence and uniqueness result was proved by K. Itô (1946).

(1) If for all i, j, x, and y we have $|\sigma_{ij}(x) - \sigma_{ij}(y)| \leq K|x - y|$ and $|b_i(x) - b_i(y)| \leq K|x - y|$, then the stochastic differential equation

$$X_t = x + \int_0^t \sigma(X_s) \, dB_s + \int_0^t b(X_s) \, ds$$

has a unique solution.

Proof We construct the solution by successive approximation. Let $X_t^0 \equiv x$ and define:

$$X_t^n = x + \int_0^t \sigma(X_s^{n-1}) \, dB_s + \int_0^t b(X_s^{n-1}) \, ds \quad \text{for } n \geq 1.$$

9.3 Uniqueness of Solutions to SDE's with Lipschitz Coefficients

Let $\Delta_n(t) = E\left(\sup_{0 \le s \le t} |X_s^n - X_s^{n-1}|^2\right)$.

We estimate Δ_n by induction. The first step is easy:

$$X_s^1 = x + \sigma(x)B_s + b(x)s,$$

so

$$|X_s^1 - X_s^0| \le |\sigma(x)B_s| + |b(x)|s.$$

Squaring and using the fact that

$$\sup_{0 \le s \le t} |\sigma(x)B_s| \stackrel{d}{=} t^{1/2} \sup_{0 \le s \le 1} |\sigma(x)B_s|,$$

gives

$$\Delta_1(t) \le C_1 t + C_2 t^{3/2} + |b|^2 t^2 \le C(t + t^2)$$

where

$$C = C_1 + C_2 + |b|^2.$$

To bound $\Delta_m(t)$ for $m \ge 2$, we recall that $|a + b|^2 \le 2a^2 + 2b^2$ and observe that this implies that for $n \ge 1$,

$$\Delta_{n+1}(T) \le 2E \sup_{0 \le t \le T} \left|\int_0^t \sigma(X_s^n) - \sigma(X_s^{n-1}) \, dB_s\right|^2$$

$$+ 2E \sup_{0 \le t \le T} \left|\int_0^t b(X_s^n) - b(X_s^{n-1}) \, ds\right|^2.$$

To bound the second term, we observe that the Cauchy-Schwarz inequality implies that

$$\sum_{i=1}^d \left(\int_0^t b^i(X_s^n) - b^i(X_s^{n-1}) \, ds\right)^2 \le \sum_{i=1}^d \left(\int_0^t (b^i(X_s^n) - b^i(X_s^{n-1}))^2 \, ds\right)\left(\int_0^t 1 \, ds\right),$$

so we have

(a) $2E \sup_{0 \le t \le T} \left|\int_0^t b(X_s^n) - b(X_s^{n-1}) \, ds\right|^2 \le 2TE \int_0^T |b(X_s^n) - b(X_s^{n-1})|^2 \, ds$

$$\le 2dTK^2 E \int_0^T |X_s^n - X_s^{n-1}|^2 \, ds.$$

To bound the other term, let σ_i be the ith row of σ and observe that Doob's inequality implies that

$$E \sup_{0 \le t \le T} \left(\int_0^t (\sigma_i(X_s^n) - \sigma_i(X_s^{n-1})) \cdot dB_s\right)^2 \le 4E\left(\int_0^T (\sigma_i(X_s^n) - \sigma_i(X_s^{n-1})) \cdot dB_s\right)^2$$

$$= 4E \int_0^T |\sigma_i(X_s^n) - \sigma_i(X_s^{n-1})|^2 \, ds,$$

so we have

(b) $2E \sup_{0 \leq t \leq T} \left| \int_0^t (\sigma(X_s^n) - \sigma(X_s^{n-1})) \cdot dB_s \right|^2$

$$\leq 2E \sum_{i=1}^d \sup_{0 \leq t \leq T} \left| \int_0^t (\sigma_i(X_s^n) - \sigma_i(X_s^{n-1})) \cdot dB_s \right|^2$$

$$\leq 8E \sum_{i=1}^d \int_0^T |\sigma_i(X_s^n) - \sigma_i(X_s^{n-1})|^2 \, ds$$

$$\leq 8d^2 K^2 E \int_0^T |X_s^n - X_s^{n-1}|^2 \, ds.$$

Combining the last two inequalities shows that

(c) $\Delta_{n+1}(T) \leq BE \int_0^T |X_s^n - X_s^{n-1}|^2 \, ds$

$$\leq B \int_0^T \Delta_n(s) \, ds$$

(where $B = 2dTK^2 + 8d^2K^2$ depends on T). Since $\Delta_1(t) \leq C(t + t^2)$, iterating (c) gives that if $t \leq T$, then

$$\Delta_2(t) \leq B \int_0^t C(s + s^2) \, ds = BC\left(\frac{t^2}{2} + \frac{t^3}{3}\right)$$

$$\Delta_3(t) \leq B \int_0^t BC\left(\frac{s^2}{2} + \frac{s^3}{3}\right) ds = B^2 C\left(\frac{t^3}{3!} + \frac{2t^4}{4!}\right),$$

and it follows by induction that if $t \leq T$, then

(d) $\Delta_n(t) \leq B^{n-1} C\left(\dfrac{t^n}{n!} + \dfrac{2t^{n+1}}{n+1!}\right).$

With this estimate established, the rest of the proof is routine. Chebyshev's inequality shows that

$$P\left(\sup_{0 \leq t \leq T} |X_t^n - X_t^{n-1}| > 2^{-n}\right) \leq 2^{2n} \Delta_n(T).$$

Since the right-hand side is summable, the Borel-Cantelli lemma implies that

$$P\left(\sup_{0 \leq t \leq T} |X_t^n - X_t^{n-1}| > 2^{-n} \text{ i.o.}\right) = 0,$$

so with probability 1, $X_t^n \to$ a limit X_t uniformly on $[0, T]$, and it follows from the estimates above that

(e) for all $m \leq n < \infty$,

$$E\left(\sup_{0 \leq s \leq T} |X_s^m - X_s^n|^2\right) \leq \left(\sum_{k=m+1}^n \Delta_k(T)^{1/2}\right)^2.$$

If we let $X_t^\infty = X_t$, then the result also holds for $n = \infty$.

9.3 Uniqueness of Solutions to SDE's with Lipschitz Coefficients

Proof If $n < \infty$, then it follows from the triangle inequality that

$$\left\| \sup_{0 \le s \le T} |X_s^m - X_s^n|^2 \right\|_2 \le \sum_{k=m+1}^{n} \left\| \sup_{0 \le s \le T} |X_s^m - X_s^{m-1}| \right\|_2 \equiv \sum_{k=m+1}^{n} \Delta_k(T)^{1/2}.$$

Letting $n \to \infty$ and using Fatou's lemma proves the second claim.

At this point, we have assembled all the ingredients. The rest of the proof consists of applying what we have learned to prove (1). To see that X_t is a solution, we observe that if we let

$$\tilde{Y}_t = x + \int_0^t \sigma(Y_s) \, dB_s + \int_0^t b(Y_s) \, ds,$$

then it follows from the proof of (c) above that

(c') $\quad E\left(\sup_{0 \le t \le T} |\tilde{Y}_t - \tilde{Z}_t|^2 \right) \le BE \int_0^T |Y_s - Z_s|^2 \, ds.$

Letting $Y_t = X_t^n$ and $Z_t = X_t$ shows that

$$E\left(\sup_{0 \le t \le T} |X_t^{n+1} - \tilde{X}_t|^2 \right) \le BE \int_0^T |X_s^n - X_s|^2 \, ds$$

$$\le BTE\left(\sup_{0 \le s \le T} |X_s^n - X_s|^2 \right) \to 0$$

by (e), so $\tilde{X}_t = \lim X_t^{n+1} = X_t$.

To prove uniqueness, observe that (e) implies that

$$E\left(\sup_{0 \le s \le T} |x - X_s|^2 \right) \le \left(\sum_{m=1}^{\infty} \Delta_m(T)^{1/2} \right)^2 < \infty,$$

so if Y is another solution with $E\sup_{0 \le s \le T} |Y_s|^2 < \infty$ and we let

$$\varphi(t) = E\left(\sup_{0 \le s \le t} |X_s - Y_s|^2 \right),$$

then $\varphi(0) = 0$, (c') implies that

$$\varphi(t) \le B \int_0^t \varphi(s) \, ds,$$

and it follows from an easy argument that $\varphi(t) \le \varepsilon e^{Bt}$ for any $\varepsilon > 0$, so $\varphi \equiv 0$ and, hence, $X = Y$. To remove the integrability condition on Y, observe that for any $R < \infty$, we can modify σ and b outside $|x| > R$ in such a way that σ and b still satisfy $|\sigma_{ij}(x) - \sigma_{ij}(y)| \le K|x - y|$ and $|b_i(x) - b_i(y)| \le K|x - y|$, but σ and b are $\equiv 0$ off some compact set. When we do this, any solution has $E(\sup_{0 \le s \le T} |Y_s|^2) < \infty$, so we can conclude that any solution of the original equation agrees with the solution we have constructed until it leaves $|x| \le R$ and then since (e) shows that our solution does not explode, we can conclude that ours is the only solution of (∗).

Before the reader forgets the proof given above, we would like to observe that the argument above gives a continuity result for $x \to X_t^x$ (the solution starting at $X_0 = x$). Let X and Y denote solutions of (*) with $X_0 = x$ and $Y_0 = y$, and let X^n and Y^n be the sequence of processes generated by the construction above when $X_t^0 \equiv x$ and $Y_t^0 \equiv y$. If we let

$$\Delta_n'(t) = E\left(\sup_{0 \le s \le t} |X_s^n - Y_s^n|^2\right),$$

observe that $\Delta_0'(t) \equiv |x - y|^2$, and iterate (c'), we see that if $t \le T$, then

$$\Delta_1'(t) \le B \int_0^t |x - y|^2 \, dt = B|x - y|^2 t$$

$$\Delta_2'(t) \le B \int_0^t B|x - y|^2 s \, ds = B^2 |x - y|^2 \frac{t^2}{2}.$$

It follows by induction that if $t \le T$, then

(d') $\Delta_n'(t) \le B^n |x - y|^2 \dfrac{t^n}{n!}$,

so summing as we did in the proof of (e) gives

(e') $E\left(\sup\limits_{0 \le s \le T} |X_y - Y_t|^2\right) \le \left(\sum\limits_{i=0}^{\infty} \Delta_k(T)^{1/2}\right)^2$
$\le |x - y|^2 C_T^2,$

where

$$C_T = \sum_{k=0}^{\infty} \left(\frac{(BT)^n}{n!}\right)^{1/2} \le \infty$$

and

$$BT = 2dK^2 T^2 + 8d^2 K^2 T.$$

Remark: If $X_0 = X$ and $Y_0 = Y$ are random, then the same result holds with $|x - y|^2$ replaced by $E|X - Y|^2$.

At this point, we have completed Itô's construction of solutions to (*). Before we proceed, however, the reader should note that we are not in the usual Markov process setup (cf. Section 1.1). We started with a Brownian motion defined on some probability space $(\Omega, \mathscr{F}, P)$, and by successive approximation we defined for each $x \in R^d$ a process X_t^x that satisfies

$$X_t^x = x + \int_0^t \sigma(X_s^x) \, dB_s + \int_0^t b(X_s^x) \, ds.$$

In other words, we have one probability measure (P) and a family of stochastic processes ($X^x, x \in R^d$), rather than one set of random variables (the coordinate maps on $(C, \mathscr{C})$) and a family of measures ($P^x, x \in R^d$) (which was what we got from Skorohod's construction in Section 9.2 and from the Cameron-Martin transformation in Section 8.4).

Note: The material in this section has appeared previously in a number of places. The proof of (1) above is a hybrid of the proofs in Friedman (1975), pages 98–102, and Stroock and Varadhan (1979), pages 124–126, with one small improvement: We estimate $\Delta_n(t) = E\sup\{|X_s^n - X_s^{n-1}|^2 : 0 \le s \le t\}$, rather than $E|X_t^n - X_t^{n-1}|^2$, and this simplifies the argument somewhat.

9.4 Some Examples

Having solved our stochastic differential equation using two methods under two different sets of conditions, it is time to compare the results by looking at some examples. We begin with a "trivial" one.

Example 1 Let $A(x) \equiv 0$. Then the stochastic differential equations become deterministic:

$$dX_s = b(X_s)\,ds.$$

Skorohod's approach allows us to construct solutions by successive approximation whenever b is continuous. In contrast, Itô's approach requires that b be Lipschitz continuous but implies that for such b, there is, for each $x \in R^d$, a unique (deterministic) process X_t that has $X_0 = x$ and that solves (∗).

A simple family of examples shows that uniqueness need not hold if the assumption of Lipschitz continuity is replaced by Hölder continuity of any order < 1.

Example 2 Let $d = 1$, $a \equiv 0$, $b(x) = |x|^\delta$, $0 < \delta < 1$. Since $b(0) = 0$, $X_t \equiv 0$ is a solution. We will now show, by guessing and verifying, that there is a nonzero solution when $\delta < 1$. Let $X_s = Cs^p$. Then

$$dX_s = Cps^{p-1}\,ds$$
$$b(X_s) = C^\delta s^{p\delta},$$

so to make $dX_s = b(X_s)\,ds$, we first pick p so that $p\delta = p - 1$ (i.e., $p = (1 - \delta)^{-1}$) and then pick C such that $Cp = C^\delta$ (i.e., $p = C^{1-\delta}$ or $C = 1/(1-\delta)^{1-\delta}$).

Example 3 Let $d = 1$, $a(x) = |x|^\delta$, $b \equiv 0$. Since a is continuous, Skorohod's approach allows us to construct solutions by successive approximation for any $\delta > 0$. In contrast, Itô's approach requires that $\sigma(x) = |x|^{\delta/2}$ be Lipschitz continuous, so we are restricted to $\delta \ge 2$, but in this range we can conclude that the solution is unique.

Although the condition required by Itô's approach is not sharp in this case, there is a good reason why Itô's approach, or any other approach that proves uniqueness, does not work for small δ—the solution of (∗) is not unique when $\delta < 1$. To prove this statement observe that by using the approach in the second part of Section 9.2, we can think of $a(x) = 1/g(x)$ where $g(x) = |x|^{-\delta}$ and solve the equation by time-changing a Brownian motion. If $\delta < 1$, then

(a) $\displaystyle E_0 \int_0^t |B_s|^{-\delta} ds = \int_0^t s^{-\delta/2} E_0 |B_1|^{-\delta} ds$

$\displaystyle = \left(\int_0^t s^{-\delta/2} ds\right) E_0 |B_1|^{-\delta} < \infty$

(since the first factor is $< \infty$ for $\delta < 2$ and the second for $\delta < 1$), so we can construct a solution of $dX_s = |X_s|^{\delta/2} dB_s$ that starts at 0 and does not stay there.

Once we have two solutions starting from 0, it is easy to see that there are two solutions starting from $x \neq 0$. If $x > 0$ and $\tau = \inf\{t : B_t = 0\}$, then by results in Section 1.9,

(b) $\displaystyle E_x \int_0^\tau |B_s|^{-\delta} 1_{(B_s \leq x)} ds = \int_0^x |y|^{-\delta}(x \wedge y) dy$

$\displaystyle = \int_0^x y^{1-\delta} dy < \infty$

(whenever $\delta < 2$), and we have

$\displaystyle \int_0^\tau |B_s|^{-\delta} 1_{(B_s > x)} ds \leq |x|^{-\delta} \tau < \infty$

(for any $\delta < \infty$). Combining the last two results gives

$\displaystyle T = \int_0^\tau |B_s|^{-\delta} ds < \infty \quad \text{a.s.,}$

that is, starting at x, we hit 0 at a time $T < \infty$. Once we hit zero, we have two choices: We can stop, or we can continue by using the time substitution (or, if you are more sophisticated, you can stop the first time the local time of B_s at zero exceeds a fixed or exponentially distributed level, or ...).

At this point, we have settled the uniqueness question when $\delta \geq 2$ (unique) and when $\delta < 1$ (not unique). Looking at the proofs of (a) and (b) more carefully reveals that you cannot escape from 0 when $\delta \geq 1$ and you cannot reach 0 when $\delta \geq 2$ (see Exercises 1 and 2 below for converses). On the basis of this conclusion, you might guess that the solution is unique when $1 \leq \delta < 2$ and that all solutions stop when they hit 0. This conjecture is indeed true. The first fact is a consequence of a theorem of Yamada and Watanabe (1971).

(1) Suppose that
(i) there is a strictly increasing function ρ with $|\sigma(x) - \sigma(y)| \leq \rho(|x - y|)$ that has $\rho(0) = 0$ and

$\displaystyle \int_0^\varepsilon \rho^{-2}(u) du = \infty \text{ for all } \varepsilon > 0$

9.4 Some Examples

(ii) there is an increasing and concave function λ with $|b(x) - b(y)| \leq \lambda(|x - y|)$ that has $\lambda(0) = 0$ and

$$\int_0^\varepsilon \lambda^{-1}(u)\, du = \infty \quad \text{for all } \varepsilon > 0.$$

Then (∗) has a unique solution.

Proof See pages 168–170 of Ikeda and Watanabe (1981). (1) implies that our equation has a unique solution when $\delta \geq 1$, so we now have a complete picture of Example 3.

$$\text{The solution of } (*) \text{ is } \begin{cases} \text{unique for } \delta \geq 1 \\ \text{not unique for } \delta < 1. \end{cases}$$

While (1) helps us with Example 3; Example 2, on the other hand, helps us to understand the reason for the difference between the assumptions about σ and b in (1). When $b(x) = |x|^\delta$ and $a \equiv 0$, (1) implies that

$$\text{the solution of } (*) \begin{cases} \text{unique for } \delta \geq 1 \\ \text{not unique for } \delta < 1, \end{cases}$$

which is the same as the conclusion for Example 3, except for the fact that it pertains to $\sigma(x) = |x|^{\delta/2}$.

Exercise 1 To justify my remark that "you cannot escape when $\delta \geq 1$," show that for all $\varepsilon > 0$,

$$\int_0^\varepsilon |B_s|^{-1}\, ds = \infty \quad P_0 \text{ a.s.}$$

Proof By scaling and monotonicity, it suffices to prove the result where ε is replaced by $T_1 = \inf\{t : |B_t| = 1\}$, and to do this, it suffices to show that

$$\limsup_{n \to \infty} \int_0^{T_1} |B_s|^{-1} 1_{(|B_s| \leq 2^{-n})}\, ds > 0 \quad P_0 \text{ a.s.}$$

To prove the last result, let $R_0 = 0$, $S_n = \inf\{t > R_n : |B_t| = 2^{-n}\}$, $R_{n+1} = \inf\{t > S_n : B_t = 0\}$ for $n \geq 0$, and $N = \sup\{n : S_n < T_1\}$. Observe that

$$\int_0^{T_1} \geq \sum_{m=0}^N \int_{R_m}^{S_m} \geq \sum_{m=0}^N 2^n(S_m - R_m)$$

where $EN \geq 2^n$ and the $S_m - R_m$ are i.i.d. with $E(S_m - R_m) = C2^{-2n}$. A simple argument (compute variances) shows that the lim inf $\geq C$ a.s.

Exercise 2 To justify my remark that "you cannot reach 0 when $\delta \geq 2$," show that if $\tau = \inf\{t : B_t = 0\}$, then for all $x \neq 0$

$$\int_0^\tau |B_s|^{-2}\, ds = \infty \quad P_x \text{ a.s.}$$

Proof The game is the same as in the last example, but the stopping times are different and don't work as well. Let $S_{-1} = 0$, $R_n = \inf\{t > S_{n-1} : |B_t| = 2^{-n-1}\}$, $S_n = \inf\{t > R_n : |B_t| = 2^{-n}\}$ for $n \geq 0$, and $N = \sup\{n : S_n < T_1\}$. Again, we have

$$\int_0^{T_1} \geq \sum_{m=0}^{N} \int_{R_m}^{S_m} \geq \sum_{m=0}^{N} 2^{2n}(S_m - R_m),$$

but this time $EN = 1$ and $E2^{2n}(S_m - R_m) = C$, so an even simpler argument shows that the lim inf $= \infty$ a.s.

9.5 Solutions Weak and Strong, Uniqueness Questions

Having solved our stochastic differential equation twice and seen some examples, our next step is to introduce some terminology that allows us to describe in technical terms what we have done. The solution constructed in Section 9.3 using Itô's method is called a strong solution. Given a Brownian motion B_t and an $x \in R^d$, we constructed a process X_t on the same probability space in such a way that

$$(*) \quad X_t = x + \int_0^t A(X_s) \, dB_s + \int_0^t b(X_s) \, ds.$$

In contrast, the solution constructed in Section 9.2 by discretizing and taking limits is called a weak solution. Its weakness is that we first defined X_t on some probability space and then constructed a Brownian motion such that $(*)$ holds.

With each concept of solution there is associated a concept of uniqueness.

We say that pathwise uniqueness holds, or that there is a unique strong solution, if whenever B_t is a Brownian motion (defined on some probability space $(\Omega, \mathcal{F}, P)$) and X and $\bar{X}$ are two strong solutions of $(*)$, it follows that, with probability 1, $X_t = \bar{X}_t$ for all $t \geq 0$.

We say that distributional uniqueness holds, or that there is a unique weak solution, if all solutions of $(*)$ give rise to the same probability law on $(C, \mathcal{C})$ when we map $\omega \in \Omega \to X(\omega) \in C$.

Itô's theorem implies that pathwise uniqueness holds when the coefficients are Lipschitz continuous. It is easy to show that in this case there is also distributional uniqueness.

(1) If the coefficients σ and b are Lipschitz continuous, then distributional uniqueness holds.

Proof Let B_t and $\bar{B}_t$ be two Brownian motions, and let X_t^n and $\bar{X}_t^n$ be the sequence of processes defined in the proof of Itô's theorem when $X_t^0 = \bar{X}_t^0 \equiv x$ and the Brownian motions are B_t and $\bar{B}_t$, respectively. An easy induction shows that for each n, X^n and $\bar{X}^n$ have the same distribution, so letting $n \to \infty$ proves (1).

9.5 Solutions Weak and Strong, Uniqueness Questions

With two notions of solution, it is natural, and almost inevitable, to ask about the relationship between the two concepts. A simple example due to Tanaka (see Yamada and Watanabe (1971)) shows that, contrary to the naive idea that it is easier to be weak than to be strong, you may have a unique weak solution but several strong solutions.

Example 1 Let $\sigma(x) = 1$ for $x \geq 0$ and $= -1$ for $x < 0$. Let W be a Brownian motion starting at 0 and let

$$B_t = \int_0^t \sigma(W_s) \, dW_s.$$

Since $B = \sigma \cdot W$ is a local martingale with $\langle B \rangle_t \equiv t$, B is a Brownian motion. The associative law implies that

$$\sigma(W) \cdot B = \sigma(W)^2 \cdot W = W,$$

so we have

$$W_t = \int_0^t \sigma(W_s) \, dB_s.$$

Since $\sigma(-x) = -\sigma(x)$ for all $x \neq 0$, we also have

$$-W_t = \int_0^t \sigma(-W_s) \, dB_s.$$

The last two equations show that there is more than one strong solution of $dX_s = \sigma(x_s) \, dB_s$. To prove that there is a unique weak solution, we observe that if $dX_s = \sigma(X_s) \, dB_s$, then X is a local martingale with $\langle X \rangle_t = t$ and, hence, a Brownian motion.

In the other direction, we have the following result of Yamada and Watanabe (1971):

(2) Pathwise uniqueness implies distributional uniqueness.

We will not prove this because we are lazy and this is not important for the developments below. The reader can find a discussion of this result in Williams (1981) and a proof in either Ikeda and Watanabe (1981), pages 149–152, or Stroock and Varadhan (1979), Section 8.1.

The last result and Tanaka's example are the basic facts about the differences between pathwise and distributional uniqueness. The best results about distributional uniqueness are due to Stroock and Varadhan (1969). To avoid the consideration of explosion, we will state their result only for bounded coefficients.

(3) Suppose that

(i) A is bounded, continuous, and positive definite at each point
(ii) b is bounded and Borel measurable. Then there is a unique weak solution.

Much of Chapters 6 and 7 in Stroock and Varadhan (1979) is devoted to preparing for and proving (3), so we will not go into the details here. The key step is to prove the result when $b \equiv 0$ and $|A_{ij}(x) - \delta_{ij}| \leq \varepsilon$ for all $x \in R^d$. The reader can find a nice exposition of this part in Ikeda and Watanabe (1981), pages 171–176.

In the material that follows, we will develop the theory of stochastic differential equations only for coefficients that are Lipschitz continuous. We do this not only because we have omitted the proof of (3), but also because in the developments in Section 9.6, the proofs of the Markov and Feller properties, we will use the fact that our processes are constructed by Itô's iteration scheme. You can also prove these results in the generality of (3) by knowing that there is a unique solution to the martingale problem, but for this you have to read Stroock and Varadhan (1979).

9.6 Markov and Feller Properties

Having constructed solutions of (∗) and having considered their nature and number at some length, we finally turn our attention to finding conditions that guarantee that $v(t, x) = Ef(X_t^x)$ is a solution of (1) of Section 9.1. In Section 8.1, when we dealt with $X_t^x = B_t$, a Brownian motion, the first step was to observe that if f is bounded, then the Markov property implies that

$$E_x(f(B_t)|\mathscr{F}_s) = v(t - s, B_s),$$

so $v(t - s, B_s)$ is a martingale. To generalize this proof to our new setting, we need to show that the X_t^x have the Markov property, that is,

(1) If f is bounded and continuous and $v(t, x) = Ef(X_t^x)$, then

$$E(f(X_t^x)|\mathscr{F}_s) = v(t - s, X_s^x).$$

Proof Let $X_{s,x}(t)$ (the process starting at x at time s) be defined as the solution of

$$X_t = x + \int_s^t \sigma(X_r)\, dB_r + \int_s^t b(X_r)\, dr$$

for $t \geq s$, and $X_{s,x}(t) = x$ for $t \leq s$. It follows from uniqueness that if $s < t < u$, then

$$X_{s,x}(u) = X_{t, X_{s,x}(t)}(u) \quad \text{a.s.}$$

(recall that all the random variables $X_{s,x}(u)$ are defined on the same probability space, $(C, \mathscr{C}, P)$). From the last result, it follows immediately that if $0 \leq s_1 < \cdots < s_n \leq t < u$, if $A_1, \ldots, A_n$ are Borel sets, and if f is bounded, then

$$E(f(X_u^x); X^x(s_1) \in A_1, \ldots, X^x(s_n) \in A_n)$$
$$= E(f(X_{t, X^x(t)}(u)); X^x(s_1) \in A_1, \ldots, X^x(s_n) \in A_n)$$

(recall that $X_t^x = X_{0,x}(t)$). To prove (1), it is enough to prove the following:

9.6 Markov and Feller Properties

(2) $$E(f(X_{t,X^x(t)}(u))|\mathcal{F}_t) = v(u - t, X_t^x).$$

To this end, observe that for any $y \in R^d$,

$$X_{t,y}(u) \in \sigma(B_{t+s} - B_t, s \geq 0)$$

and is, therefore, independent of $\mathcal{F}_t$, so

$$E(f(X_{t,y}(u))|\mathcal{F}_t) = Ef(X_{t,y}(u)) = v(u - t, y).$$

Now if $Y: \Omega \to R^d$ is $\mathcal{F}_t$ measurable and takes on only a finite number of values $y_1, \ldots, y_n$, then

$$X_{t,Y}(u) = \sum_{i=1}^n 1_{(Y=y_i)} X_{t,Y_i}(u) \quad \text{a.s.}$$

It follows from the last result that

$$E(f(X_{t,Y}(u))|\mathcal{F}_t) = v(u - t, Y).$$

To prove this equality, let $A \in \mathcal{F}_t$ with $A \subset \{Y = y_i\}$ and observe that it follows from the first result that

$$E(f(X_{t,Y}(u)); A) = E(f(X_{t,y_i}(u)); A)$$
$$= v(u - t, y_i)P(A) = E(v(u - t, Y); A).$$

To extend our results to a general $Y \in \mathcal{F}_t$ (and, hence, to prove (1')), pick a sequence Y_n of random variables that take on only finitely many values and have $Y_n \to Y$ a.s. and $E|Y_n - Y|^2 \to 0$ as $n \to \infty$. From the continuity result (e') in Section 9.3 (or, to be more precise, the remark afterwards), we have that

$$X_{t,Y_n}(u) \to X_{t,Y}(u) \quad \text{in } L^2$$

and, hence,

$$E(f(X_{t,Y_n}(u))|\mathcal{F}_t) \to E(f(X_{t,Y}(u))|\mathcal{F}_t) \quad \text{in } L^2.$$

By the result above for simple Y, the left-hand side is $v(u - t, Y_n)$. To complete the proof of (2) (and, hence, of (1)), it suffices to prove

(3) Suppose that f is bounded and continuous. Then for fixed t, $x \to v(t, x) = Ef(X_t^x)$ is continuous.

Proof The continuity result (e') in Section 9.3 implies that

$$E\left(\sup_{0 \leq s \leq T} |X_s^x - X_s^y|^2\right) \leq |x - y|^2 C_T^2,$$

where C_T is a constant whose value depends only on T. From this result, it follows immediately that if $x_n \to x$, $X_t^{x_n} \to X_t^x$ in probability and, hence,

$$v(t, x_n) = Ef(X_t^{x_n}) \to Ef(X_t^x) = v(t, x).$$

Remark: The proof of (1) given above follows Stroock and Varadhan (1979), pages 128–130. I think it is a good example of the power of the idea from measure theory that to prove an equality for the general variable Y, it is enough to prove the result for an indicator function (for then you can extend by linearity and take limits to prove the result).

With the Markov property established, it is now easy to prove

(4) Suppose that f is bounded. If v is smooth, then it satisfies part (a) of (1) in Section 9.1.

Proof The Markov property implies that

$$E(f(X_t^x)|\mathcal{F}_s) = v(t-s, X_s^x).$$

Since the left-hand side is a martingale, $v(t-s, X_s^x)$ is also a martingale. If v is smooth, then repeating the computation in the proof of (2) (at the beginning of this chapter) shows that

$$v(t-s, X_s^x) - v(t, x) = \int_0^s (-v_t + Lv)(t-r, X_r^x)\,dr + \text{a local martingale},$$

where

$$Lf = \frac{1}{2}\sum_{ij} A^{ij} D_{ij} f + \sum_i b^i D_i f,$$

so it follows that the integral on the right-hand side is a local martingale. Since this process is continuous and locally of bounded variation, it must be $\equiv 0$ and, hence, $(-v_t + Lv) = 0$ in $(0, \infty) \times R^d$.

With (4) established, the next step is to show the following:

(5) If f is bounded and continuous, then v satisfies part (b) of (1) of Section 9.1.

Proof From the continuity result used in the last proof, it follows that if $x_n \to x$ and $t_n \to 0$, then $X^{x_n}(t_n) \to x$ in probability and, hence,

$$v(t_n, x_n) = Ef(X^{x_n}(t_n)) \to f(x).$$

9.7 Conditions for Smoothness

In this section, we finally confront the problem of finding conditions that guarantee that $v(t, x) = Ef(X_t^x)$ is smooth and, hence, satisfies

(1) (a) $u_t = Lu$ in $(0, \infty) \times R^d$
(b) u is continuous in $[0, \infty) \times R^d$ and $u(0, x) = f(x)$.

In this category, the probabilistic approach has not been very successful. By purely analytical methods (i.e., the parametrix method; see Friedman (1964), Chapter 1), one can show

(2) Suppose that $A_{ij}(x)$ and $b_i(x)$ are bounded for each i and j and that

9.7 Conditions for Smoothness

(a) there is an $\alpha > 0$ such that for all $x, y \in R^d$,
$$\sum_{ij} y_i A_{ij}(x) y_j \geq \alpha |y|^2$$

(b) there is a $\beta > 0$ and $C < \infty$ such that for all i, j, x, and y,
$$|A_{ij}(x) - A_{ij}(y)| \leq C|x-y|^\beta$$
$$|b_i(x) - b_i(y)| \leq C|x-y|^\beta.$$

Then there is a positive function $p_t(x, y)$ that is jointly continuous in all of its variables and such that if f is a bounded continuous function, then
$$v(t, x) = \int p_t(x, y) f(y) \, dy.$$

Remark: This result is a combination of several theorems in Friedman (1964); see Friedman (1975), pages 141–142. For analysts, $p_t(x, y)$ is the fundamental solution with pole at x (i.e., $p_t(x, \cdot) \Rightarrow \delta_x$ at $t \to 0$); for probabilists, $p_t(x, y)$ is the transition probability
$$p_t(x, y) = P(X_t^x = y).$$

On the other hand, the best result I know of, which can be proved by purely probabilistic means, is on page 122 of Friedman (1975).

(3) Suppose that b, σ, and f are C^2 and that these functions and their derivatives of order ≤ 2 are bounded by $C(1 + |x|^\gamma)$ for some C, $\gamma < \infty$. Then v is smooth and, hence, satisfies (1).

Remark: The reader should observe that although (3) requires more smoothness for the coefficients, it does not require the "strict ellipticity," (a) in (2), and hence can be applied in situations where σ degenerates. In this context, the results obtained from (3) are almost the same as those obtained by Olenik (1966) using purely analytical methods. Probabilists (and anyone who does not read Italian) can find this result in Stroock and Varadhan (1979), Theorem 3.2.6.

Proof Since the proof is rather lengthy, we content ourselves with simply giving an idea of what is involved by assuming that everything is bounded and indicating why $D_i v$ exists. (In our defense, we would like to observe that not even Friedman (1975) spells out the details for the second derivatives; see page 123.)

To deal with derivatives with respect to x_i, we will show that $x \to X_t^x$ is, in the "L^2 sense," a differentiable function of x. To explain this statement, we need a definition.

A function $g(x, \omega)$ on $R^d \times \Omega$ is said to have $\partial g / \partial x_i = f$ in the L^2 sense if
$$E \left(\frac{g(x + he_i, \omega) - g(x, \omega)}{h} - f(x, \omega) \right)^2 \to 0$$

as $h \to 0$.

With this definition introduced, we can state our first differentiability result as

(4) If $D_j\sigma$ and $D_j b$ exist for all j and are bounded and continuous, then $\partial X_t^x/\partial x_i$ exists in the L^2 sense; furthermore, if we let $\partial_i(t) = (\partial_i^1(t), \ldots, \partial_i^d(t)) = \partial X_t^x/\partial x_i$, then ∂_i satisfies

$$\partial_i(t) = e_i + \sum_j \int_0^t \partial_i^j(s) D_j b(X_s^x) \, ds + \sum_j \int_0^t \partial_i^j(s) D_j \sigma(X_s^x) \, dB_s,$$

where e_i is the ith unit vector (i.e., $(e_i)_j = \delta_{ij}$).

Proof There is only one way to start the proof of a result like this. Let $h > 0$ and write

$$\frac{1}{h}(X_t^{x+he_i} - X_t^x) = e_i + \int_0^t \frac{1}{h}(b(X_s^{x+he_i}) - b(X_s^x)) \, ds$$
$$+ \int_0^t \frac{1}{h}(\sigma(X_s^{x+he_i}) - \sigma(X_s^x)) \, dB_s.$$

To change the first integral on the right-hand side into something that looks like the first integral in the desired answer, we write

$$\frac{1}{h} \int_0^t b(X_s^{x+he_i}) - b(X_s^x) \, ds$$
$$= \frac{1}{h} \int_0^t ds \int_0^1 d\theta \, \frac{d}{d\theta} b(X_s^x + \theta(X_s^{x+he_i} - X_s^x))$$
$$= \int_0^t ds \int_0^1 d\theta \sum_j D_j b(X_s^x + \theta(X_s^{x+he_i} - X_s^x)) \frac{X_s^{x+he_i,j} - X_s^{x,j}}{h},$$

where $X_s^{x,j}$ is the jth coordinate of X_s^x. The same trick works for the second term, with the result that

$$\frac{1}{h} \int_0^t \sigma(X_s^{x+he_i}) - \sigma(X_s^x) \, dB_s$$
$$= \frac{1}{h} \int_0^t dB_s \int_0^1 d\theta \, \frac{d}{d\theta} \sigma(X_s^x + \theta(X_s^{x+he_i} - X_s^x))$$
$$= \int_0^t dB_s \int_0^1 d\theta \sum_j D_j \sigma(X_s^x + \theta(X_s^{x+he_i} - X_s^x)) \frac{X_s^{x+he_i,j} - X_s^{x,j}}{h}$$

gives a stochastic integral equation for

$$\Delta_h(s) = \frac{X_s^{x+he_i} - X_s^x}{h}$$

in which the coefficients are almost the ones given in (4). The last step in the proof of (4), then, is to prove a result that says that if the coefficients of the equation converge in a suitable way, then so do the solutions. There is a large body of literature on this subject, which goes under the heading "stability of solutions" (e.g., Jacod and Memin (1981)). A result that is sufficient for our purposes is given on pages 118–119 of Friedman (1975); the desired conclusion follows immediately from that result.

(4) shows that if the coefficients σ and b are C^1, then so is the solution X_t^x when viewed as a function of x. Once this result is shown, it is not hard to show that $v(t, x) = Ef(X_t^x)$ is C^1. One does this by proving the "chain rule":

(5) $$D_i E f(X_t^x) = \sum_j ED_j f(X_t^x) \partial_i^j(t).$$

Proof The proof is based on the trick used to prove (3). We write

$$E(f(X_s^{x+he_i}) - f(X_s^x))$$

$$= E \int_0^1 d\theta \frac{d}{d\theta} f(X_s^x + \theta(X_s^{x+he_i} - X_s^x))$$

$$= D \int_0^1 d\theta \sum_j D_i f(X_s^x + \theta(X_s^{x+he_i} - X_s^x)) \frac{X_s^{x+he_i, j} - X_s^{x, j}}{h}$$

and let $h \to 0$. Further details are left to the reader. A complete discussion of the results in this section can be found in Section 5.5 of Friedman (1975).

Notes on Chapter 9

To steal a line from somebody, this book ends "not with a bang, but with a whimper." The results in this chapter are but a small sample of the results known about SDE's and their relationships with PDE's, and even worse, in many cases we have thrown up our hands and referred the reader to Friedman (1975), or Stroock and Varadhan (1979), or Ikeda and Watanabe (1981) for the details. In our defense, we can only say that the book had to end somewhere and that the three sources of which we have referred are all good places to learn more about the subject.

Appendix A Primer of Probability Theory

A.1 Some Differences in the Language

For an analyst, reading the probability literature must be like being an American in England. The language that is spoken is basically the same, but some of the words are different or have slightly different meanings. My first task, then, is to explain some of the colloquialisms that probabilists use. For convenience of exposition, we will begin at the very beginning.

A probability space is a triple $(\Omega, \mathscr{F}, P)$ where Ω is a set, $\mathscr{F}$ is a σ-field of subsets of Ω, and P is a probability measure, that is, a nonnegative, countably additive function on $\mathscr{F}$ that has $P(\Omega) = 1$.

Let $\mathscr{R}$ be the set of all Borel subsets of R. A function $X : \Omega \to R$ is said to be measurable if for each $B \in \mathscr{R}$ we have that $\{\omega : X(\omega) \in B\} \in \mathscr{F}$. For convenience, the phrase "X is measurable with respect to $\mathscr{F}$" is often abbreviated $X \in \mathscr{F}$, and measurable functions are commonly referred to as random variables.

In measure theory, one often talks about a sequence of functions f_n converging to a limit f "in measure" or "almost everywhere." These concepts are also used in probability, but they go by different names.

A sequence of random variables X_n is said to converge in probability to a limit X if for all $\varepsilon > 0$, $P(|X_n - X| > \varepsilon) \to 0$ as $n \to \infty$.

X_n is said to converge to X almost surely if $P(\omega : X_n(\omega) \to X(\omega) \text{ as } n \to \infty) = 1$. The last conclusion is usually abbreviated as $X_n \to X$ a.s.

The words *almost surely* and their abbreviation *a.s.* are used throughout probability as substitutes for *almost everywhere* and *a.e.* For instance, if $P(\omega : X(\omega) = Y(\omega)) = 1$, then we say that $X = Y$ a.s.

As in measure theory, $X_n \to X$ a.s. implies that $X_n \to X$ in probability, and the converse is false, but $X_n \to X$ in probability implies that there is a subsequence $X_{n_k} \to X$ a.s. We will prove the last statement, because the proof gives us an excuse to state some more definitions.

A.1 Some Differences in the Language

The indicator of a set A is the function

$$1_A(\omega) = \begin{cases} 1 & \omega \in A \\ 0 & \omega \notin A. \end{cases}$$

The notation is meant to suggest that this function is 1 on A. We do not use χ, because it looks too much like X, our favorite letter for random variables, and we do not call this a characteristic function, because that term is reserved for something else (see Chapter 6 of Chung (1974)).

If A_n is a sequence of sets, then

$$\limsup A_n = \{\omega : \limsup_{n \to \infty} 1_{A_n} = 1\}$$

$$= \bigcap_{N=1}^{\infty} \bigcup_{n=N}^{\infty} A_n.$$

The set defined above is usually referred to as $\{\omega : \omega \in A_n \text{ i.o.}\}$, where i.o. is short for infinitely often. As the next result indicates, we often make $\{\omega : \omega \in A_n \text{ i.o.}\}$ even shorter by dropping the ω's.

(1) **Borel-Cantelli Lemma.** If $\sum_n P(A_n) < \infty$, then $P(A_n \text{ i.o.}) = 0$.

Proof For any N,

$$P(\limsup A_n) \le P\left(\bigcup_{n=N}^{\infty} A_n\right) \le \sum_{n=N}^{\infty} P(A_n).$$

Letting $N \to \infty$ gives the desired result.

With (1) established, it is trivial to prove that $X_n \to X$ in probability implies that there is a subsequence $X_{n_k} \to X$ a.s. Let $\varepsilon_k \downarrow 0$, pick $n_k \to \infty$ such that $P(|X_{n_k} - X| > \varepsilon_k) \le 2^{-k}$, and then apply (1). This proof is, of course, nothing more than the standard proof from measure theory translated into the language of probability theory.

Up to this point, all of the changes have been semantic. When we turn our attention to integration, we encounter our first serious differences in notation. What an analyst would write as

$$\int_\Omega X \, dP \quad \left(\text{assuming that } \int_\Omega |X| \, dP < \infty \right)$$

a probabilist writes as EX (assuming that $E|X| < \infty$) and calls the expected value of X, or the mean of X. One clear advantage of the probabilistic notation is indicated by the typography of the last sentence—EX consumes less space and does not have to be displayed. There is also one clear disadvantage. To steal a quip from Dynkin, "If you use E to denote expectation with respect to P, then what do you use for expectation with respect to Q?" The obvious answer, F, is obviously unacceptable. Dynkin's remedy is to write PX instead of EX. Although this suggestion has considerable merit and would be useful at several points in the text, we will stick with the traditional notation.

Extending the notation above to integration over sets, we will let

$$E(X; A) = \int_A X\, dP.$$

Again, the notation is for typographical convenience and is motivated by the fact that the set A often has a complicated description. The proof of the next result illustrates the use of this notation.

(2) *Chebyshev's Inequality.* Let $Y \geq 0$ and let $\varphi \geq 0$ be a function that is increasing on $[0, \infty)$. Then

$$\varphi(a) P(Y \geq a) \leq E\varphi(Y).$$

Proof Since φ is increasing and ≥ 0,

$$\varphi(a) P(Y \geq a) \leq E(\varphi(Y); Y \geq a) \leq E\varphi(Y).$$

This result is trivial but useful. The following is a typical application:

(3) $$P(|X| \geq \varepsilon) \leq \frac{EX^2}{\varepsilon^2}.$$

A.2 Independence and Laws of Large Numbers

I have heard it said that "probability is just measure theory plus the notion of independence." Although I think that this statement is about as accurate as saying that "complex analysis is just real analysis plus $\sqrt{-1}$," there is no doubt that independence is one of the most important concepts in probability. We begin with what is hopefully a familiar definition and then work our way up to a definition that is appropriate for our current setting.

Two sets A and B are said to be independent if

$$P(A \cap B) = P(A) P(B).$$

Two random variables X and Y are said to be independent if for all Borel sets A and B,

$$P(X \in A, Y \in B) = P(X \in A) P(Y \in B).$$

Two σ-fields $\mathscr{F}$ and $\mathscr{G}$ are said to be independent if for all $A \in \mathscr{F}$ and $B \in \mathscr{G}$,

$$P(A \cap B) = P(A) P(B).$$

The third definition is a generalization of the second: Let $\mathscr{F} = \sigma(X) =$ the σ-field generated by X ($=$ the smallest σ-field $\mathscr{F}$ such that $X \in \mathscr{F}$), let $\mathscr{G} = \sigma(Y)$, and observe that $A \in \sigma(X)$ if and only if $A = \{\omega : X(\omega) \in C\}$ where $C \in \mathscr{R}$. The second definition is, in turn, a generalization of the first: Let $X = 1_A$, let $Y = 1_B$, and observe that if A and B are independent, then so are A^c and B, A^c and B^c, A and Ω, A and $\emptyset$, and so on.

In view of the last two remarks, when we define what it means for several things to be independent, we take things in the opposite order.

σ-fields $\mathscr{F}_1, \ldots, \mathscr{F}_n$ are said to be independent if whenever $A_i \in \mathscr{F}_i$ for $i = 1, \ldots, n$ we have that

$$P\left(\bigcap_{i=1}^n A_i\right) = \prod_{i=1}^n P(A_i).$$

Random variables $X_1, \ldots, X_n$ are said to be independent if whenever $A_i \in \mathscr{R}$ for $i = 1, \ldots, n$ we have that

$$P\left(\bigcap_{i=1}^n \{X_i \in A_i\}\right) = \prod_{i=1}^n P(X_i \in A_i).$$

Sets $A_1, \ldots, A_n$ are said to be independent if whenever $I \subset \{1, \ldots, n\}$ we have that

$$P\left(\bigcap_{i \in I} A_i\right) = \prod_{i \in I} P(A_i).$$

If you think about it for a minute, you will see that the third definition is what we get when we specialize the second to $X_i = 1_{A_i}$. It is important to note that the last definition is *not* equivalent to requiring that $P(A_i \cap A_j) = P(A_i)P(A_j)$ whenever $i \ne j$ (this is called pairwise independence).

Example 1 Let X_1, X_2, and X_3 be independent random variables that have $P(X_i = 1) = P(X_i = -1) = 1/2$, and let $A_1 = \{X_2 = X_3\}$, $A_2 = \{X_3 = X_1\}$, and $A_3 = \{X_1 = X_2\}$. These events are pairwise independent, since $P(A_i \cap A_j) = P(X_1 = X_2 = X_3) = 1/4 = P(A_i)P(A_j)$, but they are not independent, since $P(A_1 \cap A_2 \cap A_3) = P(X_1 = X_2 = X_3) = 1/4 \ne 1/8 = P(A_1)P(A_2)P(A_3)$.

Of the three definitions in the second list above, the first is the most important, so if it is unfamiliar, it would be a good idea to spend a minute and prove the next three results to get acquainted with the ideas involved.

(1) If $\mathscr{F}_1, \ldots, \mathscr{F}_n$ are independent and $X_i \in \mathscr{F}_i$ have $E|X_i| < \infty$, then

$$E\left(\prod_{i=1}^n X_i\right) = \prod_{i=1}^n EX_i.$$

Note that $E|\prod_{i=1}^n X_i| < \infty$ is part of the conclusion. The proof of this result follows a plan of attack that is standard in measure theory: Prove the result first for $X_i = 1_{A_i}$, use linearity to extend the result to simple random variables, monotone convergence to extend it to $X_i \ge 0$, and write $X_i = X_i^+ - X_i^-$ to prove the result in general.

(2) If $X_1, \ldots, X_n$ are independent random variables, then the σ-fields $\mathscr{F}_i = \sigma(X_i)$ are independent, and, consequently, if $f_1, \ldots, f_n$ are Borel functions, then $f_1(X_1), \ldots, f_n(X_n)$ are independent.

(3) Generalize the proof of (2) to conclude that if $1 \le n_1 < n_2 < \cdots < n_k = n$ and the $f_i : \Omega \to R^{n_i - n_{i-1}}$ are Borel measurable, then

$$f_1(X_1, \ldots, X_{n_1}), \ldots, f_k(X_{n_{k-1}+1}, \ldots, X_{n_k})$$

and independent.

Hint: Start with f_i that are of the form $\prod_j 1_{A_j}(X_j)$ and then use the approach described for (1) to work your way up to the general result.

Sequences of independent random variables are very important in probability theory, because they are (hopefully) what is generated when an experiment is repeated or a survey is taken. Motivated by this example, one of the main problems of the subject is to give conditions under which the "sample mean" $(X_1 + \cdots + X_n)/n$ approaches the "true mean" as $n \to \infty$, and to estimate the rate at which this occurs. Much of the first quarter of a graduate probability course is devoted to developing these results, but since they are not essential for what we will do, we will just prove one sample result to illustrate some of the concepts in this section and then state two more results in order to give you a taste of the theory.

(4) Let $X_1, X_2, \ldots$ be a sequence of independent and identically distributed random variables (i.e., $P(X_i \le x)$ is independent of i) that have $EX_i^2 < \infty$. As $n \to \infty$,

$$\frac{1}{n}(X_1 + \cdots + X_n) \to EX_1 \text{ in probability.}$$

Proof Let $\mu = EX_i$, $Y_i = X_i - \mu$. Now

$$\frac{1}{n}(X_1 + \cdots + X_n) - \mu = \frac{1}{n}(Y_1 + \cdots + Y_n),$$

so it suffices to show that the right-hand side $\to 0$ in probability. The key to doing this is to observe that

$$E\left(\sum_{i=1}^n Y_i\right)^2 = E\left(\sum_{i=1}^n \sum_{j=1}^n Y_i Y_j\right)$$

$$= \sum_{i=1}^n \sum_{j=1}^n EY_i Y_j = \sum_{i=1}^n EY_i^2 = nEY_1^2$$

(since if $i \ne j$, $EY_i Y_j = EY_i EY_j = 0$). If we let $S_n = Y_1 + \cdots + Y_n$ and $C = EY_1^2$, then

$$ES_n^2 = Cn, \quad \text{i.e.,} \quad E\left(\frac{S_n}{n}\right)^2 = \frac{C}{n}$$

and Chebyshev's inequality implies that

$$P\left(\left|\frac{S_n}{n}\right| \ge \varepsilon\right) \le \varepsilon^{-1} E\left(\frac{S_n}{n}\right)^2 = \frac{C\varepsilon^{-1}}{n} \to 0.$$

A.2 Independence and Laws of Large Numbers

The result above is a weak form of the weak law of large numbers. The strongest form of the strong law is

(5) Let $X_1, X_2, \ldots$ be a sequence of i.i.d. random variables (for a translation, see (4)) with $E|X_i| < \infty$. As $n \to \infty$,

$$\frac{1}{n}(X_1 + \cdots + X_n) \to EX_1 \quad \text{a.s.}$$

Analysts may recognize this result as being a consequence of Birkhoff's ergodic theorem. It is easy to show (see Exercise 2 below) that $E|X_i| < \infty$ is necessary for $(X_1 + \cdots + X_n)/n$ to converge to a finite limit so the condition in (5) is "sharp." The weak law holds in a little greater generality:

(6) Let $X_1, X_2, \ldots$ be a sequence of i.i.d. random variables. There is a sequence of constants a_n such that

$$\frac{1}{n}(X_1 + \cdots + X_n) - a_n \to 0 \text{ in probability}$$

if and only if $nP(|X_i| > n) \to 0$. In this case, we can take

$$a_n = E(X_i; |X_i| \le n).$$

Proof We leave the proof as an exercise for the reader. Once someone tells you to look at $X_{i,n} = X_i 1_{(|X_i| \le n)}$ and use Chebyshev's inequality, the rest is not hard. See Feller (1971), page 235, for a solution.

The next two exercises are much easier, but the first is much more important.

Exercise 1 *The Second Borel-Cantelli Lemma.* If $A_1, A_2, \ldots$ are independent and $\sum P(A_n) = \infty$, then $P(A_n \text{ i.o.}) = 1$.

Hint: If $M \le N < \infty$, then

$$P\left(\bigcap_{n=M}^{N} A_n^c\right) = \prod_{n=M}^{N} (1 - P(A_n)),$$

and, for any M, the right-hand side $\to 0$ as $N \to \infty$.

Exercise 2 Let $X_1, X_2, \ldots$ be a sequence of i.i.d. random variables with $E|X_i| = \infty$. Then

$$P(|S_n - S_{n+1}| > n \text{ i.o.}) = 1,$$

so S_n/n cannot converge to a finite limit on a set of positive probability.

Hint:

$$\sum_{n=0}^{\infty} P(|X_i| > n) \ge \int_0^{\infty} P(|X| > x)\, dx = E|X|.$$

A.3 Conditional Expectation

Given a probability space $(\Omega, \mathcal{F}_0, P)$, a σ-field $\mathcal{F} \subset \mathcal{F}_0$, and a random variable $X \in \mathcal{F}_0$ (i.e., $\mathcal{F}_0$ measurable) with $E|X| < \infty$, we define $E(X|\mathcal{F})$ to be any random variable Y that has

(i) $Y \in \mathcal{F}$

(ii) for all $A \in \mathcal{F}$, $\int_A X \, dP = \int_A Y \, dP$.

Y is said to be a version of $E(X|\mathcal{F})$. Any two versions of $E(X|\mathcal{F})$ are equal almost surely.

Interpretation: We think of $\mathcal{F}$ as describing the information we have at our disposal: For each $A \in \mathcal{F}$, we know whether or not A has occurred. $E(X|\mathcal{F})$ is, then, our "best guess" of the value of X given the information at our disposal. Some examples should help to clarify this. In each case, you should check that the answer we have given satisfies (i) and (ii).

Example 1 If $X \in \mathcal{F}$, then $E(X|\mathcal{F}) = X$, that is, if we know X, then our "best guess" is X itself. In general, the only thing that can keep X from being $E(X|\mathcal{F})$ is condition (i).

Example 2 Suppose that $\Omega_1, \ldots, \Omega_n$ is a partition of Ω into disjoint sets, each of which has positive probability, and that $\mathcal{F} = \sigma(\Omega_1, \ldots, \Omega_n)$, the σ-field generated by these sets. Then on Ω_i,

$$E(X|\mathcal{F}) = \frac{E(X;\Omega_i)}{P(\Omega_i)}.$$

In words, the information in $\mathcal{F}$ tells us the element of the partition which contains our outcome, and given this information, our best guess for X is the average value of X over Ω_i.

Example 3 Suppose that X is independent of $\mathcal{F}$, that is, $P(\{X \in A\} \cap B) = P(X \in A)P(B)$ for all $A \in \mathcal{R}$ and $B \in \mathcal{F}$. In this case, $E(X|\mathcal{F}) = EX$, that is, the information in $\mathcal{F}$ is of no help in guessing the value of X.

Example 4 Let $X \geq 0$ and let Q be the measure that has density X with respect to P, that is, $Q(A) = \int_A X \, dP$. Let P' and Q' be the restrictions of P and Q to $\mathcal{F}$. Then $Q' \ll P'$, and $E(X|\mathcal{F})$ is the Radon-Nikodym derivative dQ'/dP'. This is, in fact, how we show that the conditional expectation exists.

Conditional expectation has many of the same properties as ordinary expectation:

(a) linearity

$$E(aX + Y|\mathcal{F}) = aE(X|\mathcal{F}) + E(Y|\mathcal{F})$$

A.3 Conditional Expectation

(b) order preserving

if $X \le Y$, then $E(X|\mathcal{F}) \le E(Y|\mathcal{F})$

(c) monotone convergence

if $X_n \uparrow X$, then $E(X_n|\mathcal{F}) \uparrow E(X|\mathcal{F})$

(d) Jensen's inequality

if φ is convex and $E|X|, E|\varphi(X)| < \infty$, then
$\varphi(E(X|\mathcal{F})) \le E(\varphi(X)|\mathcal{F})$

(e) L^p convergence

if $X_n \to X$ in $L^p, p \ge 1$, then
$E(X_n|\mathcal{F}) \to E(X|\mathcal{F})$ in L^p

(f) dominated convergence

if $X_n \to X$ a.s. and $|X_n| \le Y$ with $EY < \infty$, then
$E(X_n|\mathcal{F}) \to E(X|\mathcal{F})$ a.s.

These properties are not very hard to prove using the definition of conditional expectation. To prove (a), we simply check that the right-hand side $\in \mathcal{F}$ and has the same integral as $aX + Y$ over all $A \in \mathcal{F}$. To prove (b), we observe that if $A \in \mathcal{F}$,

$$\int_A E(X|\mathcal{F})\,dP = \int_A X\,dP \le \int_A Y\,dP = \int_A E(Y|\mathcal{F})\,dP,$$

and applying this result to $A = \{E(X|\mathcal{F}) > E(Y|\mathcal{F})\}$, we conclude that $A = \emptyset$, that is, $E(X|\mathcal{F}) \le E(Y|\mathcal{F})$.

With these two examples as a guide, you should be able to prove (c) through (e), but, for the moment, I want to discourage you from doing this. It is more important to understand the following properties of conditional expectation, which have no analogue for ordinary expectation, and I leave the proofs of these properties as recommended exercises.

(1) $E(E(X|\mathcal{F})) = E(X)$.

(2) If $\mathcal{F}_1 \subset \mathcal{F}_2$, then

   (i) $E(E(X|\mathcal{F}_1)|\mathcal{F}_2) = E(X|\mathcal{F}_1)$
   (ii) $E(E(X|\mathcal{F}_2)|\mathcal{F}_1) = E(X|\mathcal{F}_1)$.

   In words, the smaller σ-field always wins out.

(3) If $A \in \mathcal{G}$ and $E|Y| < \infty$, then

$$E(Y1_A|\mathcal{G}) = 1_A E(Y|\mathcal{G}).$$

By using linearity and taking limits, we can easily extend this to

(4) If $X \in \mathcal{G}$ and $E|Y|, E|XY| < \infty$, then

$$E(XY|\mathcal{G}) = XE(Y|\mathcal{G}).$$

From (4) (and (d)), we get a geometric interpretation of the conditional expectation.

(5) Suppose that $EX^2 < \infty$. Then $L^2(\mathcal{F}_0) = \{Y \in \mathcal{F}_0 : EY^2 < \infty\}$ is a Hilbert space, and $L^2(\mathcal{F})$ is a closed subspace. In this case, $E(X|\mathcal{F})$ is the projection of X onto $L^2(\mathcal{F})$, or, in statistical terms, $E(X|\mathcal{F})$ is the random variable $Y \in \mathcal{F}$ that minimizes the mean square error $E(X - Y)^2$.

A.4 Martingales

Let $\mathcal{F}_n$ be an increasing sequence of σ-fields. A sequence X_n is said to be adapted to $\mathcal{F}_n$ if $X_n \in \mathcal{F}_n$ for all n. If X_n is adapted to $\mathcal{F}_n$ and, for all n, $E|X_n| < \infty$ and $E(X_{n+1}|\mathcal{F}_n) = X_n$, then X is said to be a martingale. If, in the last definition, $=$ is replaced by $\leq$ or $\geq$, then X_n is said to be a supermartingale or submartingale, respectively.

To give an example of a martingale, consider successive tosses of a fair coin, and let $\xi_n = 1$ if the nth toss is heads and $\xi_n = -1$ if the nth toss is tails. Let $S_n = \xi_1 + \cdots + \xi_n$. S_n is called the (symmetric) simple random walk. It represents the amount of money a gambler has after the nth toss if each time the coin is tossed he bets one dollar on the coin coming up heads. Let $\mathcal{F}_n = \sigma(\xi_1, \ldots, \xi_n)$. As we mentioned before, we think of $\mathcal{F}_n$ as giving the information we have at time n, which in this case is the outcomes of the first n tosses. S_n is a martingale with respect to $\mathcal{F}_n$. To prove this, we observe that $S_n = \xi_1 + \cdots + \xi_n \in \mathcal{F}_n$, $E|S_n| < \infty$, and, by (3) in Section A.2, ξ_{n+1} is independent of $\mathcal{F}_n$, we have

$$\begin{aligned} E(S_{n+1}|\mathcal{F}_n) &= E(S_n + \xi_{n+1}|\mathcal{F}_n) \\ &= E(S_n|\mathcal{F}_n) + E(\xi_{n+1}|\mathcal{F}_n) \\ &= S_n + E\xi_{n+1} \\ &= S_n. \end{aligned}$$

If the successive tosses have $P(\xi_n = 1) \leq 1/2$, then the computation above shows that

$$E(S_{n+1}|\mathcal{F}_n) \leq S_n,$$

that is, S_n is a supermartingale. Since S_n corresponds to betting on an unfavorable game, we see that there is nothing "super" about a supermartingale. The name comes, instead, from the fact that superharmonic functions, when composed with Brownian motion, give rise to supermartingales.

For what follows, we will need a few simple facts about martingales, the proofs of which we leave as exercises for the reader.

(1) If X_n is a martingale and $m < n$, then
$$E(X_n|\mathcal{F}_m) = X_m.$$

(2) If X_n is a martingale and φ is a convex function with $E|\varphi(X_n)| < \infty$ for all n, then $\varphi(X_n)$ is a submartingale.

(3) If X_n is a submartingale and φ is an increasing convex function with $E\varphi(X_n) < \infty$ for all n, then $\varphi(X_n)$ is a submartingale.

(4) *Orthogonality of Martingale Increments.* If X_n is a martingale with $EX_n^2 < \infty$ for all n and $l \leq m \leq n$, then
$$E((X_n - X_m)X_l) = 0.$$

Since the proof of (4) is a classic example of manipulating conditional expectations, we will give the proof and let the reader justify the steps

$$\begin{aligned} E((X_n - X_m)X_l) &= EE((X_n - X_m)X_l|\mathcal{F}_m) \\ &= E[X_l E(X_n - X_m|\mathcal{F}_m)] \\ &= E[X_l(E(X_n|\mathcal{F}_m) - X_m)] \\ &= E[X_l \cdot 0] = 0. \end{aligned}$$

From the proof above, it is immediate that we have

(5) Under the hypotheses of (4),
$$E((X_n - X_m)^2|\mathcal{F}_m) = E(X_n^2|\mathcal{F}_m) - X_m^2.$$

A.5 Gambling Systems and the Martingale Convergence Theorem

Let $\mathcal{F}_n$ be an increasing sequence of σ-fields. H_n is said to be predictable if $H_n \in \mathcal{F}_{n-1}$ for all $n \geq 1$.

In words, the value of the process at time n may be predicted (with certainty) from the information available at time $n-1$. You should think of H_n as the amount of money a gambler bets at time n. This amount can be based on the outcomes at times $1, \ldots, n-1$, but it cannot depend on the outcome at time n.

Once we start thinking of H_n as a gambling system, it is natural to ask how much money we would win if we used it. For concreteness, let us suppose that the game consists of flipping a coin and that for each dollar bet we win one dollar when the coin comes up heads and lose one dollar when the coin comes up tails (most games in casinos reduce to this situation when you ignore all the ritual). Let S_n be the net amount of money we would have won at time n if we had bet one dollar each time. If we bet according to a gambling system H, then our net winnings at time n would be

$$(H \cdot S)_n = \sum_{m=1}^n H_m(S_m - S_{m-1}),$$

since $S_m - S_{m-1} = +1$ or -1 when the mth toss results in a win or loss, respectively.

The next result is the most basic fact about gambling systems and is, apparently, little known. It says that there is no system for beating an unfavorable game.

(1) Let X_n be a supermartingale. If $H_n \geq 0$ is predictable and each H_n is bounded, then $(H \cdot X)_n$ is a supermartingale.

Proof

$$\begin{aligned} E((H \cdot X)_{n+1} | \mathscr{F}_n) &= (H \cdot X)_n + E(H_{n+1}(X_{n+1} - X_n) | \mathscr{F}_n) \\ &= (H \cdot X)_n + H_{n+1} E(X_{n+1} - X_n | \mathscr{F}_n) \\ &\leq (H \cdot X)_n, \end{aligned}$$

since $E(X_{n+1} - X_n | \mathscr{F}_n) \leq 0$ and $H_{n+1} \geq 0$.

Remark: The same result is obviously also valid for submartingales and for martingales (and in the second case without the restriction $H_n \geq 0$). To keep from being repetitious, we will state our results for only one type of process and leave it to the reader to translate the result to the other two.

In my remark preceding (1), I did not mean to dismiss gambling systems as worthless. There is one system that allows us to prove the martingale convergence theorem. Let $h > 0$, let $N_0 = 0$, and for $k \geq 1$ let

$$\begin{aligned} N_{2k-1} &= \inf\{m > N_{2k-2} : X_m \leq a\} \\ N_{2k} &= \inf\{m > N_{2k-1} : X_m \geq a + h\} \\ H_m &= \begin{cases} 1 & \text{if } N_{2k-1} \leq m - 1 < N_{2k} \text{ for some } k \\ 0 & \text{otherwise} \end{cases} \\ U_n &= \sup\{k : N_{2k} \leq n\}. \end{aligned}$$

Since $X(N_{2k-1}) \leq a$ and $X(N_{2k}) \geq a + h$, between times N_{2k-1} and N_{2k}, X crosses from $\leq a$ to $\geq a + h$. H_n is a gambling system that tries to take advantage of the "upcrossings." In stock-market terms, we buy when $X_m \leq a$ and sell when $X_m \geq a + h$. In this way, every time an upcrossing is completed, we make a profit $\geq h$. Last but not least, U_n is the number of upcrossings completed by time n.

(2) *The Upcrossing Inequality.* If X_m is a submartingale, then

$$hEU_n \leq E(X_n - a)^+ - E(X_0 - a)^+.$$

Proof Let $Y_n = (X_n - a)^+$. Since Y_n is a submartingale that upcrosses $[0, h]$ the same number of times that X_n upcrosses $[a, a+h]$, it suffices to prove the result when $a = 0$ and $X_n \geq 0$. In this case, we have that

A.5 Gambling Systems and the Martingale Convergence Theorem

$$hU_n \le (H \cdot X)_n,$$

since a final incomplete upcrossing (if there is one) makes a nonnegative contribution to the right-hand side. Let $K_m = 1 - H_m$.

$$X_n - X_0 = (H \cdot X)_n + (K \cdot X)_n,$$

and it follows from (1) that $E(K \cdot X)_n \ge E(K \cdot X)_0 = 0$, so $E(H \cdot X)_n \le E(X_n - X_0)$, proving (2) in the special case and, hence, in general.

Remark: We have proved the result above in its classical form even though this approach is a little misleading. The key fact is that $E(K \cdot X)_n \ge 0$, that is, even by buying high and selling low, we cannot lose money on a submartingale, or in other words, it is the reluctance of submartingales to go from above $a + h$ to below a that limits the number of upcrossings.

From the upcrossing inequality, we easily obtain

(3) *The Martingale Convergence Theorem.* If X_n is a submartingale with $\sup_n EX_n^+ < \infty$, then as $n \to \infty$, X_n converges almost surely to a limit X with $E|X| < \infty$.

Proof Since $(X - a)^+ \le X^+ + |a|$, (2) implies that $EU_n \le (|a| + E|X_n|)/h$, so if we let $U = \lim U_n$ be the number of upcrossings of $[a, a + h]$ by the whole sequence, then $EU < \infty$ and, hence, $U < \infty$ a.s. Since this result holds for all rational a and h,

$$\bigcup_{a, h \in Q} \left\{ \liminf_{n \to \infty} X_n < a < a + h < \limsup_{n \to \infty} X_n \right\}$$

has probability 0, and

$$\limsup_{n \to \infty} X_n = \liminf_{n \to \infty} X_n \quad \text{a.s.},$$

which implies $X = \lim X_n$ exists a.s. Fatou's lemma guarantees that $EX^+ \le \liminf EX_n^+ < \infty$, so $X < \infty$ a.s. To see that $X > -\infty$, we observe that $EX_n^- = EX_n^+ - EX_n \le EX_n^+ - EX_0$ (since X_n is a submartingale), and another application of Fatou's lemma shows that $EX^- \le \liminf EX_n^- < \infty$.

From (3), it follows immediately that we have

(4) If $X_n \le 0$ is a submartingale, then as $n \to \infty$, $X_n \to X$ a.s.

The last two results are easy to rationalize. Submartingales are like increasing sequences of real numbers—if they are bounded above, they must converge almost surely. The next example shows that they need not converge in L^1.

Example 1 (Double or nothing) Suppose that we are betting on a symmetric simple random walk and we use the following system:

$$H_n = \begin{cases} 2^{n-1} & \text{on } S_{n-1} = n - 1 \\ 0 & \text{otherwise.} \end{cases}$$

In words, we start by betting one dollar on heads. If we win, we add our winnings to our original bet and bet everything again. When we lose, we lose everything and quit playing. Let $X_n = 1 + (H \cdot S)_n$. From (1), it follows that X is a martingale. The definition implies that $X_n \geq 0$, so $E|X_n| = EX_n = EX_0 = 1$, but it is easy to see that $P(X_n > 0) = 2^{-n}$, so $X_n \to 0$ a.s. as $n \to \infty$.

This is a very important example to keep in mind as you read the next three sections.

A.6 Doob's Inequality, Convergence in L^p, $p > 1$

A random variable N is said to be a stopping time if $\{N = n\} \in \mathcal{F}_n$ for all $n < \infty$. If you think of N as the time a gambler stops gambling, then the condition above says that the decision to stop at time n must be measurable with respect to the information available at that time. The following is an important property of stopping times:

(1) If X_n is a submartingale and N is a stopping time, then $X_{n \wedge N}$ is a submartingale.

Proof Let $H_n = 1_{(N \geq n)}$. Since $\{N \geq n\} = \{N \leq n-1\}^c \in \mathcal{F}_{n-1}$, H_n is predictable, and it follows from (1) of the last section that $(H \cdot X)_n = X_{n \wedge N}$ is a submartingale.

(2) If X_n is a submartingale and N is a stopping time with $P(N \leq k) = 1$, then
$$EX_0 \leq EX_N \leq EX_k.$$

Proof Since $X_{N \wedge n}$ is a submartingale, it follows that $EX_0 = EX_{N \wedge 0} \leq EX_{N \wedge k} = EX_N$. To prove the right-hand inequality, let $H_m = 1_{(N < m)}$. Now $\{N < m\} = \{N \leq m - 1\} \in \mathcal{F}_{m-1}$, so $(H \cdot X)_n = X_n - X_{N \wedge n}$ is a submartingale and $EX_k - EX_N = E(H \cdot X)_k \geq E(H \cdot X)_0 = 0$.

Remark: Let X_n be the martingale described in the last section and let $N = \inf\{n : X_n = 0\}$. Then $EX_0 = 1 > 0 = EX_N$, so the first inequality need not hold for unbounded stopping times. In Section A.8, we will consider conditions that guarantee that $EX_0 \leq EX_N$ for unbounded N.

From (2), we immediately get

(3) *Doob's Inequality.* If X_m is a submartingale and $A = \left\{\max_{0 \leq m \leq n} X_m \geq \lambda\right\}$, then
$$\lambda P(A) \leq EX_n 1_A \leq EX_n^+.$$

Proof Let $N = \inf\{m : X_m \geq \lambda \text{ or } m \geq n\}$. $X_N \geq \lambda$ on A and $N = n$ on A^c, so it follows from (2) that
$$\lambda P(A) \leq E(X_N 1_A) \leq EX_n 1_A$$
(observe that $X_N = X_n$ on A^c). The second inequality in (3) is trivial.

A.7 Uniform Integrability and Convergence in L^1

Integrating the inequality in (3) gives

(4) If $\bar{X}_n = \max_{0 \le m \le n} X_m$, then for $p > 1$,

$$E(\bar{X}_n^p) \le \left(\frac{p}{p-1}\right)^p E(X_n^+)^p.$$

Proof Fubini's theorem implies that

$$E\bar{X}_n^p = \int_0^\infty p\lambda^{p-1} P(\bar{X}_n > \lambda)\, d\lambda$$

$$\le \int_0^\infty p\lambda^{p-1} \left(\lambda^{-1} \int_{\{\bar{X}_n \ge \lambda\}} X_n^+\, dP\right) d\lambda$$

$$= \int_\Omega X_n^+ \left(\int_0^{\bar{X}_n} p\lambda^{p-2}\, d\lambda\right) dP$$

$$= \frac{p}{p-1} \int_\Omega X_n^+ \bar{X}_n^{p-1}\, dP.$$

If we let $q = p/(p-1)$ be the exponent conjugate to p and apply Hölder's inequality, we see that the above

$$\le q(E|X_n^+|^p)^{1/p}(E|\bar{X}_n|^p)^{1/q}.$$

At this point, we would like to divide both sides of the inequality above by $(E|\bar{X}_n|^p)^{1/q}$ to prove (4). Unfortunately, the laws of arithmetic do not allow us to divide by something that may be ∞. To remedy this difficulty, we observe that $P(\bar{X}_n \wedge N > \lambda) \le P(\bar{X}_n > \lambda)$, so repeating the proof above shows that

$$(E(\bar{X}_n \wedge N)^p)^{1/p} \le q(E|X_n^+|^p)^{1/p},$$

and letting $N \to \infty$ proves (4).

From (4), we get the following L^p convergence theorem:

(5) If X_n is a martingale, then for $p > 1$, $\sup_n E|X_n|^p < \infty$ implies that $X_n \to X$ in L^p.

Proof From the martingale convergence theorem, it follows that $X_n \to X$ a.s. Since $|X_n|$ is a submartingale, (4) implies that $\sup_m |X_m| \in L^p$, and it follows from the dominated convergence theorem that $E|X_n - X|^p \to 0$.

Remark: Again, the martingale described at the end of the last section shows that this result is false for $p = 1$.

A.7 Uniform Integrability and Convergence in L^1

In this section, we will give conditions that guarantee that a martingale converges in L^1. The key to this is the following definition:

A collection of random variables $\{X_i, i \in I\}$ is said to be uniformly integrable if

$$\lim_{M \to \infty} \left(\sup_{i \in I} E(|X_i|; |X_i| > M) \right) = 0.$$

Uniformly integrable families can be very large.

(1) If $X \in L^1$, then $\{E(X|\mathscr{F})\}$ is uniformly integrable.

Proof If A_n is a sequence of sets with $P(A_n) \to 0$, then the dominated convergence theorem implies that $E(|X|; A_n) \to 0$. From the last result, it follows that if $\varepsilon > 0$, we can pick $\delta > 0$ such that if $P(A) \le \delta$, then $E(|X|; A) \le \varepsilon$.

Pick M large enough so that $E|X|/M \le \delta$. Jensen's inequality and the definition of conditional expectation imply that

$$E(|E(X|\mathscr{F})|; |E(X|\mathscr{F})| > M) \le E(E(|X||\mathscr{F}); E(|X||\mathscr{F}) > M)$$
$$= E(|X|; E(|X||\mathscr{F}) > M),$$

and we have that

$$P(E(|X||\mathscr{F}) > M) \le \frac{E(E(|X||\mathscr{F}))}{M} = \frac{E|X|}{M} \le \delta,$$

so for this choice of M,

$$\sup_{\mathscr{F}} E(|E(X|\mathscr{F})|; |E(X|\mathscr{F})| > M) \le \varepsilon,$$

and since ε was arbitrary, the collection is uniformly integrable.

Another common example is

Exercise 1 Let φ be any function with $\varphi(x)/x \to \infty$ as $x \to \infty$, for example, $\varphi(x) = x^p$ or $\varphi(x) = x \log x$. If $E\varphi(|X_i|) \le C$ for all $i \in I$, then $\{X_i, i \in I\}$ is uniformly integrable.

The relevance of uniform integrability to L^1 convergence is explained by

(2) If $X_n \to X$ a.s., then the following statements are equivalent:
(a) $\{X_n, n \ge 0\}$ is uniformly integrable
(b) $X_n \to X$ in L^1
(c) $E|X_n| \to E|X| < \infty$.

Proof (a) $\Rightarrow$ (b): Let

$$\varphi_M(x) = \begin{cases} M & M \le x \\ x & -M \le x \le M \\ -M & x \le -M \end{cases}$$

and observe that by patiently checking the nine possible cases,

$$E|X_n - X| \le E|\varphi_M(X_n) - \varphi_M(X)| + E(|X_n|; |X_n| > M) + E(|X|; |X| > M).$$

As $n \to \infty$, the first term $\to 0$. If $\varepsilon > 0$ and M is large, then the second term $\le \varepsilon$.

To bound the third term, we observe that uniform integrability implies that $\sup_n E|X_n| < \infty$, so Fatou's lemma implies that $E|X| < \infty$, and by picking M larger, we can make the third term $\leq \varepsilon$. Combining the last three facts shows that $\limsup E|X_n - X| \leq 2\varepsilon$, proving (b).

(b) $\Rightarrow$ (c): $|E|X_n| - E|X|| \leq E||X_n| - |X|| \leq E|X_n - X|$.

(c) $\Rightarrow$ (a): Let $\psi_M(x) = x$ on $[0, M-1]$, $\psi_M = 0$ on $[M, \infty)$, and let ψ_M be linear on $[M-1, M]$. If M is large, $E|X| - E\psi_M(|X|) < \varepsilon/2$. The bounded convergence theorem implies that $E\psi_M(|X_n|) \to E\psi_M(|X|)$, so using (c) we get that if $n \geq n_0$,

$$E(|X_n|; |X_n| > M) \leq E|X_n| - E\psi_M(|X_n|) < \varepsilon.$$

By choosing M larger, we can make $E(|X_n|; |X_n| > M) < \varepsilon$ for $0 \leq n < n_0$, so X_n is uniformly integrable.

We are now ready to state the main theorems of this section. Since we have already done all the work, the proofs are short.

(3) For a submartingale, the following statements are equivalent:

(a) it is uniformly integrable
(b) it converges in L^1.

Proof (a) $\Rightarrow$ (b): Uniform integrability implies that $\sup E|X_n| < \infty$, which by (3) of Section A.5 implies that $X_n \to X$ a.s., which by (2) above implies that $X_n \to X$ in L^1. The converse, (b) $\Rightarrow$ (a), is a corollary of (2).

(4) For a martingale, the following statements are equivalent:

(a) it is uniformly integrable
(b) it converges in L^1
(c) there is an integrable random variable X such that $X_n = E(X|\mathcal{F}_n)$.

Proof (a) $\Rightarrow$ (b): This result follows from the proof given in (3).

(b) $\Rightarrow$ (c): Let $X = \lim X_n$. If $m > n$, then $E(X_m|\mathcal{F}_n) = X_n$, so if $A \in \mathcal{F}_n$, $E(X_n; A) = E(X_m; A)$. As $m \to \infty$, $X_m 1_A \to X 1_A$ in L^1, so we have that $E(X_n; A) = E(X; A)$ for all $A \in \mathcal{F}_n$, and it follows that $X_n = E(X|\mathcal{F}_n)$.

(c) $\Rightarrow$ (a): This result follows from (1) above.

A.8 Optional Stopping Theorems

In this section, we will prove a number of results that allow us to conclude that if X_n is a submartingale and $M \leq N$ are stopping times, then $EX_M \leq EX_N$. The first step is to show

(1) If X_n is a uniformly integrable submartingale, then for any stopping time N, $X_{N \wedge n}$ is uniformly integrable.

Proof We begin by observing that $X_{N\wedge n}$ is a submartingale with $EX_{N\wedge n}^+ \leq EX_n^+$ (since X_n^+ is a submartingale), so

$$\sup_n EX_{N\wedge n}^+ \leq \sup_n EX_n^+ < \infty$$

(since $X_n \to X_\infty$ in L^1), and it follows from the martingale convergence theorem that $X_{N\wedge n} \to X_N$ a.s. and $E|X_N| < \infty$. With this result established, the rest is easy, since

$$E(|X_{N\wedge n}|;|X_{N\wedge n}| > M) \leq E(|X_N|;|X_N| > M) + E(|X_n|;|X_n| > M)$$

and X_n is uniformly integrable.

From (1) it follows immediately that we have:

(2) If X_n is a uniformly integrable submartingale, then for any stopping time $N \leq \infty$,

$$EX_0 \leq EX_N \leq EX_\infty.$$

Proof (2) in Section A.6 implies that $EX_0 \leq EX_{N\wedge n} \leq EX_n$. Letting $n \to \infty$ and observing that (1) above and results in Section A.7 imply that $X_{N\wedge n} \to X_N$ in L^1 and $X_n \to X_\infty$ in L^1, gives the desired result.

From (2), we get the following useful corollary:

(3) *The Optional Stopping Theorem.* If $X_{N\wedge n}$ is a uniformly integrable submartingale, then for any stopping time $M \leq N$,

$$EX_M \leq EX_N.$$

We have given (3) a name since it is the basic result in this section and it is usually the result we are referring to when we use the words "optional stopping theorem" in the text. In applying (3), the following fact is useful:

(4) If X_n is a submartingale and $N < \infty$ is a stopping time with

(a) $E|X_N| < \infty$
(b) $E(|X_n|; N > n) \to 0$,

then $X_{N\wedge n}$ is uniformly integrable.

Proof

$$E(|X_{N\wedge n}|;|X_{N\wedge n}| > M) \leq E(|X_N|;|X_N| > M) + E(|X_n|; N > n, |X_n| > M).$$

Let $\varepsilon > 0$. If we pick n_0 large enough, then $E(|X_n|; N > n) < \varepsilon/2$ for all $n > n_0$. Having done this, we can pick M large enough so that $E(|X_N|;|X_N| > M) < \varepsilon/2$ and, for $0 \leq n \leq n_0$, $E(|X_n|;|X_n| > M) < \varepsilon/2$, so it follows from the first inequality that

$$E(|X_{N\wedge n}|;|X_{N\wedge n}| > M) < \varepsilon$$

for all n, and hence $X_{N\wedge n}$ is uniformly integrable.

Finally, there is one stopping theorem that does not require uniform integrability:

A.8 Optional Stopping Theorems

(5) If X_n is a nonnegative supermartingale and $N \leq \infty$ is a stopping time, then $EX_0 \geq EX_N$, where $X_\infty = \lim X_n$ (which exists by (4) of Section A.5).

Proof $EX_0 \geq EX_{N \wedge n}$. The monotone convergence theorem implies that

$$E(X_N, N < \infty) = \lim_{n \to \infty} E(X_N; N \leq n),$$

and Fatou's lemma implies that

$$E(X_\infty; N = \infty) \leq \liminf_{n \to \infty} E(X_n; N > n),$$

so adding the last two lines gives the desired result.

References

In the list of references below, we have for convenience shortened the standard abbreviations for the sources referred to most often:

AMS	American Mathematical Society
BAMS	*Bulletin of the American Mathematical Society*
CMP	*Communications on Mathematical Physics*
CPAM	*Communications on Pure and Applied Mathematics*
JFA	*Journal of Functional Analysis*
LMS	London Mathematical Society
LNM	Lecture Notes in Mathematics, Springer-Verlag, New York
MAA	Mathematical Association of America
PAMS	*Proceedings of the American Mathematical Society*
PJM	*Pacific Journal of Mathematics*
Sem.	*Seminar de Probabilités* (Strasbourg)
TAMS	*Transactions of the American Mathematical Society*
TPA	*Theory of Probability and Its Applications*
ZfW	*Zeitschrift für Wahrscheinlichkeitstheorie und Verwandte Gebiete*

Agmon, S. (1965). *Lectures on elliptic boundary value problems*. Van Nostrand, New York.
───── (1982). *Lectures on exponential decay of solutions of second-order elliptic equations*. Mathematical Notes. Princeton Univ. Press, Princeton.
Ahlfors, L. (1966). *Complex analysis*. 2d ed. McGraw-Hill, New York.
Aizenman, M., and B. Simon (1982). Brownian motion and a Harnack inequality for Schrödinger operators. *CPAM* 35:209–273.
Ash, J. M., ed. (1976). *Studies in harmonic analysis*. MAA Studies in Mathematics, vol. 13. MAA, Washington, D.C.
Austin, D. G. (1966). A sample function property of martingales. *Ann. Math. Stat.* 37:1396–1397.
Bachelier, L. (1900). Théorie de las speculation. *Ann. Sci. École Norm. Sup.* 17:21–86.
Baernstein, A., II (1978). Some sharp inequalities for conjugate functions. *Indiana Math. J.* 27:833–852.

Bary, N. K. (1964). *A treatise on trigonometric series.* Pergamon Press, New York.
Bernard, A., and B. Maisoneuve (1977). Decomposition atomique de martingales de la classe H^1. In *Sem. XII*, pp. 303–323. Springer LNM 649.
Bers, L., F. John, and M. Schechter (1964). *Partial differential equations.* AMS, Providence, RI.
Berthier, A., and B. Gaveau (1978). Critere de convergence des fonctionelles de Kac et applications en mechanique et en géométrie. *JFA* 29:416–424.
Billingsley, P. (1968). *Convergence of probability measures.* John Wiley, New York.
Blumenthal, R. M., and R. K. Getoor (1968). *Markov processes and their potential theory.* Academic Press, New York.
Breiman, L. (1968). *Probability.* Addison-Wesley, New York.
Brelot, M., and J. L. Doob (1963). Limites angulaires et limites fines. *Ann. Inst. Fourier* (Grenoble) 13:395–415.
Brossard, J. (1975). Thèse de troisième cycle.
———— (1976). Comportement nontangentiel et comportement Brownien des fonctions harmonique dans un demi-espace: Demonstration probabiliste d'un théorème de Calderon et Stein. In *Sem. XII*, pp. 378–397. Springer LNM 649.
Burkholder, D. L. (1962). Successive conditional expectation of an integrable function. *Ann. Math. Stat.* 33:887–893.
———— (1964). Maximal inequalities as necessary conditions for almost everywhere convergence. *ZfW* 3:75–88.
———— (1966). Martingale transforms. *Ann. Math. Stat.* 37:1494–1505.
———— (1970). Martingale inequalities. In *Martingales*, ed. H. Dinges, pp. 1–8. Springer LNM 190.
———— (1973). Distribution function inequalities for martingales. *Ann. Prob.* 1:19–42.
———— (1975). One-sided maximal functions and H^p. *JFA* 18:429–454.
———— (1976). Harmonic analysis and probability. In *Studies in harmonic analysis*, ed. Ash (1976), pp. 136–149.
———— (1977a). Brownian motion and classical analysis. *Proc. Symp. Pure Math.* 31:5–14.
———— (1977b). Exit times of Brownian motion, harmonic majorization and Hardy spaces. *Adv. in Math.* 26:182–205.
———— (1978). Boundary value estimation of the range of an analytic function. *Michigan Math. J.* 25:197–211.
———— (1979a). Martingale theory and harmonic analysis in Euclidean spaces. In *Harmonic analysis in Euclidian space*, ed. Weiss and Wainger (1979).
———— (1979b). A sharp inequality for martingale transforms. *Ann. Prob.* 7:858–863.
———— (1981). A geometrical characterization of Banach spaces in which martingale difference sequences are unconditional. *Ann. Prob.* 9:997–1011.
———— (1982). A nonlinear partial differential equation and the unconditional constant of the Haar system in L^p. *BAMS* (New Series) 7:591–595.
———— (in press). Boundary value problems and sharp inequalities for martingale transforms. *Ann. Prob.*
Burkholder, D. L., B. Davis, and R. F. Gundy (1972). Integral inequalities for convex functions of operators and martingales. In *Proceedings of the Sixth Berkeley Symposium*, vol. II, pp. 223–240.
Burkholder, D. L., and R. F. Gundy (1970). Extrapolation and interpolation of quasilinear operators on martingales. *Acta Math.* 124:249–304.
———— (1972). Distribution function inequalities for the area integral. *Studia Math.* 44:527–544.

_____ (1973). Boundary behavior of harmonic functions in a half space and Brownian motion. *Ann. Inst. Fourier* (Grenoble) 23:195–212.
Burkholder, D. L., R. F. Gundy, and M. L. Silverstein (1971). A maximal function characterization of the class H^p. *TAMS* 157:137–153.
Calderon, A. P. (1950a). On the behavior of harmonic functions at the boundary. *TAMS* 68:47–54.
_____ (1950b). On a theorem of Marcinkiewicz and Zygmund. *TAMS* 68:55–61.
_____ (1950c). On the theorems of M. Riesz and Zygmund. *PAMS* 1:533–535.
_____ (1966). Singular integrals. *BAMS* 72:426–465.
Calderon, A. P., and A. Zygmund (1964). On higher gradients of harmonic functions. *Studia Math.* 24:211–226.
Cameron, R. H., and W. T. Martin (1944a). Transformation of Wiener integrals under translations. *Ann. Math.* 45:386–396.
_____ (1944b). The Wiener measure of Hilbert neighborhoods in the space of real continuous functions. *J. Math. Phys.* 23:195–209.
_____ (1945a). Evaluations of various Wiener integrals by use of Sturm-Liouville differential equations. *BAMS* 51:73–90.
_____ (1945b). Transformation of Wiener integrals under a general class of linear transformations. *TAMS* 58:184–219.
_____ (1949). The transformation of Wiener integrals by nonlinear transformations. *TAMS* 66:253–283.
Carleson, L. (1958). An interpretation problem for bounded analytic functions. *Amer. J. Math.* 80:921–930.
_____ (1961). On the existence of boundary values for harmonic functions in several variables. *Archiv für Math.* 4:393–399.
_____ (1962). Interpolations by bounded analytic functions and the corona problem. *Ann. Math.* 76:547–559.
Carmona, R. (1978). Pointwise bounds for Schrödinger eigenstates. *CMP* 62:97–106.
_____ (1979a). Processus de diffusion gouverné par la formé de Dirichlet de l'operateur de Schrödinger. In *Sem. XIII*, pp. 557–569. Springer LNM 721.
_____ (1979b). Regularity projects of Schrödinger and Dirichlet semigroups. *JFA* 33:259–296.
Carmona, R., and B. Simon (1981). Pointwise bounds on eigenfunctions and wave packets in n-body quantum systems V. *CMP* 80:59–98.
Chao, J. A., and M. H. Taibleson (1973). A subregularity inequality for conjugate systems on local fields. *Studia Math.* 46:249–257.
Chung, K. L. (1974). *A course in probability theory.* 2d ed. Academic Press, New York.
_____ (1976). Excursions in Brownian motion. *Archiv für Math.* 14:155–177.
_____ (1979). On stopped Feynman-Kac functionals. In *Sem. XIV*, pp. 347–356. Springer LNM 784.
_____ (1981). Feynman-Kac functional and the Schrödinger equation. In *Seminar on Stochastic Processes I*, ed. E. Cinlar, K. L. Chung, and R. Getoor. Birkhauser, Boston.
_____ (1982). *Lectures from Markov processes to Brownian motion.* Springer-Verlag, New York.
Chung, K. L., R. Durrett, and Z. Zhongxin (1983). Extension of domains with finite gauge. *Math. Annalen* 264:78–79.
Chung, K. L., and P. Li (1983). Comparison of probability and eigenvalue methods for the Schrödinger equation. *Adv. in Math.*
Chung, K. L., and K. M. Rao (1980). Sur la théorie du potentiel avec la fonctionelle de

Feynman-Kac. *C. R. Acad. Sci. Paris* 290A:629–631.

———— (1981). Potential theory with the Feynman-Kac functional. *Prob. Math. Stat.* 1.

Chung, K. L., and S. R. S. Varadhan (1979). Positive solutions of the Schrödinger equation in one dimension. *Studia Math.* 68:249–260.

Ciesielski, Z., and S. J. Taylor (1962). First passage times and sojurn times for Brownian motion in space and exact Hausdorff measure. *TAMS* 103:434–450.

Coifman, R. R. (1974). A real variable characterization of H^p. *Studia Math.* 51:269–274.

Coifman, R. R., and C. Fefferman (1974). Weighted norm inequalities for maximal functions and singular integrals. *Studia Math.* 51:241–250.

Coifman, R. R., and R. Rochberg (1980). Another characterization of *BMO*. *PAMS* 79:249–254.

Coifman, R. R., and G. Weiss (1977). Extensions of Hardy spaces and their use in analysis. *BAMS* 83:569–645.

Conway, J. B. (1978). *Functions of one complex variable.* 2d ed. Springer-Verlag, New York.

Courant, R., and D. Hilbert (1953, 1962). *Methods of mathematical physics.* Vols. I, II. Interscience, New York.

Davis, B. (1968). Comparison tests for martingale convergence. *Ann. Math. Stat.* 39:2141–2144.

———— (1969). A comparison test for martingale inequalities. *Ann. Math. Stat.* 40:505–508.

———— (1970). On the integrability of the martingale square function. *Israel J. Math.* 8:187–190.

———— (1973a). On the distributions of conjugate functions of nonnegative measures. *Duke Math. J.* 40:695–700.

———— (1973b). An inequality for the distribution of the Brownian gradient function. *PAMS* 37:189–194.

———— (1974). On the weak type (1, 1) inequality for conjugate functions. *PAMS* 44:307–311.

———— (1975). Picard's theorem and Brownian motion. *TAMS* 213:353–362.

———— (1976). On Kolmogorov's inequalities $\|\tilde{f}\|_p \leq C_p \|f\|_1, 0 < p < 1$. *TAMS* 222:179–192.

———— (1979a). Applications of the conformal invariance of Brownian motion. In *Harmonic analysis in Euclidean space*, ed. Weiss and Wainger (1979).

———— (1979b). Brownian motion and analytic functions. *Ann. Prob.* 7:913–932.

———— (1980). Hardy spaces and rearrangements. *TAMS* 261:211–233.

Deift, P., W. Hunziker, B. Simon, and E. Vock (1978). Pointwise bounds on eigenfunctions and wave packets in n-body quantum systems IV. *CMP* 64:1–34.

Dellacherie, C. (1980). Un survol de la théorie de l'intégrale stochastique. *Stoch. Proc. Appl.* 10:115–144.

Dellacherie, C., and P. A. Meyer (1978). *Probabilities and potential.* English translation of *Probabilités et potentiel.* North Holland, Amsterdam.

Doleans-Dade, C. (1970). Quelques applications de la formule de changements de variables pour les semimartingales. *ZfW* 16:181–194.

Donsker, M. D., and M. Kac (1950). A sampling method for determining the lowest eigenvalue and the principal eigenfunction of Schrödinger's equation. *J. Res. Nat. Bureau of Standards* 44:551–557.

Donsker, M. D., and S. R. S. Varadhan (1974). Asymptotic evaluation of certain Wiener integrals for large time. In *Proceedings of the International Conference on Function Space Integration.* Oxford Univ. Press.

———— (1975a). Asymptotics for the Wiener sausage. *CPAM* 28:525–566; errata, p. 677.

―――― (1975b). On a variational formula for the principal eigenvalue for operators with a maximum principle. *Proc. Nat. Acad. Sci., USA* 72:780–783.

―――― (1975–1976). Asymptotic evaluation of certain Markov process expectations for large time. *CPAM* (I) 28:1–47; (II) 28:279–301; (III) 29:389–461.

―――― (1976). On the principal eigenvalue of second-order elliptic differential operators. *CPAM* 29:595–622.

Doob, J. L. (1953). *Stochastic processes*. John Wiley, New York.

―――― (1954). Semimartingales and subharmonic functions. *TAMS* 77:86–121.

―――― (1955a). Martingales and one-dimensional diffusion. *TAMS* 78:168–208.

―――― (1955b). A probability approach to the heat equation. *TAMS* 80:216–280.

―――― (1956). Probability methods applied to the first boundary value problem. In *Proceedings of the Third Berkeley Symposium*, vol. II, pp. 49–80.

―――― (1957). Conditional Brownian motion and the boundary limits of harmonic functions. *Bull. Soc. Math. France* 85:431–458.

―――― (1958a). Boundary limit theorems for a half-space. *J. Math. Pures Appl.* 37:385–392.

―――― (1958b). Probability theory and the first boundary value problem. *Illinois J. Math.* 2:19–36.

―――― (1959). A nonprobabilistic proof of the relative Fatou theorem. *Ann. Inst. Fourier* (Grenoble) 9:293–300.

―――― (1960). Relative limit theorems in analysis. *J. Analyse Math.* 8:289–306.

―――― (1961). Conformally invariant cluster value theory. *Illinois J. Math.* 5:521–549.

―――― (1962). Boundary properties of functions with finite Dirichlet integrals. *Ann. Inst. Fourier* (Grenoble) 12:573–622.

―――― (1963). One-sided cluster-value theorems. *Proc. LMS* 13:461–470.

―――― (1964). Some classical function theory theorems and their modern versions. *Ann. Inst. Fourier* (Grenoble) 15:113–136.

―――― (1966). Remarks on the boundary limits of harmonic functions. *J. SIAM—Numerical Analysis* 3:229–235.

Dubins, L. E., and D. Gilat (1978). On the distribution of maxima of martingales. *PAMS* 68:337–338.

Dubins, L. E., and G. Schwarz (1965). On continuous martingales. *Proc. Nat. Acad. Sci., USA* 53:913–916.

Dunford, N., and J. T. Schwarz (1957). *Linear operators, Part I: General theory*. Interscience, New York.

Duren, P. L. (1970). *The theory of H^p spaces*. Academic Press, New York.

Duren, P. L., B. W. Romberg, and A. L. Shields (1969). Linear functionals on H^p spaces with $0 < p < 1$. *J. Reine Angew. Math.* 238:32–60.

Durrett, R. (1982). A new proof of Spitzer's result on the winding of two-dimensional Brownian motion. *Ann. Prob.* 10:244–246.

Dvoretsky, A., P. Erdös, and S. Kakutani (1961). Nonincreasing everywhere of the Brownian motion process. In *Proceedings of the Fourth Berkeley Symposium*, vol. II, pp. 103–116.

Dynkin, E. B. (1960). Markov processes and related problems of analysis. *Russian Math. Surveys* 15, no. 2, pp. 1–24.

―――― (1963). Markov processes and problems in analysis. In *Proceedings of the International Congress, Stockholm, 1962*. AMS Translations, series 2, vol. 31, pp. 1–24.

―――― (1981). *Markov processes and related problems of analysis: Selected papers*. LMS Lecture Notes. Cambridge Univ. Press, Cambridge.

Einstein, A. (1905). On the movement of small particles suspended in a stationary liquid demanded by the molecular kinetic theory of heat. *Ann. Phys.* 17.

——— (1926). *Investigations on the theory of the Brownian movement.* Reprinted by Dover Books, New York, 1956.

Fatou, P. (1906). Séries trigonométriques et séries de Taylor. *Acta Math.* 30:335–400.

Fefferman, C. (1971). Characterization of bounded mean oscillation. *BAMS* 77:587–588.

——— (1976). Harmonic analysis and H^p spaces. In *Studies in harmonic analysis*, ed. Ash (1976), pp. 38–75.

Fefferman, C., and E. M. Stein (1972). H^p spaces in several variables. *Acta Math.* 129:137–193.

Feller, W. (1971). *An introduction to probability theory and its applications.* Vol. II. John Wiley, New York.

Feynman, R. J. (1948). Space-time approach to nonrelativistic quantum mechanics. *Rev. Mod. Phys.* 20:367–387.

Folland, G. B. (1976). *Introduction to partial differential equations.* Princeton Univ. Press, Princeton.

Freedman, D. (1970). *Brownian motion and diffusion.* Holden-Day, San Francisco.

Friedman, A. (1964). *Partial differential equations of parabolic type.* Prentice-Hall, Englewood Cliffs, NJ.

——— (1969). *Partial differential equations.* Holt, Rinehart & Winston, New York.

——— (1975). *Stochastic differential equations and applications.* Academic Press, New York.

Garnett, J. (1979). Two constructions in *BMO*. In *Harmonic analysis in Euclidean space*, ed. Weiss and Wainger (1979).

——— (1980). *Bounded analytic functions.* Academic Press, New York.

Garnett, J., and P. Jones (1978). The distance in *BMO* to L^∞. *Ann. Math.* 108:373–393.

——— (1982). *BMO* from dyadic *BMO*. *PJM* 99:351–371.

Garsia, A. M. (1970). *Topics in almost everywhere convergence.* Markham, Chicago.

——— (1973a). The Burges-Davis inequality via Fefferman's inequality. *Ark. Mat.* 11:229–237.

——— (1973b). On a convex function inequality for martingales. *Ann. Prob.* 1:171–174.

——— (1973c). *Martingale inequalities: Seminar notes on recent progress.* Benjamin, Reading, MA.

Getoor, R. K., and M. J. Sharpe (1972). Conformal martingales. *Invent. Math.* 16:271–308.

——— (1979). Excursions of Brownian motion and Bessel processes. *ZfW* 47:83–106.

Gikhman, I. I., and A. V. Skorohod (1973). *Stochastic differential equations.* Springer-Verlag, New York.

Gilbarg, D., and N. S. Trudinger (1977). *Elliptic partial differential equations of second order.* Springer-Verlag, New York.

Girsanov, I. V. (1960). On transforming a certain class of stochastic processes by absolutely continuous change of measure. *TPA* 5:285–301.

——— (1961). On Itô's stochastic integral equation. *Soviet Math.* 2:506–509.

——— (1962). An example of nonuniqueness of the solution of K. Itô's stochastic integral equation. *TPA* 7:325–331.

Gundy, R. F. (1968). A decomposition for L^1 bounded martingales. *Ann. Math. Stat.* 39:134–138.

——— (1969). On the class $L\log L$, martingales, and singular integrals. *Studia Math.* 33:109–118.

——— (1980a). Inequalités pour martingales à un et deux indices: l'espace H^p. *École d'été de Probabilités de Saint-Flour, VIII.* Springer LNM 774.

———— (1980b). Local convergence of a class of martingales in multidimensional time. *Ann. Prob.* 8:607–614.

Gundy, R. F., and E. M. Stein (1979). H^p theory for the polydisc. *Proc. Nat. Acad. Sci., USA* 76:1026–1029.

Gundy, R. F., and N. Varopoulos (1976). A martingale that occurs in harmonic analysis. *Ark. Mat.* 14:179–187.

———— (1979). Les transformations de Riesz et les integrales stochastiques. *C. R. Acad. Sci. Paris* 289A:13–16.

Gustafson, K. E. (1980). *Introduction to partial differential equations.* John Wiley, New York.

Hardy, G. H. (1915). The mean value of the modulus of an analytic function. *Proc. LMS* 14:269–277.

———— (1928). Remarks on three recent notes in the *Journal. J. LMS* 3:166–169.

Hardy, G. H., and J. E. Littlewood (1926). Some new properties of Fourier constants. *Math. Ann.* 97:159–209.

———— (1930). A maximal theorem with function-theoretic applications. *Acta Math.* 54:81–116.

———— (1931). Some properties of conjugate functions. *J. Reine Angew. Math.* 167:405–423.

Helson, H., and G. Szegö (1960). A problem in prediction theory. *Ann. Mat. Pura Appl.* 51:107–138.

Herz, C. (1974a). Bounded mean oscillation and regulated martingales. *TAMS* 193:199–216.

———— (1974b). H_p spaces of martingales $0 < p \leq 1$. *ZfW* 28:189–205.

Hewitt, E., and K. Stromberg (1969). *Real and abstract analysis.* Springer-Verlag, New York.

Hoffman-Ostenhof, M. and T., and B. Simon (1980a). Brownian motion and a consequence of Harnack's inequality, nodes of quantum wave functions. *PAMS* 80:301–305.

———— (1980b). On the nodal structure of atomic eigenfunctions. *J. Phys. A.* 13:1131–1133.

Hunt, G. A. (1956). Some theorems concerning Brownian motion. *TAMS* 81:294–391.

Hunt, R. A., B. Muckenhoupt, and R. L. Wheeden (1973). Weighted norm inequalities for conjugate functions and Hilbert transforms. *TAMS* 176:227–251.

Ikeda, N., and S. Watanabe (1981). *Stochastic differential equations and diffusion processes.* North Holland, Amsterdam.

Itô, K. (1944). Stochastic integrals. *Proc. Imp. Acad. Tokyo* 20:519–524.

———— (1946). On a stochastic integral equation. *Proc. Japan Acad.* 22:2, 32–35.

———— (1950a). Brownian motions in a Lie group. *Proc. Japan Acad.* 26:8, 4–10.

———— (1950b). Stochastic differential equations in a differentiable manifold. (I) *Nagoya Math. J.* 1:35–47; (II) *Mem. Coll. Sci. U. Kyoto*, Series A, 28, 1, 81–85.

———— (1951a). On a formula concerning stochastic differentials. *Nagoya Math. J.* 3:55–65.

———— (1951b). On stochastic differential equations. *Memoirs AMS*, no. 4.

———— (1960). Wiener integral and Feynman integral. In *Proceedings of the Fourth Berkeley Symposium*, vol. II, pp. 227–238.

Itô, K., and H. McKean, Jr. (1964). *Diffusion processes and their sample paths.* Springer-Verlag, New York.

Jacod, J. (1979). *Calcul stochastique et problèmes de martingales.* Springer LNM 714.

Jacod, J., and J. Memin (1981). Weak and strong solutions of stochastic differential equations: Existence and stability. In *Stochastic integrals*, ed. Williams (1981), pp. 169–212.

Janson, S. (1977). Characterization of H^1 by singular integral transformations on martingales and R^n. *Math. Scand.* 41:140–152.

───── (1979). Singular integrals on local fields and generalizations to martingales. In *Harmonic analysis in Euclidean space*, ed. Weiss and Wainger (1979).

───── (1981). BMO and commutators of martingale transforms. *Ann. Inst. Fourier* (Grenoble) 31:265–270.

John, F. (1961). Rotation and strain. *CPAM* 14:391–413.

───── (1982). *Partial differential equations*. 4th ed. Springer-Verlag, New York.

John, F., and L. Nirenberg (1961). On functions of bounded mean oscillation. *CPAM* 14:415–426.

Jones, P. (1980). Carleson measures and the Fefferman-Stein decomposition of *BMO (R)*. *Ann. Math.* 111:197–208.

Kac, M. (1946). On the average of a certain Wiener functional and a related limit theorem in the calculus of probability. *TAMS* 59:401–414.

───── (1949). On distributions of certain Wiener functionals. *TAMS* 65:1–13.

───── (1951). On some connections between probability theory and differential and integral equations. In *Proceedings of the Second Berkeley Symposium*, pp. 189–215.

───── (1953). An application of probability theory to the study of Laplace's equation. *Ann. de Socièté Math. Polonaise* 25:122–130.

───── (1959). *Probability and related topics in physical science*. Interscience, New York.

───── (1966a). Can one hear the shape of a drum? *Amer. Math. Monthly* 73:1–23.

───── (1966b). Wiener and integration in function spaces. *BAMS* 72:52–68.

───── (1970). On some probabilistic aspects of classical analysis. *Amer. Math. Monthly* 77:586–597.

───── (1972). On applying mathematics: Reflections and examples. *Quart. J. Appl. Math.* 30:17–29.

───── (1974). A stochastic model related to the telegrapher's equation. *Rocky Mountain J.* 4:497–509.

Kac, M., and J. M. Luttinger (1974). Probabilistic methods in scattering theory. *Rocky Mountain J.* 4:511–537.

───── (1975). Scattering length and capacity. *Ann. Inst. Fourier* (Grenoble) 25:317–321.

Kahane, J. P. (1976). Brownian motion and classical analysis. *Bull. LMS* 7:145–155.

Kakutani, S. (1944a). On Brownian motion in n-space. *Proc. Imp. Acad. Tokyo* 20:648–652.

───── (1944b). Two-dimensional Brownian motion and harmonic functions. *Proc. Imp. Acad. Tokyo* 20:706–714.

───── (1945). Markov processes and the Dirichlet problem. *Proc. Imp. Acad. Tokyo* 21:227–233.

Kato, T. (1973). Schrödinger operators with singular potentials. *Israel J. Math.* 13:135–148.

Katznelson, Y. (1968). *An introduction to harmonic analysis*. John Wiley, New York.

Kellog, O. D. (1929). *Foundations of potential theory*. Springer-Verlag. Reprinted by Dover Books, New York, 1954.

Khas'minskii, R. Z. (1959). On positive solutions of the equation $Au + Vu = 0$. *TPA* 4:309–318.

Khintchine, A. (1933). Asymptotische Geseteze der Wahrscheinlichkeitsrechnung. *Ergebn. Math.*, Berlin.

Knight, F. B. (1971). A reduction of continuous square-integrable martingales to Brownian motion. In *Martingales*, ed. H. Dinges, pp. 19–31. Springer LNM 190.

———— (1981). *Essentials of Brownian motion and diffusion*. AMS, Providence, RI.

Kolmogorov, A. (1925). Sur les fonctions harmoniques conjuguées et les seriés de Fourier. *Fund. Math.* 7:24–29.

Koosis, P. (1980). *Lectures on H^p spaces*. LMS Lecture Notes. Cambridge Univ. Press, Cambridge.

Kunita, H., and S. Watanabe (1967). On square integrable martingales. *Nagoya Math. J.* 30:209–245.

Ladyzenskaya, O. A., V. A. Solonnikov, and N. N. Ural'ceva (1968). *Linear and quasilinear equations of parabolic type*. AMS Translations of Mathematical Monographs 23.

Latter, R. (1977). A decomposition of $H^p(R^n)$ in terms of atoms. *Studia Math.* 62:92–101.

———— (1979). The atomic decomposition of Hardy spaces. In *Harmonic analysis in Euclidean space*, ed. Weiss and Wainger (1979).

Lebesgue, H. (1924). Conditions de régularité, conditions d'irrégularité, conditions d'impossibilité dans le problème de Dirichlet. *C. R. Acad. Sci. Paris*, pp. 349–354.

Lenglart, E. (1977). Sur la convergence presque sûre des martingales locales. *C. R. Acad. Sci. Paris* 284A:1085–1088.

Lepingle, D. (1978). Sur le comportement asymptotique des martingales locales. In *Sem. XII*, pp. 148–161. Springer LNM 649.

Lévy, P. (1939). Sur certains processes stochastiques homogènes. *Compositio Math.* 7:283–339.

———— (1940). Le mouvement Brownien plan. *Amer. J. Math.* 62:487–550.

———— (1948). *Processus stochastiques et mouvement Brownien*. Gauthier-Villars, Paris.

———— (1951). Wiener's random function. In *Proceedings of the Second Berkeley Symposium*, pp. 171–186.

Lieb, E. (1976). Bounds on the eigenvalues of the Laplace and Schrödinger operators. *BAMS* 82:751–753.

Lieb, E., and B. Simon (1982). Pointwise bounds on eigenfunctions and wave packets in n-body quantum systems VI: Asymptotics in the two-cluster region. *Adv. Appl. Math.* 1:324–343.

Littlewood, J. E. (1926). On a theorem of Kolmogorov. *J. LMS* 1:229–231.

———— (1929). On a theorem of Zygmund. *J. LMS* 4:305–307.

Loomis, L. H. (1946). A note on the Hilbert transform. *BAMS* 52:1082–1086.

Lyons, T. J., and H. P. McKean (in press). Winding of plane Brownian motion. Preprint. Oxford Univ. Press, New York.

McKean, H. P. (1969). *Stochastic integrals*. Academic Press, New York.

———— (1977). $-\Delta$ plus a bad potential. *J. Math. Phys.* 18:1277–1279.

Marcinkiewicz, J., and A. Zygmund (1938). A theorem of Lusin. *Duke Math. J.* 4:473–485.

Messulam, P., and M. Yor (1982). On D. Williams' punching method and some applications. *J. LMS*.

Meyer, P. A. (1966). *Probability and potentials*. Blaisdel, Waltham, MA.

———— (1976). Un cours sur les intégrales stochastiques. In *Sem. X*, pp. 245–400. Springer LNM 511.

———— (1977). Le dual de $H^1(R^v)$ demonstrations probabilistes. In *Sem. XI*, pp. 132–195. Springer LNM 581.

Millar, P. W. (1968). Martingale integrals. *TAMS* 133:145–166.

Moser, J. (1961). On Harnack's theorem of elliptic differential equations. *CPAM* 14:577–591.

Nevanlinna, F. and R. (1922). Über die Eigenschaften analytischer Funktionen in der Umgebung einer singulären Stelle oder Linie. *Acta Soc. Sci. Fenn.* 50, no. 5.

Neveu, J. (1975). *Discrete parameter martingales.* North Holland, Amsterdam.

Olenik, O. A. (1966). Alcuni risultati sulle equazioni lineari e quasilineari ellitico-paraboliche a derivate parziali del second ordine. *Rend. Classe Sci. Nat. Acad. Naz. Lincei,* Series 8, 40:775–784.

Paley, R., N. Wiener, and A. Zygmund (1933). Note on random functions. *Math. Z.* 37:647–668.

Paley, R. E. A. C., and A. Zygmund (1932). A note on analytic functions in the unit circle. *P. Camb. Phil. Soc.* 28:266–272.

Petersen, K. E. (1977). *Brownian motion, Hardy spaces, and bounded mean oscillation.* LMS Lecture Notes. Cambridge Univ. Press, Cambridge.

Pichorides, S. K. (1972). On the best values of the constants in the theorems of M. Riesz, Zygmund, and Kolmogorov. *Studia Math.* 46:165–179.

Pitman, J. (1979). A note on L^2 maximal inequalities. In *Sem. XV,* pp. 251–258. Springer LNM 850.

Plessner, A. (1928). Über das Verhalten analytischen Funktionen auf dem Rande des Definitions-bereiches. *J. Reine Angew. Math.* 158:219–227.

Port, S., and C. Stone (1978). *Brownian motion and classical potential theory.* Academic Press, New York.

Privalov, I. I. (1916). Sur les fonctions conjuguées. *Bull. Soc. Math. France* (1916), pp. 100–103.

_____ (1924). Generalization of a theorem of Fatou. *Mat. Sbornik* 31:232–235.

Protter, M. H., and H. F. Weinberger (1966). On the spectrum of general second-order equations. *BAMS* 72:251–255.

_____ (1967). *Maximum principles in differential equations.* Prentice-Hall, New York.

Rao, K. M. (1969). On decomposition theorems of Meyer. *Math. Scand.* 24:66–78.

Ray, D. (1954). On the spectra of second-order differential operators. *TAMS* 77:299–321.

Riesz, F. (1923). Über die Randwerte einer analytischen Funktion. *Math. Z.* 18:87–95.

Riesz, M. (1927). Sur les fonctions conjuguées. *Math. Z.* 27:218–244.

Rosenblatt, M. (1951). On a class of Markov processes. *TAMS* 71:120–135.

Royden, H. (1968). *Real analysis.* 2d ed. Macmillan, New York.

Sharpe, M. J. (1980). Local times and singularities of continuous local martingales. In *Sem. XIV,* pp. 76–101. Springer LNM 784.

Simon, B. (1974–1975). Pointwise bounds on eigenfunctions and wave packets in n-body quantum systems. (I) *PAMS* 42:395–401; (II) *PAMS* 45:454–456; (III) *TAMS* 208:317–329.

_____ (1979). *Functional integration and quantum physics.* Academic Press, New York.

_____ (1980). Brownian motion, L^p properties of Schrödinger operators and the localization of binding. *JFA* 35:215–229.

_____ (1981). Large time behavior of the L^p norm of Schrödinger semigroups. *JFA* 40:66–83.

_____ (1982). Schrödinger semigroups. *BAMS* (new series) 7:447–526.

_____ (1983). Instantons, double wells, and large deviations. *BAMS* (new series) 8:323–326.

Sion, M. (1958). On general minmax theorems. *PJM* 8:171–176.

Skorohod, A. V. (1956). Limit theorems for stochastic processes. *TPA* 1:261–290.

_____ (1961). Existence and uniqueness of solutions to stochastic differential equations. *Sibirsk. Mat. Zur.* 2:129–137.

_____ (1965). *Studies in the theory of random processes.* Addison-Wesley, Reading, MA.
Spanne, S. (1966). Sur l'interpolation entre les espaces $\mathscr{L}_k^{p,\varphi}$. *Ann. Scusla Norm Sup. Pisa* 20:625–648.
Spencer, D. C. (1943). A function theoretic identity. *Amer. J. Math.* 65:147–160.
Spitzer, F. (1958). Some theorems concerning two-dimensional Brownian motion. *TAMS* 87:187–197.
Spivak, M. (1979). *A comprehensive introduction to differential geometry.* Publish or Perish, Inc., Berkeley, CA.
Stein, E. M. (1961). On the theory of harmonic functions of several variables II. *Acta Math.* 106:137–174.
_____ (1962). Conjugate harmonic functions in several variables. In *Proceedings of the International Congress of Mathematics, Stockholm*, pp. 414–420.
_____ (1967). Singular integrals, harmonic functions and differentiability properties of functions of several variables. *Proc. Symp. Pure Math.* 10:316–335.
_____ (1969). A note on $L \log L$. *Studia Math.* 31:305–310.
_____ (1970a). *Singular integrals and differentiability properties of functions.* Princeton Univ. Press, Princeton.
_____ (1970b). *Topics in harmonic analysis related to the Littlewood-Paley theory.* Annals of Mathematical Studies, no. 63. Princeton Univ. Press, Princeton.
Stein, E. M., and G. Weiss (1959). An extension of a theorem of Marcinkiewicz and its applications. *J. Math. Mech.* 8:263–284.
_____ (1960). On the theory of harmonic functions of several variables I. *Acta Math.* 103:25–62.
_____ (1968). Generalization of the Cauchy-Riemann equations and representations of the rotation group. *Amer. J. Math.* 90:163–196.
_____ (1971). *Introduction to Fourier analysis on Euclidian spaces.* Princeton Univ. Press, Princeton.
Stein, P. (1933). On a theorem of M. Riesz. *J. LMS* 8:242–247.
Stroock, D. W. (1973). Application of Fefferman-Stein type interpolation to probability theory and analysis. *CPAM* 26:477–496.
_____ (1981). The Malliavin calculus and its applications. In *Stochastic integrals*, ed. D. Williams (1981), pp. 394–432.
_____ (1983). The Malliavin calculus and its application to second-order parabolic equations. *Math. Systems Theory* 14:25–65.
Stroock, D. W., and S. R. S. Varadhan (1969). Diffusion processes with continuous coefficients. *CPAM* 22: (I) 345–400, (II) 479–530.
_____ (1970). On the support of diffusion processes with applications to the strong maximum principle. In *Proceedings of the Sixth Berkeley Symposium*, vol. III, pp. 333–360.
_____ (1972a). On degenerate elliptic parabolic operators of second order and their associated diffusions. *CPAM* 25:651–713.
_____ (1972b). Diffusion processes. In *Proceedings of the Sixth Berkeley Symposium*, vol. III, pp. 361–368.
_____ (1974). A probabilistic approach to $H^p(R^d)$. *TAMS* 192:245–260.
_____ (1979). *Multidimensional diffusion processes.* Springer-Verlag, New York.
Taibleson, M. H. (1975). *Fourier analysis on local fields.* Mathematical Notes. Princeton Univ. Press, Princeton.
_____ (1979). An introduction to Hardy spaces on local fields. In *Harmonic analysis in Euclidean space*, ed. Weiss and Wainger (1979).

Taibleson, M. H. and G. Weiss (1979). The molecular characterization of Hardy spaces. In *Harmonic analysis in Euclidean space*, ed. Weiss and Wainger (1979).
Tanaka, H. (1964). Existence of diffusions with continuous coefficients. *Mem. Fac. Sci. Kyushu*, Series A 18:89–103.
Tsirel'son, B. S. (1976). An example of a stochastic differential equation having no strong solution. *TPA* 20:416–418.
Uchiyama, A. (1982). A constructive proof of the Fefferman-Stein decomposition of $BMO(R^n)$. *Acta Math.* 148:215–241.
———— (1983). A constructive proof of the Fefferman-Stein decomposition of BMO on simple martingales. In *Conference on harmonic analysis in honor of Antoni Zygmund*, vol. II, ed. Beckner, Calderon, Fefferman, and Jones, pp. 495–505. Wadsworth, Belmont, CA.
Varopoulos, N. (1977a). BMO functions and the $\bar{\partial}$ equation. *PJM* 71:221–273.
———— (1977b). A remark on BMO and bounded harmonic functions. *PJM* 73:257–259.
———— (1979). A probabilistic proof of the Garnett-Jones theorem on BMO. *PJM* 90:201–221.
———— (1980a). Aspects of probabilistic Littlewood-Paley theory. *JFA* 38:25–60.
———— (1980b). The Helson-Szegö theorem and A_p functions for Brownian motion and several variables. *JFA* 39:85–121.
Walsh, J. B. (1977). A property of conformal martingales. In *Sem. XI*, pp. 490–501. Springer LNM 581.
Walsh, J. L. (1929). The approximation of harmonic functions by harmonic polynomials and harmonic rational functions. *BAMS* 35:499–544.
Weiss, G. (1970). Complex method in harmonic analysis. *Amer. Math. Monthly* 77:465–474.
Weiss, G., and S. Wainger, eds. (1979). *Harmonic analysis in Euclidean space*. Proceedings of Symposia in Pure Mathematics, vol. 35. AMS, Providence, RI.
Widder, D. V. (1975). *The heat equation*. Academic Press, New York.
Wiener, N. (1923). Differential space. *J. Math. Phys.* 2:131–174.
———— (1924a). Certain notions in potential theory. *J. Math. Phys.* 3:24–51.
———— (1924b). The Dirichlet problem. *J. Math. Phys.* 3:127–146.
———— (1924c). Un problème de probabilitése énombrables. *Bull. Soc. Math. de France* 52:569–578.
———— (1925). Note on a paper of O. Perron. *J. Math. Phys.* 4:21–32.
———— (1930a). Generalized harmonic analysis. *Acta Math.* 55:117–258.
———— (1930b). The homogeneous chaos. *Amer. J. Math.* 80:897–936.
Williams, D., ed. (1981). *Stochastic integrals*. Proceedings LMS Durham Symposium. Springer LNM 851.
———— (ND). A simple geometric proof of Spitzer's winding number formula for 2-dimensional Brownian motion. Unpublished paper. University College of Swansea, Wales.
Yamada, T., and S. Watanabe (1971). On the uniqueness of solutions to stochastic differential equations. *J. Math. Kyoto* 11: (I) 155–167, (II) 553–563.
Zygmund, A. (1929). Sur les fonctions conjuguées. *Fund. Math.* 13:284–303; correction, pp. 18, 312.
———— (1949). On the boundary values of several complex variables. *Fund. Math.* 36:207–235.
———— (1959). *Trigonometric series*. 2d ed. Cambridge Univ. Press, London.

Index of Notation

R^d	1	D	36	$\mathcal{N}_a$	106		
$\mathscr{R}^d$	3	k_y	36	$\mathcal{N}$	106		
C	5	h_θ	38	$A_a u$	106		
$\mathscr{C}$	5	$G_D(x,y)$	39	$\mathscr{A}_a$	106		
P_x	5	Λ	44	$\mathscr{A}$	106		
P_μ	5	Π	49	$\mathscr{L}^*$	107		
$\mathscr{F}_s^0$	8	$\langle X \rangle_t$	52	$\mathcal{N}^*$	107		
$p_t(x,y)$	8	$\langle X, Y \rangle_t$	54	$\mathscr{A}^*$	107		
θ_s	10	Π_0	55	P^θ a.s.	107		
$\mathscr{F}_s^+$	12	$b\Pi_0$	55	$\mathscr{P}f$	119		
$\mathscr{F}_s'$	17	$(H \cdot X)_t$	55	$S_\alpha(\theta)$	126		
$\mathscr{A}$	17	Π_1	56	$N_\alpha u$	127		
T_a	17	$\Pi_2(X)$	57	h^p	144		
$\mathscr{F}_s$	19	$\|H\|_X$	57	$d_p(u)$	144		
T_A	19	$\mathscr{M}^2$	57	H^p	144		
θ_S	20	$\Pi_3(X)$	58	π	144		
$\mathscr{F}_S$	20	$\ell b \Pi$	59	X^*	147		
$D(x,\delta)$	23	J	80	$\mathscr{M}^p$	147		
i.o.	28	P_z^θ	96	$\|X\|_p$	148		
D_i	28	$\mathscr{I}$	102	X_t^*	149		
Δ	28, 71	V_a^θ	105	$\mathscr{L}^p$	150		
$G(x,y)$	31	$\mathscr{L}_a$	105	$\mathscr{K}^p$	150		
H	32	$\mathscr{L}$	105	$\|X\|_{\mathscr{K}}$	151		
$G_H(x,y)$	35	$N_a u$	106	$\|X\|_{\mathscr{L}}$	151		

Index of Notation

$\langle\!\langle X \rangle\!\rangle_p$	152	$\mathscr{BMO}_h$	192	σ	272
H_0^p	155	BMO	193	X_t^x	282
$(A * X)_t$	163	$\|\varphi\|_*$	193	$X_{s,x}$	288
$\bar{u}$	170	$\langle\!\langle \varphi \rangle\!\rangle_*$	198	a.s.	294
$(\mathscr{M}^p)^*$	184	M	203	1_A	295
$\mathscr{BMO}$	186	N	203	i.o.	295
$\|X\|_*$	186	$\alpha_0(X)$	211	$\sigma(X)$	296
$\mathscr{BMO}_2$	188	K_d	228	$E(X	\mathscr{F})$	300
$\langle\!\langle X \rangle\!\rangle_*$	188	K_d^{loc}	229		
$\mathscr{M}_h^1$	192	L	271		

Subject Index

adapted process, 44
almost surely, 294
associative law, 62
asymptotic σ-field, 17
atomic decomposition
  for $p = 1$, 184
  for $p < 1$, 216

Blumenthal's 0-1 law, 14
Borel Cantelli lemmas, 295, 299
bounded mean oscillation, 186, 193, 199, 205
Brownian motion, 1
Brownian paths
  nondifferentiability, 5
  quadratic variation, 6
  Holder continuity, 7, 15
Burkholder-Gundy
  inequalities, 155
Burkholder's weak type
  inequality, 164

Cameron-Martin
  transformation, 234
Cauchy process, 2, 33
Chebyshev's inequality, 296
Ciesielski and Taylor's
  "paradox," 260
conditional expectation, 300
conditioned Brownian
  motions
  in H, 94, 97
  in $R^d - \{0\}$, 100
  in D, 126
conjugate harmonic
  functions, 162, 170
covariance
  of two local martingales, 54
  of two stochastic
    integrals, 61
  of two semimartingales, 68

Dirichlet problem, 43, 246
Donsker and Varadhan, 266

Doob's inequality, 147, 306
  sharp constant in, 151
Doob-Meyer
  decomposition, 52
distributive law, 68
duality theorem
  for $\mathcal{M}^1$, 184
  for H^1, 199
Dynkin's $\pi - \lambda$ theorem, 9

explosions
  from too much drift, 240
  from too much variance, 277
  Feller's test, 242
exponential martingale, 27, 70

Fefferman's inequality, 187, 190
Fefferman-Stein
  decomposition, 202
Feynman-Kac formula, 229
filtration, 50
fine topology, 108

Garnett-Jones theorem, 211, 214
Girsanov's formula, 82
good λ-inequalities, 153
Green's function, 39
Gundy's $L \log L$ inequality, 148

Hardy space H^p, 144
Harnack's inequality (for
  solutions of Schrödinger's equation), 261
heat equation, 220
Helson-Szegö theorem, 215
h-transform, 92, 94

independence, 296
integration by parts, 68
Itô's formula

  for a local martingale, 64
  for several
    semimartingales, 68

Janson's theorem, 167
Jensen's formula, 132
Jensen's inequality (for
  conditional
  expectations), 301
John-Nirenberg inequality, 208, 209

Kelvin's transformations, 80, 98
Khasmin'skii's lemma, 231
Kolmogorov's
  extension theorem, 3
  continuity criterion, 6
  three series theorem, 180
  (other) weak type
    inequality, 173
Kunita-Watanabe
  inequality, 59

law of the iterated
  logarithm
  for Brownian motion, 15
  for continuous local
    martingales, 77
Lebesgue's thorn, 248
Lévy's theorem, 75
Lévy's arc sin law, 261
local martingale, 50

Markov property
  of Brownian motion, 7
  of diffusions, 288
martingale, 302
martingale convergence
  theorem, 305
martingale transforms, 162
maximum principle, 72, 115
monotone class theorem, 10

natural scale, 74, 93, 241

Nevanlinna class, 128
nontangential convergence, 105

optional projection, 191
optional σ-field, 44
optional stopping theorem, 310
orthogonality of martingale increments, 303

Pac-Man, 249
Picard's theorems, 139, 143
Poincaré's cone condition, 249
Poisson's equation, 43, 251
Poisson integral
   representation of harmonic functions
   nonnegative functions, 98, 99
   bounded functions, 119
   functions in H^p, 158
potential kernels, 31
predictable σ-field, 49

predictable sequence, 48, 303

quadratic variation
   of Brownian motion, 7
   of continuous local martingales, 65

Rayleigh-Ritz formula, 270
regular point, 247
Riesz's inequality, 170
   sharp constant in, 172
Riesz transforms, 169

Schrödinger equation, 255, 263, 264
Schrödinger semigroups, 232
semimartingale, 58
shift, 10
shift invariant events, 102
simple random walk, 48, 91, 303
Spitzer's theorem (on the winding of Brownian motion), 134

stable subspace, 87
stochastic differential equations
   existence of solutions, 274, 278
   uniqueness, 278, 286
   examples, 283
stopping time, 18, 306
stretching function, 51
strong Markov property
   for Brownian motion, 21
   for conditioned Brownian motion, 102
submartingale, 302
subordination (of analytic functions), 176
supermartingale, 302

uniform integrability, 308
usual domination argument, 62

variance process, 52
Varopoulos' staircase, 213
volume potential, 225

Date Due

JAN 29 1987			

BRODART, INC.   Cat. No. 23 233   Printed in U.S.A.